Introduction to
SNA Networking

Introduction to SNA Networking

A Professional's Guide to VTAM/NCP

Jay Ranade

George C. Sackett

Second Edition

McGraw-Hill, Inc.

New York   San Francisco   Washington, D.C.   Auckland   Bogotá
Caracas   Lisbon   London   Madrid   Mexico City   Milan
Montreal   New Delhi   San Juan   Singapore
Sydney   Tokyo   Toronto

Library of Congress Cataloging-in-Publication Data

Ranade, Jay.
    Introduction to SNA networking : a professional's guide to
VTAM/NCP / Jay Ranade, George C. Sackett. — 2nd ed.
        p.   cm. — (McGraw-Hill series on computer communications)
    Includes index.
    ISBN 0-07-051506-9
    1. SNA (Computer network architectural) 2. Virtual computer
systems. I. Sackett, George C. II. Title. III. Series.
    TK5105.5.R447 1994
    004.6'5—dc20   94-27533
                                                                          CIP

1 2 3 4 5 6 7 8 9 0   DOC/DOC   9 0 0 9 8 7 6 5

ISBN 0-07-051506-9

*The sponsoring editor for this book was Jerry Papke, the editing super-
visor was Jim Halston, and the production supervisor was Pamela A.
Pelton. This book was set in Century Schoolbook. It was composed by
McGraw-Hill's Professional Book Group composition unit.*

Printed and bound by R. R. Donnelley & Sons Company.

This book is dedicated to all the people around the world who conceived, developed, improved, and maintain IBM's SNA.

George
Jay

Contents

Part 4   NCP

Part 5   Operations and Network Management

Preface to the First Edition

SNA is IBM's grand architecture for designing and implementing computer networks. VTAM and NCP are communications software products that are SNA-compliant and form the basis of SNA implementation. The purpose of this book is to teach you how to use VTAM and NCP to create single-host/single-domain networks. In a companion volume, entitled *Advanced SNA Networking*, you will learn about large multi-domain networks.

The 80/20 Rule

The SNA and VTAM/NCP literature provided by IBM constitutes about 20,000 pages of reading material. This text is not a summary of the IBM manuals nor is its purpose to replace those manuals. We have applied the 80/20 rule for material to be included in this text. The material which is used 80 percent of the time constitutes 20 percent of the material available. This is what is covered in this text; the rest has been discarded. While this approach gives you sufficient knowledge to feel comfortable with the software, it does not qualify this book as a reference manual.

Who This Book Is For

This book has been written with many types of readers in mind. It can be used as a primer for those who want to be VTAM/NCP systems programmers. It provides the practical aspects of SNA network implementation for those who have only a theoretical knowledge of the subject. It is ideal for CICS systems programmers who do not have a clear picture of networks and data communications. It can be easily understood by network operators who would like to gain background information on SNA networks. Network engineers will need this book to complement their hardware knowledge with communications software information.

Because the book provides sufficient information to understand logical units, InterSystems Communications (ISC), distributed systems, and a lot more, we would even go so far as to say that application programmers working in a COBOL/CICS environment would benefit from this text. As data processing evolves more and more toward distributed application/data, it will be beneficial to all levels of personnel to understand how various components of a network are interconnected.

PC users will find Parts 1 and 2 of this book especially useful. Since OS/2 Extended Edition Communication Manager has become available, PC communications are no longer limited to asynchronous protocols. The PC community will have to learn how they fit into the large SNA environments and gain insight into PC connectivity to LANs, communications controllers, and cluster controllers. They will also have to learn about APPC and LU 6.2 in the very near future.

What Is the Prerequisite

If you have been in data processing for a couple of years, you can understand this book. If you are only a COBOL programmer, that's fine too. If you know CICS, so much the better. If you are a systems programmer, this book will be a breeze.

The authors do not expect you to know anything about data communications; it is discussed in Part 1. The book has been structured in such a way that it may be used as a self-teaching guide. It may be used as a textbook for a 3- to 5-day in-house or public seminar on VTAM/NCP.

What Environment This Book Is For

This book covers the MVS, MVS/XA, DOS/VSE, and VM operating systems environments. It is applicable to large mainframes (e.g., IBM 3090, 3084, 4381) as well as not-so-large mainframes (e.g., IBM 9370). It includes both CICS and IMS/DC installations.

A Word on the Style Used

The authors have made extensive use of diagrams, figures, examples, and illustrations. The style has purposely been kept simple. We hope to alleviate readers' fears about the intricacies of networks. As you move from one chapter to the next, you will begin to appreciate the structure and underlying simplicity behind small and large networks.

We do not go into too much theory. This is a practical book. As a matter of fact, this is the first and only practical guide on SNA. If you are more interested in the theory, please consult the bibliography for additional sources.

Why This Book Is Complete

You name it and it's in there. You will learn about hardware such as 3174s and 3745s; we include various protocols such as SDLC, BSC, half-duplex, and full-duplex. Telecommunications concepts and various physical transmission media are discussed, and a discussion of the various layers of SNA has been incorporated. It's only after we give you the necessary background information that we begin to talk about VTAM and NCP. In addition, this book is up-to-date as of the most recent releases of VTAM and NCP. The authors intend to revise the book if a significant IBM announcement makes it necessary to add, delete, or modify parts of the text.

What Is Included

Part 1 begins with the concepts of data and telecommunications. It moves on to details about the communications hardware, software, and protocols. We also talk about SNA and its various layers. This part lays the foundation for understanding networks and their components.

Part 2 concentrates on SNA domains and networks. We also discuss defining a network topography.

After finishing Part 2, you will be ready to move onto defining the network to VTAM and NCP. This takes place in Part 3. Various VTAM start options, major nodes, path tables, user tables, and the macros needed to define them are discussed. A separate discussion of channel attached VM and VSE configurations is also included.

Part 4 illustrates the use of various NCP macros such as PCCU, BUILD, HOST, PATH, GROUP, LINE, CLUSTER, TERMINAL, etc.

What Is the Next Step

This text concentrates on single-domain networks only. Single-domain networks typically involve the use of a single host. If you work in such an environment, that's all you will need to know. However, most of the computer installations that need to communicate over a network have two or more mainframes. A sequel to this text, entitled *Advanced SNA Networking: A Professional's Guide to*

VTAM/NCP, is in the final stages of completion. It will include IBM 3174, IBM 3745, multi-domain networks, SNI, VTAM initialization, Low Entry Networking, IBM 9370 connectivity, network operations, NetView, NetView/PC, and performance management. We hope you find it a suitable sequel to the current text.

After reading this book, you will not only know how to use VTAM/NCP macros, but also how various networking hardware, software, and protocols fit into the whole picture. This will give you a global view of the networking components. The authors expect that after finishing this book, you will be able to design and code for single-domain networks with a high degree of confidence and competence.

Preface to the Second Edition

The success of the first edition has proven the viability of publishing a text that provides networking architecture and implementation in a concise and useful format. The first edition of this text provided up-to-date information on IBM's strategic networking architecture, Systems Network Architecture (SNA). SNA is implemented on IBM mainframes and front end processors by two program offerings provided by IBM. These are Virtual Telecommunications Access Method (VTAM) and Network Control Program (NCP).

SNA has gone through a metamorphosis since the publication of the first edition. Traditional SNA covered in the first edition implements a hierarchical architecture. The "new" SNA is moving away from this master–slave relationship to a peer-to-peer relationship. This new architecture is called Advanced Peer-to-Peer Networking (APPN). APPN is now implemented on every major platform used in enterprise computing.

This second edition concentrates on the new versions of VTAM and NCP that implement not only traditional SNA but also are required for implementing APPN. These new versions, VTAM V4.1 and NCP V6.2, are covered in detail throughout this edition as well as the new controller offerings by IBM that can participate in APPN.

This second edition covers new internetworking functionality of IBM 3174 Establishment Controllers and IBM 3745 Communications Controllers. These controllers have not only been enhanced to communicate using APPN but also non-IBM architecture such as Transmission Control Protocol/Internet Protocol (TCP/IP).

Finally this second edition covers connectivity enhancements to mainframe applications using token-ring and Ethernet local area network communications.

The above cited enhancements warrant a second edition to this best-selling text.

Jay Ranade

Introduction and Concepts

Data Communications and Telecommunications

Communications is the ability to exchange information between two or more entities. When such entities are human beings and the distances involved are insignificant, communications is not a problem. Human beings have evolved over millennia to develop different language patterns and speech formats so that their brains can comprehend and analyze such information. When the distances involved were significant, the necessity to communicate resulted in the invention of other methods of information exchange. Throughout history, there have been examples of information transfer through the use of smoke signals, drum beats, mirror reflections, etc.

As the world advanced into the industrial era, the invention of wireless exchange of information was a giant step toward expediting the information transfer. The invention of telephones further made it possible to communicate quickly and conveniently over long distances. Recently, there has been widespread use of other telecommunications media such as microwave, satellite, and fiber optics. One of the most astonishing examples of communications is the transmission of images of the planet Uranus by Voyager II from billions of miles away. After the shortest possible sample of communications history, let's get to the relevant topic—the computer communications.

1.1 Typical IBM Communications Environments

To understand computer communications in an IBM environment, we will make extensive use of examples and sample environments. The hardware and software we will be mentioning will be discussed in

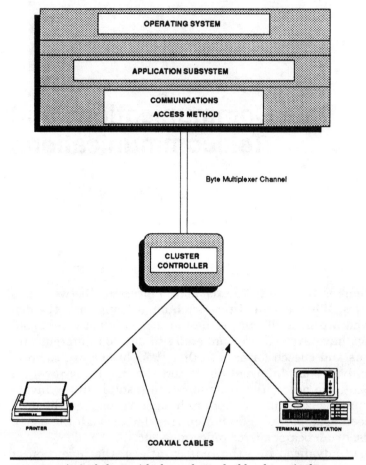

Figure 1.1   A single host with channel-attached local terminals.

subsequent chapters; this section will give you a global view of what it's all about.

1.1.1   Single-host, local terminals environment

Refer to Figure 1.1 for the simplest possible communications environment. It consists of a single mainframe computer, which is also called a "host processor." All the users who need to access the host applications are in the same building in which the host processor is installed. What we need in this case is a cluster controller which can be attached to the byte multiplexer channel of the host. A cluster controller is a device which controls a cluster of terminals. Terminals cannot be directly attached to a channel; they have to be hooked to this device, which is further attached to the channel. The byte multi-plexer channel of a host processor is a communications link between the host and the peripheral devices such as cluster controllers, communications controllers, operating system printers, etc. Another channel type, the block multiplexer channel, is generally used to connect to Direct Access Storage Devices (DASDs), tape drives, etc.

A cluster controller can control terminal (e.g., IBM 3278) and printer devices (e.g., IBM 3268 printer). Depending upon the model of the cluster controller, you can attach anywhere from 8 to 32 printers or terminals. The link between the cluster controller and the terminal or printer device is usually a coaxial cable with proper connectors on either side which provide the physical interface. Be aware that this environment is possible only if the host, the cluster controller, and the terminals or printers are in the same building. A channel-attached cluster controller cannot be more than 100 to 200 feet from the channel itself. The terminals or printers cannot be more than 500 to 1000 feet from the cluster controller.

To give you a taste of the real environment, let's look at the examples of software and hardware involved in this configuration:

- A host processor is an IBM 3090, IBM 4381, or IBM 9370.

- A channel-attached cluster controller is an IBM 3274 Model 41D or IBM 3274 Model 41A.

- A display terminal is an IBM 3278, an IBM 3290, or an IBM 3270 PC.

- A printer device is an IBM 3268 or an IBM 3287.

- The operating system is MVS/XA, MVS/ESA, or DOS/VSE.

- An application subsystem is CICS/VS, IMS/DC, or TSO.

- The communications access method is ACF/VTAM or ACF/TCAM.

If you are overwhelmed by the introduction of so many acronyms, it will console you to know that you are not expected to know or remember them. Just try to get a global picture of different environments. Then, when we start discussing the individual components, you will know where they fit in the whole puzzle.

1.1.2 Single-host local and remote terminals

Figure 1.2 shows an environment in which you have a single mainframe, but not all the attached terminals are in the same building. Suppose that the business requirements dictate that a regional office in Chicago must access the host applications in New York. While you still need local clus-

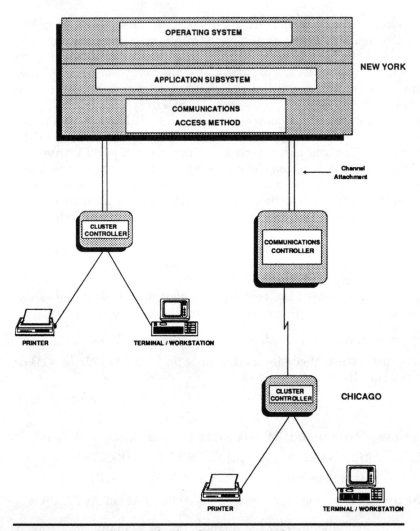

Figure 1.2   A single host with local and remote terminals.

ter controllers to support terminals and printers in the same building, you cannot attach terminals in Chicago to the same cluster controller. What you need is a "remote cluster controller" in Chicago to which all the printer and terminal devices in Chicago will be attached.

A remote cluster controller is different from a local cluster controller. A local cluster controller attaches to a computer channel, but a remote one has a link only to the communications lines. In our case, the communications lines will be the telephone link between the Chicago and New York offices. A remote cluster controller uses the same physical connectivity media, such as coaxial cables, to attach to the terminal and printer devices. Therefore, as far as the end-user devices are concerned, their connectivity is the same for remote and local cluster controllers.

Furthermore, a remote cluster controller cannot be directly attached to the computer channels over the communications lines. You need a device in the middle called a "communications controller." In our sample environment, while the cluster controller is attached to the communications controller over a telephone link, the communications controller itself is hooked to the mainframe through a channel. A communications controller can control a number of cluster controllers located at multiple geographical locations.

The following are examples of the new hardware and software introduced in this section:

- A remote cluster controller is an IBM 3274 Model 41C.

- A communications controller is an IBM 3745, IBM 3720, IBM 3725, or IBM 3705.

- The software running in a communications controller is the Network Control Program, or NCP.

While the communications software running in the mainframe and the communications controller is named ACF/VTAM and ACF/NCP, respectively, the software running in the cluster controller does not have a specific name. For our purposes, we will call the cluster controller software the "Cluster Controller Load" program.

1.1.3   Single-host local and remote communications controllers

In the previous section we talked about local and remote cluster controllers. In this section we introduce local and remote communications controllers. In our hypothetical company, the Chicago office has grown so rapidly that now it has a staff of 1500 people and an inventory of 600 terminals. All the terminals are attached to 20 different remote cluster controllers located in Chicago. The 20 cluster controllers have 20 different communications links to the channel-attached communications controller in New York. It is determined that it will be more cost-effective to install a communications controller locally in the Chicago office and attach the cluster controllers to it through modem eliminators. The connectivity would be through high-bandwidth communications lines between New York and Chicago. Such communications lines can be 56-kilobits-per-second (kbps) leased lines, which are usually referred to

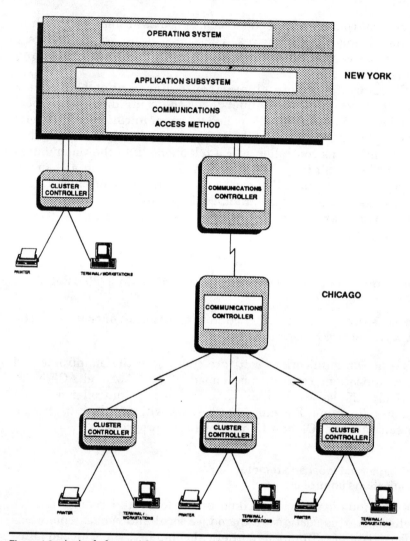

Figure 1.3   A single host with channel- and link-attached communications controllers.

as Digital Dataphone Service (DDS) lines. Figure 1.3 shows such a configuration.

Notice that the communications controller located in Chicago is *not* channel-attached to the New York host. It is link-attached (through communications links) to the channel-attached communications controller located in New York. The link-attached communications controller has almost the same characteristics as the channel-attached one. It is used for concentrating data from multiple cluster controllers at a remote location. Examples of such communications controllers are IBM 3745, IBM 3720, IBM 3725, and IBM 3705. The software running in a link-attached communications controller is also NCP.

1.1.4   Local host connectivity to a remote host

In our next scenario, there is a business requirement to open a new zonal office in Los Angeles. The processing requirements are such that a host is required to run new applications in Los Angeles. At the same time, the Los Angeles host would like to communicate with the New York office to exchange and access its data. The configuration would look as shown in Figure 1.4.

Notice that the Los Angeles host requires a channel-attached communications controller to have connectivity to the New York host through its own channel-attached communications controller.

Although we have not shown it in the figure, there will be channel-attached cluster controllers and communications controller–attached cluster controllers at various locations.

The communications access methods running in New York and Los Angeles hosts can be ACF/VTAM, and their applications subsystems can be CICS/VS.

1.1.5   Token-ring LAN connectivity to a remote host

IBM introduced token-ring networking in the middle 1980s as a local area network medium. Token-ring networking is based on IEEE 802.5 specifications. These specifications provide the framework for networking with a token. A token frame is used to determine if a LAN adapter can send data. A LAN adapter is nothing more than an interface card that attaches the LAN device to the wires that make up the LAN. If the adapter receives a free token frame and has data to send, it will take the token frame off the LAN and send a data frame. When the adapter has completed its transmission of data, it sends out a new free token frame onto the LAN for another adapter to use. In this way, token-ring networking ensures that only one adapter can be sending data at a time on the LAN. Token-ring has proven to be a stable LAN architecture and has been adopted by many corporations. For more information on token-ring LANs, see the bibliography for *IBM's Token-Ring Networking Handbook*.

Several different types of devices can be attached to a token-ring LAN. For example, IBM 3725 and 3745 Communications Controllers can be attached to provide LAN connectivity to the SNA communications network for token-ring–attached devices that require VTAM

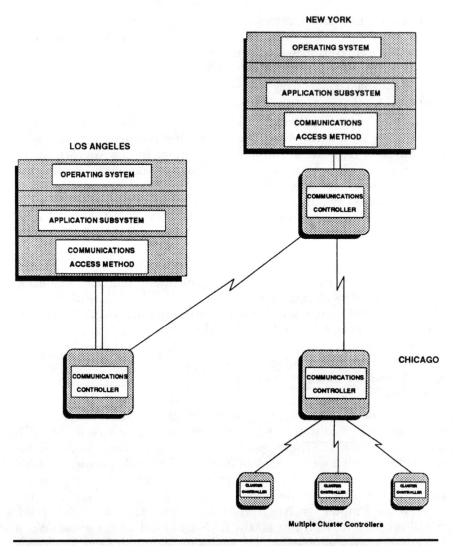

Figure 1.4 Two remote hosts having connectivity through communications controllers.

application access to the mainframe, as shown in Figure 1.5. The mainframe itself can be indirectly attached to the token-ring LAN through the IBM 3172 Interconnect Controller and by using an IBM 3174 Establishment Controller. All these configurations will allow end-user LAN access to VTAM applications residing on the mainframe computer.

End-user workstations (i.e., personal computers or midrange devices) can also be attached to token-ring LANs. In fact, this type of configuration was the original intent for token-ring LANs. However,

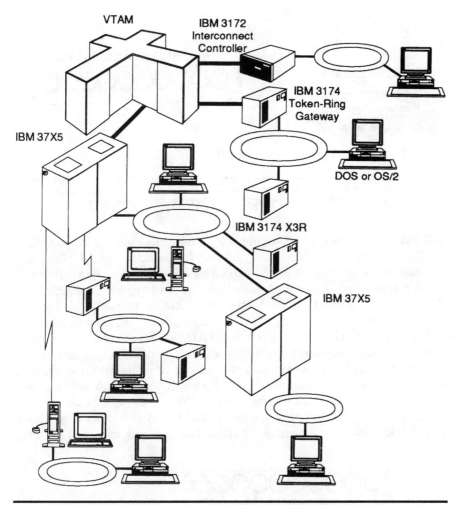

Figure 1.5   SNA/VTAM token-ring connectivity options using IBM 37x5, 3172, and 3174.

vital corporate data on the mainframe was needed by these end-user stations, which fostered the development of SNA software for these devices. This software provides connectivity to VTAM and its applications residing on the mainframe.

The basic token-ring LAN configuration is composed of a token-ring LAN adapter card, a wiring scheme, multistation access units, and bridges, as shown in Figure 1.6. The token-ring LAN adapter cards provide the interface from the end-user station to the token-ring medium. The adapters can clock the rate of data for token-ring at either 4 megabits per second (Mbps) or 16 Mbps. The clocking rate

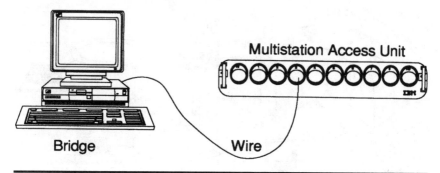

Figure 1.6 Basic token-ring LAN configuration is composed of a token-ring LAN adapter card, a wiring scheme, multistation access units, and bridges.

will determine the type of wiring (i.e., media) needed. As an example, 4 Mbps can run on all media types for token-ring, but 16 Mbps can operate properly only on either type 5 or type 9 media. The difference is that type 9 wiring is a shielded wiring and type 5 is unshielded. Type 3 wiring is used for a 4-Mbps rate, since it is unshielded twisted pair (i.e., telephone wiring), which is inexpensive and easy to install.

The Multistation Access Unit (MAU) is a passive device that derives its electric current from the electric current presented on the wire by the LAN adapter. Since it does not have its own power, it is often referred to as a nonintelligent device, a "dumb" device, or a passive device. However, as can be seen in Figure 1.7, the MAU is what actually makes up the ring on which the token is transmitted. Note that a token-ring configuration is not actually a ring or circle; in reali-

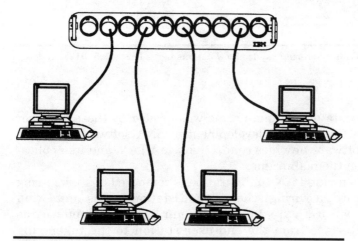

Figure 1.7 Multistation Access Unit (MAU) makes up the ring on which the token is transmitted.

ty, it is a star-wired configuration. The protocol is called token-ring because the token, if not removed from the ring, will return to the original sending station and thus has traveled in a logical loop or ring on the LAN. The LAN adapter attaches itself to the ring by sending a "phantom" current of ± 5 V on the wire to the MAU. The phantom current makes the MAU close its internal connection, effectively closing the loop, thus creating a ring.

MAUs have been usurped by the IBM 8230 Controlled Access Unit (CAU) and third-party intelligent hub devices, as shown in Figure 1.8. Unlike the MAU, the CAU is an intelligent device and provides some LAN management by automatically shutting down an interface if the attached device is not acting properly or by setting filters that allow only certain stations on specific ports. The IBM CAUs must be used with Lobe Attachment Modules (LAMs). These devices respond to CAU commands and requests and are effectively manageable MAUs. Several vendors other than IBM have designed and provide products for token-ring that match or exceed the CAUs capabilities.

A bridge is a device that attaches two LAN segments. The bridge contains software that implements token-ring network bridging functions. As seen in Figure 1.9, two LAN segments are bridged together to create a larger token-ring LAN environment. IBM provides local bridges and remote bridge pairs. Local bridges connect two LAN segments within the same building. Remote bridge pairs connect two LAN segments over distances that require communications lines. This type of configuration extends the LAN over a wide area network

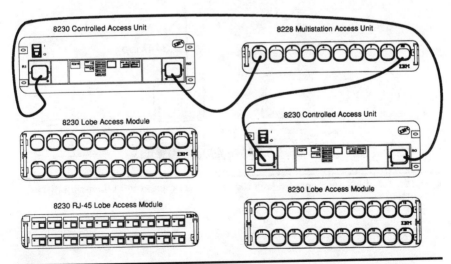

Figure 1.8 IBM 8230 Controlled Access Unit (CAU) components and configuration.

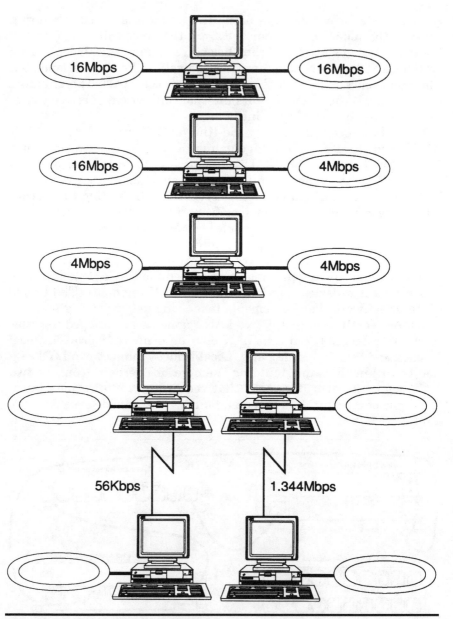

Figure 1.9 Two LAN segments bridged together to create a larger token-ring LAN environment.

(WAN). The ability to span long distances between LAN segments using bridges needs careful attention when implementing token-ring because of a concept called hops.

Token-ring architecture bases session establishment on route broadcasting. A token-ring frame contains a Routing Information Field (RIF). The RIF is composed of a bridge and a segment number. In defining bridges, a unique number is given to a LAN segment and a number is assigned to a bridge. This creates a unique RIF for access to a LAN segment. These bridge and segment numbers are placed in the RIF as a station's route broadcast passes through the bridge from one segment to another. This is called a hop. The length of the RIF field in token-ring is designed so that a maximum of 8 segments or 7 bridges can be used before the broadcast can be exhausted. Careful design and planning must be used to avoid the hop count restriction.

1.1.6 Ethernet LAN connectivity to a remote host

Ethernet was developed by Xerox and became the first de facto LAN standard. It was quickly adopted by research and development institutions as well as educational institutions. Ethernet is based on a bus topology, as shown in Figure 1.10. This is in contrast to token-ring, where the frame flows from one station to the next until the destination has been reached. Ethernet bus topology broadcasts every frame. This is often referred to as Carrier Sense Multiple Access/Collision Detection (CSMA/CD). Each station on an Ethernet receives the frame and reads it. If the frame's destination address matches the LAN adapter address, the data field is copied. If no match is found, then the frame is discarded by the adapter. In a token-ring topology,

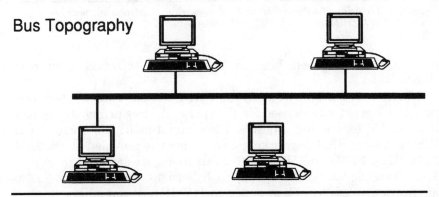

Bus Topography

Figure 1.10   Ethernet is based on a bus topology.

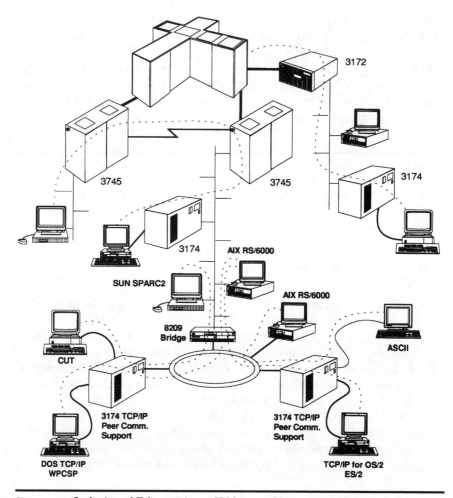

Figure 1.11 Inclusion of Ethernet in an IBM networking environment is provided by the IBM 3172, IBM 3174, and IBM 3745 controllers.

the frame would be sent back out onto the ring to travel to the next station.

Ethernet was primarily used for UNIX workstations at its inception, but has since been used for all types of devices primarily because it is simple to use and requires a low investment. The inclusion of Ethernet in an IBM networking environment is provided by the IBM 3172, IBM 3174, and IBM 3745 controllers, as depicted in Figure 1.11. These devices can now support Ethernet connectivity for VTAM application access as well as access to TCP/IP host computers.

The IBM 3172 provides mainframe access for Ethernet stations, allowing them to issue TCP/IP remote procedure calls to TCP/IP executing on the mainframe computer. The IBM 3174 provides Ethernet connectivity, which allows Ethernet stations to access both TCP/IP hosts on token-ring and the mainframe and VTAM applications on the mainframe. The IBM 3745 Ethernet capability provides for TCP/IP connections over the SNA WAN through other IBM 3745s to Ethernet-attached devices. These capabilities provide a means for IBM to expand its network connectivity offerings while preserving customer investment in networking and information technologies.

1.1.7 Remote printing

In our final scenario, we have a requirement to print enormous amounts of reports in our Houston office. In the past the requirements were met by printing the reports in New York and then mailing them to Houston. But the urgency of these reports requires that they be printed on a real-time basis. Figure 1.12 shows a configuration in which we have installed a Remote Job Entry (RJE) device in Houston. Such a device can be attached to a printer, a card reader, a

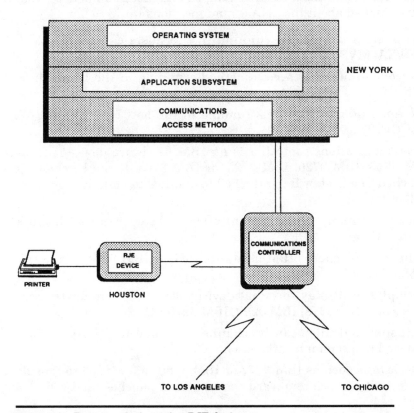

Figure 1.12   Remote printing using RJE devices.

card punch, and a console. Examples are IBM 2770, IBM 2780, IBM 3777, and IBM 3780 devices. An RJE device communicates with the host through a communications link to the communications controller.

A printer attached to an RJE device is usually a high-speed printer such as an IBM 3203 or IBM 3211. An RJE device is controlled by a host subsystem such as JES2 or JES3. Don't confuse high-speed RJE printers with low-speed VTAM printers, which are attached to a cluster controller. VTAM printers are used by application subsystems such as CICS/VS or IMS/DC, not JES2 or JES3.

1.1.8   Components of IBM communications world

In the previous examples, we have seen the global picture of different IBM communication environments. We will cover each component in detail in the following chapters. When you learn about individual hardware and software components, keep in mind the global view so that you can relate them to other components. Let's summarize what we have learned about the SNA communications world so far:

- The operating systems running in the hosts can be MVS/SP, MVS/XA, MVS/ESA, VM, and DOS/VSE.

- The communications access methods running in the host can be ACF/VTAM or ACF/TCAM.

- The application subsystems running in the host can be CICS/VS, IMS/DC, TSO, etc.

- A communications controller is an IBM hardware device such as IBM 3745, IBM 3720, IBM 3725, or IBM 3705. It can be channel-attached to a host or link-attached to another communications controller.

- Software running in a communications controller is the Network Control Program (NCP).

- A cluster controller is IBM hardware such as IBM 3274, IBM 3276, IBM 3174, etc.

- Examples of display terminal devices attached to a cluster controller are IBM 3278, IBM 3290, IBM 3270 PC, etc.

- A cluster controller can be channel-attached to a host or link-attached to a communications controller.

- RJE devices such as IBM 3777 or IBM 3780 can be used to provide connectivity to printers, card readers, card punches, and consoles at a remote location. Without RJE, such devices are channel-attached to a host.

Some of the commonly used names for the devices we have introduced so far are as follows. A host is also called a "mainframe" or a "Central Processing Unit" (CPU). A communications controller is also called a "box" or a "Front-End Processor" (FEP) or a "front end." You will frequently hear communications programmers saying that they have loaded the box. What they mean is that they have loaded NCP into the communications controller. A cluster controller is also called a "terminal controller."

1.2 Telecommunications Facilities and Concepts

In the previous section, we talked about connecting different hardware over communications lines. Such connectivity issues could involve the following:

- Connecting a communications controller to another communications controller.

- Connecting a cluster controller to a channel-attached or link-attached communications controller.

- Connecting a terminal device directly to the communications controller.

1.2.1 Common carriers versus value-added carriers

If the distance between the devices is less than a few hundred feet and they are in the same building, they can be connected directly using proper cables and physical interfaces. Such connectivity may involve the use of devices such as modem eliminators. More often than not, such devices are *not* located in the same building. They may be located in two or more different buildings in a metropolitan area. They can also be dispersed over long distances ranging from a few miles to several thousand miles. Under those circumstances, you have to use the services of a common carrier such as a telephone company. For connectivity requiring the use of equipment which falls in the domain of different regional telephone companies, you require the services of a long-distance carrier such as AT&T, MCI, Sprint, etc.

Common carriers. Common carriers build the actual physical networks; e.g., they lay telephone cables. They are regulated by government regulatory authorities such as the Federal Communications Commission (FCC), and they are licensed to build their physical networks on public property. Common carriers file their tariffs with the regulatory agencies, and upon approval these tariffs become legally binding upon the telephone company and the recipient of the service.

Value-added carriers.   There is another kind of service provided by Value-Added Carriers (VACs). VACs, unlike common carriers, are not regulated by the government. VACs lease physical networks from the common carriers and add some additional services to them. Such additional services are provided through the network of computers installed by the VACs connecting their leased physical networks. Examples of VACs are Telenet, Tymnet, and Datapac (in Canada). Services provided by VACs can be protocol conversion, code translation, and packet switching. Such networks are also called "Value-Added Networks" (VANs).

In the past, the physical network used by common carriers was based on copper wires and cables, and most of the switching equipment used by them was electromechanical in nature. Since the advent of computer technology, most of the switching is now done by computer-controlled electronic switches rather than electromechanical ones. Also, because of advancements in telecommunications technology, a lot of other physical media are used for transmission purposes. Such physical media could be microwave transmission, satellite communications, or the use of fiber optics.

1.2.2 Analog versus digital transmission

Before we get to know more about the physical media for data transmissions, we must understand two modes of transmission, analog and digital.

Analog transmission.   The analog mode of transmission is the older and still the most widely used mode of information transmission. Since the human voice is analog in nature, the telephone networks were designed to meet the need of human communications. Analog transmission is waveform in nature (see Figure 1.13).

Although the human voice can have a frequency anywhere from 200 to 7000 cycles per second (also called "hertz," Hz), the telephone transmission filters permit transmissions only between 300 and 3300 Hz. Analog signals can be used over copper wires, cables, and microwave and satellite transmission. Since analog signals get weak over distances, they are boosted at regular intervals using amplifiers. The boosters amplify not only the real analog signal but also the noise associated with it.

(a) ANALOG TRANSMISSION

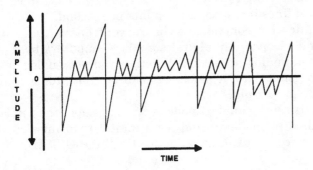

(b) DIGITAL TRANSMISSION

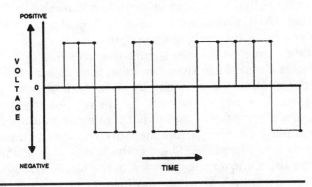

Figure 1.13   (a) Analog transmission; (b) digital transmission pulses.

Digital transmission.   While analog transmission consists of a continuous stream of different frequencies, digital transmission consists of separate pulses (see Figure 1.13b). Digital transmission is possible over copper wires and cables, microwave transmission, satellite transmission, and optical fiber facilities. Recall that analog transmission is not possible over fiber optic transmission media. Digital transmission pulses are also boosted at regular intervals using regenerative repeaters, which do not boost any extraneous signal associated with the digital pulses. Therefore, digital transmission is better in transmission quality than analog transmission and also has a lower transmission error rate. Examples of digital transmission services provided by common carriers are Dataphone Digital Service at 56 Kbps and T1 lines at 1.544 Mbps. While the capacity of an analog line is measured in hertz, for digital lines it is measured in bits per second.

Hosts, communication controllers, cluster controllers, and terminals generate and receive only digital signals. Computers by design are digital (and not analog) in nature. Whether you use analog or digital circuits to transmit data, it has to be ensured that proper conversion is done at the proper place.

When you are using the analog mode of transmission, the interface equipment is called a "modem," and for a digital transmission it is called a "Data Service Unit/Channel Service Unit" (DSU/CSU).

1.2.3   Switched versus leased lines

When you dial a phone number and carry out a conversation, the connection between your phone and the destination phone is established for the duration of that conversation. The common carrier company charges depend on the distance, the time, and the duration of the call. Such a connection is called a dial-up or switched connection. In a switched connection, the circuit used for establishing the call varies with each call. If you hang up and dial again, you may get a different circuit whose quality may be different from the previous one. You may also experience a circuit-busy signal and dial tone delays.

In a leased line, the circuit is assigned by the telephone company on a permanent basis. You may or may not use the line for any useful purpose, but you are paying for it, and it is there when you need it. The telephone company charges a flat monthly fee that primarily depends upon the distance involved. The circuits involved in the path are known and stay the same for the duration of the lease. The circuit quality is verified at the outset and does not have to be verified later because the circuit path does not change.

Switched and leased line requirements depend upon the application needs. If you need to transmit data for only a few minutes or hours, a switched line may be more economical.

1.2.4   Point-to-point versus point-to-multipoint

On a point-to-point connection, a communications link is dedicated for communication between two end objects. In our case such end objects can be a communications controller and a cluster controller. Figure 1.14a shows a point-to-point connection. In a point-to-multipoint link, multiple devices can be attached on a single link. For example, you can have up to 254 cluster controllers on a single link

(a) POINT-TO-POINT

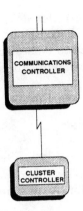

(b) POINT-TO-MULTIPOINT

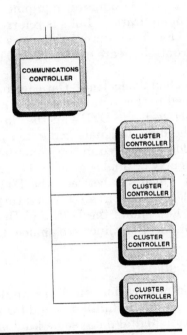

Figure 1.14 (a) A point-to-point communications link; (b) a point-to-multipoint communications link.

to a communications controller port. Figure 1.14b shows a configuration in which four cluster controllers are connected to a single link coming from a communications controller. In telecommunications terminology, point-to-multipoint is also called a "multidrop link."

In a point-to-multipoint configuration, the intelligence and control to manage the flow of information lie with the device which is at the head of the multidrop line. In our case, the communications controller controls the multiple cluster controller devices. It sends requests to each cluster controller, asking it if it has any data to send. This process is also called "polling." Each cluster controller has to wait for its turn to be polled before it can send any data to the communications controller.

1.2.5 Data terminal and data circuit-terminating equipment

Terminologies of data terminal and data circuit-terminating equipment are used quite extensively by the network engineers, so it is important that VTAM/NCP systems programmers understand them.

Data terminal equipment. Computer equipment which needs to communicate over communications links is referred to as "Data Terminal Equipment" (DTE). The communications controllers and link-attached cluster controllers are examples of DTE.

Data circuit-terminating equipment. Data terminal equipment cannot be directly attached to the communications lines. You need hardware that interfaces between the DTE and the link lines. This hardware also translates the digital signals generated by DTE and the transmission methodology employed in the communications link. The communications links may be analog or digital in nature.

The equipment which sits between the DTE and the communications links is called "Data Circuit-Terminating Equipment," or DCE for short. Figure 1.15 shows the DTE and DCE in an IBM environment. If the communications lines are analog, the DCE is a modem.

1.2.6 Modem and DSU/CSU

Modem. Most communication lines are analog in nature because these lines were required to fulfill demand for voice communications. Since DTEs generate digital data, some kind of equipment is needed which can act as a translator of signals between the digital DTEs and analog communications lines. A device that can accomplish this is called a "modem" (modem stands for *modulator/demodulator*).

A modem receives a binary bit stream of 0s and 1s from the sending DTE and modulates them to analog signals for transmission over the voice-grade lines. On the receiving end, a second modem receives the analog signals from the communications lines and demodulates them to binary data to be passed on to the receiving DTE. Modems always work in pairs because the sending and the receiving DTE both require them. In network engineering terminology, a modem is a DCE for analog lines.

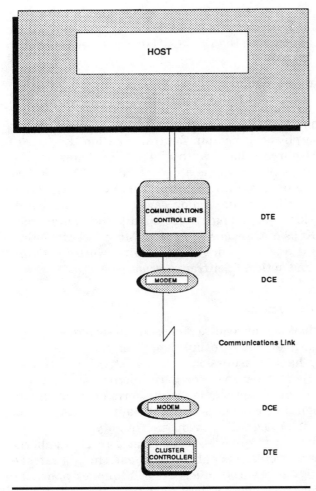

Figure 1.15 Diagram illustrating connectivity between
DTE and DCE in a network.

DSC/CSU. Quite recently, there has been a proliferation of digital circuits. Digital circuits are used not only for data communications but
also for voice. For voice communications, the voice signals are translated into digital data, transmitted over digital lines, and reconstitut-
ed on the other end into audio-frequency signals. Some of the available digital transmission services are DDS (56 Kbps) and T1 lines
(1.544 Mbps). Most probably the prominent digital service in the
future will be Integrated Services Digital Network (ISDN).

The DCE that provides an interface between the DTE and digital
communications links is called a "Data Service Unit/Channel Service
Unit" or DSU/CSU. It receives the binary data from DTE and trans-

lates it into digital pulse signals which can be transmitted over the communications links. It also provides the clocking function, which must be in synchronization with the DSU/CSU on the receiving end.

1.3   Telecommunications Media

There are different media used for telecommunications. When the voice-based telephone networks were developed in the past, copper wire was used as the physical medium for transmission. Each telephone conversation required a bandwidth of 4000 Hz. Later on, the same copper wire medium was used as a digital carrier. The method used for sending voice over digital facilities is called "Pulse Code Modulation" (PCM). Although the analog carriers require a bandwidth of 4 kilohertz (kHz) for a single voice circuit, the digital voice channel requires 64 Kbps of transmission speed. More recently, other modes of communications were developed which were based on microwave, satellite, and optical fiber transmission.

1.3.1   Microwave radio transmission

Microwave transmissions do not require any solid physical media (e.g., copper wire or optical fiber) for transmission. Therefore, this is the most widely used long-haul transmission facility in the United States. Microwave facilities require parabolic or horn microwave antennas. One of the most important requirements for this form of transmission is the need for a straight line of sight between the antennas. Any form of obstruction can cut off or degrade transmission quality.

Microwave transmission frequencies vary from 2 to 25 gigahertz (GHz). At lower frequencies, repeaters to boost the signal are required approximately every 100 miles. At a frequency above 18 GHz, you may need repeaters every mile.

At microwave frequencies, the characteristics of radio waves are close to those of light waves. Therefore heavy rain or heavy fog conditions can increase the transmission error rate and thus also increase retransmissions between DTEs.

Microwave transmission is used quite extensively in metropolitan areas. Rooftops of office buildings in every large city are crowded with microwave antennas. When data transmission between two closely spaced office buildings of a company is required, microwave transmission provides an economical alternative to common carrier services. Because of allocation of almost all frequency ranges, it is almost impossible to get a frequency assignment in large metropolitan areas.

A single radio channel, which consists of 30 megahertz (MHz) of bandwidth, can provide 6000 voice channels of 5000 Hz each.

1.3.2   T1 and fractional T1 transmission

Digital private line service is one of the fastest growing telecommunications offerings—the T1 service along with Fractional T1 (FT1). Providers of these two services are reducing the cost to the end user. These reduced costs are alluring to corporate network designers whose job is to provide the best service in a cost-efficient manner. T1 and FT1 digital services provide this characteristic to network designs.

T1 is defined as any digital transmission over any medium that supports 1.544-Mbps transmission. T1 services provide up to twenty-four 64-Kbps lines. Each line is referred to as a channel. The T1 services support time-division multiplexed digital transmission up to 1.544 Mbps. The product of the 24 channels times 64 Kbps is 1.536 Mbps. The remaining 8 Kbps supports the framing bits that provide for frame interpretation. T1 circuits are attractive to corporations with medium to large networks. The ability to multiplex the T1 has network design advantages. Figure 1.16 shows two types of applications. In Figure 1.16a, 12 of the T1 channels enter a multiplexer at

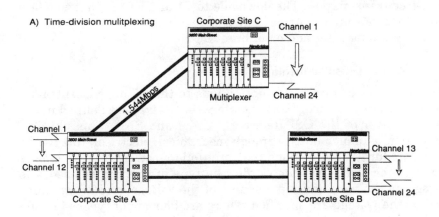

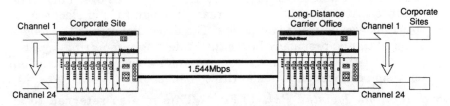

Figure 1.16   Two types of T1 application configurations.

remote site A and the second 12 continue on to remote site B. A second configuration is shown in Figure 1.16b. In this configuration, the T1 from corporate offices connects to a long-distance carrier office. The carrier then demultiplexes the T1 to its 24 individual channels and sends them over individual leased local lines to remote offices. This configuration is known as fanout.

Fractional T1 is a service offering provided by several of the long-distance carriers that employs a methodology in which the full T1 is split into smaller segments for network traffic volumes less than T1 but greater than 56 Kbps. Again, this allows for data transmission rates that are multiples of 64 Kbps, the rate of a T1 channel. Typical FT1 usage rates are 128 Kbps, 256 Kbps, 384 Kbps, and 768 Kbps. The cost difference between 768 Kbps and a full T1 usually favors the purchase of the full T1.

The advantages of T1 and FT1 services are their ability to handle large volumes of traffic for voice, data, and image. The T1 and FT1 circuit is defined for point-to-point full-duplex transmission of serial, bipolar, return-to-zero, isochronous digital traffic with a 95 percent error-free transmission. The downside to T1 and FT1 is that they are prone to potential problems that can result in network downtime, including timing jitter and error bursts.

1.3.3 LAN transmission media

The media used to connect workstations to the wiring concentrators and to connect the wiring concentrators themselves are defined in the physical layer of the OSI Reference Model and SNA. This physical connection is implemented through mechanical, electrical, functional, and procedural specifications that are understood by both endpoints of the connection. In other words, the type of connectors used, the voltage on the wire and the meaning of the voltage, and events initiated by the voltage are specified. There are four major types of media used in LAN networking: coaxial cable, telephone twisted pair, shielded twisted pair, and fiber optic.

- *Coaxial cable.* This type of cable medium has a low attenuation characteristic and drives signals at high data rates over fairly long distances. Attenuation is the decrease in magnitude of electrical or optical power between two points. The drive distance is the length a signal can be propagated over the cable before requiring a regeneration of the signal, using, for example, a repeater. Your cable television connection is made with coaxial cable.

- *Telephone Twisted Pair* (TTP). This is also referred to as "Unshielded Twisted Pair" (UTP). Unshielded means that the wires do not have a metal shield around them. A metal shield reduces interference from unwanted radio frequencies, which have a negative effect on the cable. Therefore, UTP is very susceptible to unwanted radio frequencies and emits radio frequencies itself when used for high-speed data transmission. Filters can be put in place to reduce or eliminate the unwanted radio frequencies, but the filters reduce the signal strength. UTP also suffers from cross talk, a signal generated from one pair affecting the signal on the other pair.

- *Data-Grade Media* (DGM), *shielded twisted pair.* As the name implies, this type of cable contains a shield. Within each shield there can be one or more twisted pairs. The pairs are twisted in such a way that they themselves help to alleviate radio-frequency interference within the shielded cable. Shielded twisted-pair cable is suited to handle data rates over 20 Mbps for most distances found in typical office buildings. Because of its low attenuation and high data rates over long distances, DGM can be used for both 4-Mbps and 16-Mbps token-ring segments in the IBM Token-Ring Network.

- *Fiber optic.* The previous cables all use copper wiring. Fiber optic utilizes glass and light pulses. A fiber-optic cable can contain thousands of fibers. Each fiber is thinner than the thickness of a human hair. Because fiber-optic media use light as the transmission signal, they are basically impervious to radio-frequency interference. The rate of transmission in fiber-optic media is literally the speed of light. With minor signal attenuation, fiber optic has data rates that can approach terabits per second—in other words, 1 trillion bits per second! Fiber optic provides greater security than copper wire because tapping into the fiber-optic signal will cause a loss of the signal and will be immediately detected. Copper wire, on the other hand, can allow tapping of the signal without great loss to the delivery of data.

Of the four cable types discussed, the last three are more widely used. UTP, though suffering from high attenuation and radio-frequency interference, is a well-understood medium and is inexpensive; however, it is undesirable for providing a stable token-ring network. Fiber optic possesses the highest throughput but is a relatively new technology and hence is expensive to implement. DGM shielded twisted-pair cable is the most cost-effective solution available today for providing both high-speed data rates and immunity to radio-frequency interference. Figure 1.17 contains a table outlining the pros and cons of each cable type.

Medium Type	Pros	Cons
Unshielded Copper Wire Twisted pair	• Understood technology. • Knowledgeable technicians available. • Fast and simple installation. • Least expensive medium for LAN migration. • Can use existing telephone wiring found in building.	• Affected by electromagnetic interference. • Electrical and magnetic waves can be intercepted. • Cross talk between wires can cause transmission errors. • Exterior placement needs protection from lightning and corrosion. • Low data rate.
Shielded Copper Wire Twisted pair Coaxial Twinaxial Broadband (CATV)	• Understood technology. • Knowledgeable technicians available. • Fast and simple installation (except for CATV). • Minimal emanation of electromagnetic signals. • Shield provides some protection from interference, lightning, cross talk, corrosion.	• High-grade coaxial, twinaxial, and CATV cables are fairly expensive to use for migrating to LAN. • CATV cable is thick and rigid requiring special tools to install around turns.
Fiber Optic Fiber Optic Cable	• Useful for high-speed applications. • No electromagnetic signal emanation. • Not affected by cross talk, electromagnetic interference, lightning, corrosion. • Less expensive medium than coaxial or CATV cable. • Signal transmitted over greater distances than copper wire without boosting the signal.	• Knowledgeable technicians are scarce. • Device connection more expensive than copper wire. • Bi-directional communication requires two fiber-optic lines. • High cost for installation. • Cannot be split, use for point-to-point topologies.

Figure 1.17 Pros and cons of each cable type.

1.3.4 LAN cable types

There are seven classes of cables and eight cable types currently available for use with the IBM Cabling System. Each type is used for specific installation requirements. These requirements involve location and environment. The installation requirements of cables are dictated by the local or state electrical codes in your area. Be sure to follow these codes, as they have been implemented for your safety.

Cable	Conductor	Pairs	Shielding	Outdoor	Use
Type 1	Solid Copper	Two AWG 22 Twisted	Braided or Corrugated	YES	Main paths work area to wiring closet
Type 2	Solid Copper	Two AWG 22 Twisted Four TTP (AWG 22)	Braided No	No	Work area to wiring closet
Type 3	Solid Copper	Four TTP (AWG 22/24)	No	No	Work area to wiring closet
Type 5	Fiber Optics	Two Fiber (100/140)	No	Yes	Main paths (inter bldg.)
Type 5R	Fiber Optics	Two Fiber (50/125)	No	Yes	Main paths (up to 500 m)
Type 6	Stranded Copper	Two AWG 26 Twisted	Braided	No	Jumpers in wiring closet
Type 8	Solid Copper	Two AWG 26 Parallel	Braided (each pair)	No	Work area under carpet
Type 9	Solid Copper	Two AWG 26	Braided	No	Main paths (plenum)

Figure 1.18 Characteristics of each cable class.

Figure 1.18 contains a table outlining the characteristics of each cable class.

Type 1: Twisted pair. This cable has a braided shield around two twisted pairs of #22 American Wire Gauge (AWG) solid wire for data communication. This cable type can support up to 16-Mbps data transfer rates and is used strictly for data. As with all cable types listed here, this cable can be installed in conduits. Type 1 can also come in two flavors: plenum and riser. Plenum cable has an outer fire-resistant covering, such as Teflon, to reduce the fire hazard and emitted PVC fumes should a fire start. The resistance to fire and low hazardous vapor emissions gives this cable adequate protection to allow installation in air ducts (plenums) and other spaces within a building used for air passages without the use of conduits.

Riser cable also has fire-resistant characteristics without the use of conduit. However, this cable is more suitable for use in a building riser. For example, building risers include elevator shafts and mail-drop chutes. Type 1 cable can also be used outdoors. Used in this way, the cable has a corrugated metallic cable shield around two twisted pair. Each wire of the twisted pair is made up of #22 AWG wire. This cable type can be used in an aerial installation or placed in conduit underground.

Type 2: Telephone twisted pair. This is the same as type 1 cable but with an additional four twisted pairs of #22 AWG solid copper telephone wires. The cable contains six twisted pairs, two shielded and four unshielded. This cable type has the added advantage of supporting two token-ring connections and one voice connection concurrently. The shielded pairs can carry data at speeds up to 16 Mbps, and the unshielded pairs (voice pairs) can carry data at 2, 4, and 10 Mbps.

Type 3: Telephone twisted pair. This is the recommended cable type for implementing LAN networks over unshielded telephone twisted pair. This cable type has three or more twisted pairs of #22 or #24 AWG solid copper wire and supports either voice or data. The solid copper wire must have a minimum of two twisted pairs per foot. The IBM Cabling System recommends that a maximum of 72 devices can be attached to a token-ring using type 3 cable.

Type 5: Optical fiber. Optical fiber itself is fairly inexpensive, but it is costly to install. It is not provided for plenum specifications and therefore must be installed in some kind of protective conduit. The nonplenum optical fiber type 5 cable contains two optical fiber strands, one for transmit and one for receive. This nonplenum type 5 can be used indoors, aerial, or underground, as long as it is installed using a conduit.

Fiber-optic cables known as Type 5 R can also be used in building risers. These cables will contain multiple fiber strands. They can be useful when making connections over long distances within a building. The cable may be laid horizontally and vertically. This cable does not require conduit but cannot be run through a plenum.

Outdoor fiber-optic cables also contain multiple fibers and are used primarily as backbone or campus connections between buildings. This cable requires installation in a conduit.

Fiber-optic cables can be used for both voice and data with speeds of 16 Mbps or greater. Fiber-optic cables are measured in microns instead of AWG. There are four thicknesses to a fiber, as described in Figure 1.18. The IBM Cabling System recommends the use of a 62.5-micron size for the fiber core and 125 microns for the cladding.

Type 6: Twisted pair. This shielded twisted-pair cable uses #26 AWG solid copper wire and is for data communications. The twisted pairs are surrounded by a braided shield. Type 6 cable is used only for patch or jumper cable.

Type 8: Parallel pairs. This cable type consists of two parallel pairs of #26 AWG wires for data communications. This cable is used for under-carpet installation. The cable is used in situations where the locations of the endpoints do not allow for connection through plenums, conduits, or risers. Type 8 cables support data rates up to 16 Mbps. Each wire is a solid piece of copper wire.

Type 9: Twisted pair. This cable type has two twisted pairs with a braided shield. Each wire of the pairs is #26 AWG. The wires can be made up of strands as well as a solid piece of copper. This cable is used mainly through plenum for main connections between LAN segments. It can service data rates up to 16 Mbps. This cable can also be used in building risers. Type 9 cable is a lower-cost solution than type 1 plenum cable.

1.3.5 Satellite communications

Communications satellites have become quite popular in the last 25 years. They orbit the earth over the equator at a geosynchronous altitude of 22,300 miles. At this altitude, a satellite travels at the same speed as the spin of the earth so that its location remains the same relative to any point on earth. These satellites can cover the entire earth's surface for broadcast transmission.

A satellite can have an enormous transmission capability. For example, the INTELSAT VI satellite can transmit 33,000 voice channels in addition to four TV channels.

Satellites are currently spaced at 3-degree intervals. However, it is possible that in the future they will be spaced at 2-degree intervals to accommodate more satellites.

The transmission frequency for a satellite has quite a wide range. The preferred frequency range, also called "C band," varies between 3700 and 4200 MHz on the down link and 5925 and 6425 MHz on the up link. At this frequency range, impairment of transmission quality by rain and fog is insignificant. C-band frequencies are almost unavailable these days because of congestion.

The available frequency range consists of 11.7 to 12.2 GHz on the down link and 14 to 14.5 GHz on the up link side. This frequency range is called "Ku band." Because of its high frequency (hence small wave length), you require a smaller and more economical earth station antenna to receive the signal. However, the low reliability of Ku band because of rain attenuation is a serious problem. A still higher frequency range called "Ka band" is also available, but the rain factor does not make it a popular choice.

Advantages. Satellite transmission is more desirable for certain applications for the following reasons:

- While terrestrial communication cost is dependent on the distance involved, satellite communications costs is independent of such distance.

- Satellites have enormous bandwidth. They are more suitable for applications requiring large data transfers.

- Using satellite transmission facilities, you can directly reach your destination, thus bypassing the telephone companies.

- Satellite transmission is excellent for broadcast purposes. If data has to be transmitted from a single source to multiple destinations, satellite is much more economical. For example, there could be applications in which a brokerage house's central location in New York must transmit stock prices on a real-time basis to its 200 branch offices. This could be accomplished by a single satellite broadcast, while the alternative would be to have 200 terrestrial links between the central office and the branch offices.

- Earth stations, after transmitting data to a satellite, can verify the accuracy of their transmission by listening back to the return signal.

- Satellite communications are suitable for those areas which are difficult to reach because of their rough terrain. They are also more economical for far-away areas where it is uneconomical to install and maintain terrestrial transmission facilities.

Disadvantages.   Satellite transmission comes with some inherent problems, as follows:

- A satellite transmission takes a trip of 22,300 miles from the transmitting earth station to the satellite and 22,300 miles from the satellite to the receiving earth station. Even though radio signals travel at the speed of light, the distance imposes approximately $\frac{1}{2}$ second (s) delay in transmission. This delay can cause severe performance problems for those link-level transmission protocols which have to send an acknowledgment on each transmission. Such protocols are ARQ and BSC. It is recommended that link-level protocols be changed to SDLC under those circumstances or that satellite transmission not be used.

- At the higher frequencies of the Ku and Ka bands, the rain attenuation factor can cause data errors and hence retransmission delays.

- Because of the higher reliability of C band, there is more crowding in that frequency range. This can cause interference from the terrestrial microwave transmission facilities operating on the same frequency.

1.3.6   Fiber-optic communications

Fiber-optic communications makes use of light (also called "photons") as a transmission object, as compared to the radio-frequency signals used by microwave and satellite communications. The transmission medium used by it is an optical fiber, which is a hair-thin fine strand of pure quartz glass. Optically transmitted data is transmitted at the speed of 125,000 miles per second through these strands.

Fiber optics is the fastest growing transmission technology. It seems that it will be the predominant voice and data communications medium in the near future. One of the long-distance communications carriers, US Sprint, has 100 percent of its network based on optical fiber.

Advantages. Fiber optics has the following advantages over other communications media:

- Because the technology is based on light wave transmission, it is immune from Radio-Frequency Interference (RFI) and Electromagnetic Interference (EMI). Thus optical cable can run in any environment without the risk of generating or receiving such interference from other sources.
- With the advancement of technology, optical fiber can carry enormous bandwidth.
- Optical fiber can withstand temperatures of up to 1800 degrees Fahrenheit.
- Transmission is highly reliable. Optical fiber has a Bit Error Rate (BER) of one in a trillion (a trillion is a thousand billion, or 10^{12}).
- Since there are no electromagnetic emissions involved, optical transmission is highly secure. Therefore it is particularly useful for military purposes.
- It is lightweight. While a mile of optical cable weighs about 50 pounds (lbst), a mile of other types of cable weighs about 325 lb.

Thus we see that optical fiber is superior in usefulness to the copper wire cables which have been in use over the last century or so. Moreover, copper is a limited natural resource, and its prices fluctuate. Optical fibers are made from sand or silica, which is the most abundant natural resource on earth.

The only hazard against which an optical fiber must be protected is water because ice can damage a cable. Ice can also create attenuation and thus affect transmission quality.

1.4 Summary

In this chapter we discussed various environments involving networking of IBM hardware and software. Although the detailed discussion of each of the discussed components is the subject matter for the rest of the book, it gave us a global view of typical networking environments. We also introduced a lot of commonly used telecommunications facilities and transmission media. A lot of the telecommunication terminology introduced in this chapter will be referred to in the subsequent chapters. Therefore it is important that you understand it thoroughly before proceeding further. And now, on to the next chapter on communications hardware.

2

Communications Hardware— Cluster Controllers and Communications Controllers

SNA gives the underlying architecture for connecting various communications hardware and software. Although this book is primarily oriented toward learning VTAM and NCP, it is essential to understand the hardware that makes up the physical network. First we will get a global picture of all the hardware components involved. After learning the basic function of each one, we will get more detailed information about some of them.

2.1 Global View of the Communications Hardware

Normally, an end user, whether an application programmer or a nontechnical user of the application systems, is only aware of the physical existence of a terminal. In addition, an application programmer will also be aware of the presence of a computer at the other end. The existence of the complex network hardware in the middle is usually transparent to them. Here is an overview of each component:

Terminal. A device through which the end user communicates with the host computer. Usually, a terminal is connected to a cluster controller through a coaxial cable.

Cluster controller. It sits between the terminals and the host for a channel-attached configuration and between the terminals and the communications controller for a link-attached configuration. Depending upon the model, it can control a cluster of 8 to 32 terminals.

Communications controller. It runs NCP software, which performs the network control functions. Link-attached cluster controllers, RJE/RJP devices, and asynchronous terminals are attached to the communications controller.

Network controller. An IBM 3710 network controller acts as a protocol converter and line concentrator for start/stop and BISYNC lines and devices. It is connected to a communications controller.

Network conversion unit. An IBM 3708 provides for the linking of ASCII devices to an SNA network.

RJE/RJP device. It handles batch as well as remote job entry functions for the console: printers, card readers, and card punches.

Physical interfaces. Physical interfaces define the physical connectivity and electrical interface between Data Terminal Equipment (DTE), such as a communications or cluster controller, and Data Circuit-Terminating Equipment (DCE), such as a modem. Most common examples of such interfaces are RS-232-C, V.35, and RS-449.

Modem. This is a device that acts as an interface between the DTE device, such as communications or cluster controllers, and the analog communications links, such as voice-grade telephone lines.

Patch panels. Although you can directly connect a DTE and a DCE over a physical interface (e.g., RS-232-C), usually this is done via a patch panel. Patch panels make moving DTE devices from one place to another less cumbersome. They are also used to connect terminal devices to the cluster controllers. Such patch panels have coaxial links coming from the terminals and the cluster controllers. Any time a terminal device is moved from one office to another, patch panel links make it easy to establish a connection.

Multiplexers. They are used whenever multiple links are concentrated over a single link to use the single link as a transmission medium. On the other end, the single link is further split into multiple link connections. Individual links on either side seem to have a direct connectivity between them. Multiplexers on each side of the link do the necessary conversion for the multiple links on either side. A T1 multiplexer is used to split the bandwidth of 1.544 Mbps into multiple-line-speed channels. A coaxial multiplexer (e.g., IBM 3299) is used to concentrate up to eight terminal lines on a single coaxial cable.

We have reviewed almost all the hardware devices that will be referred to in this book. Now, let's discuss the cluster controllers and communications controllers in more detail. Other hardware components will be discussed in the next chapter.

2.2   Cluster Controllers

2.2.1   Functions

As the name suggests, a cluster controller is used to control a number (cluster) of terminals. Depending upon the model, you can attach anywhere from 4 to 64 terminals or printers to it. A cluster controller can be directly attached to a host computer channel or can be link-attached to a communications controller (Figure 2.1).

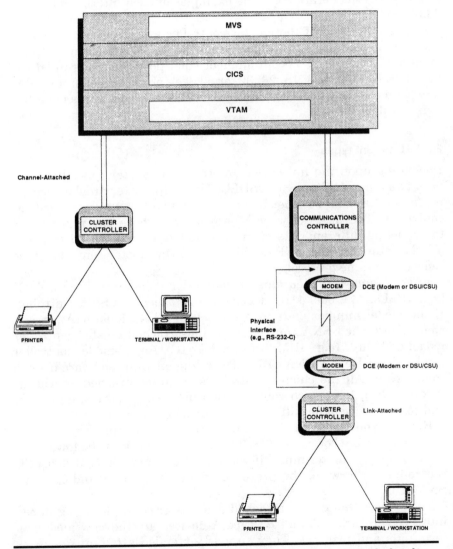

Figure 2.1   An example of cluster controller configuration—channel-attached and link-attached.

A channel-attached cluster controller is linked to the byte or block multiplexer channel of a mainframe. Its interface is the bus and tag interface to the host channel. A link-attached cluster controller is linked to the communications controller through a communications link. The hardware that links it to the communications link is called "DCE." DCE can be a modem for analog lines or a DSU/CSU for digital circuits. The physical interface between the cluster controller and the modem depends upon the line speed and communications controller port considerations. Some examples of such interfaces are RS-232-C, RS-449, V.35, etc.

At the other end of the cluster controller are a number of coaxial cable connector ports. You can attach a 3270-family terminal or a printer to each port. The number of such ports depends upon the model and can be 4, 8, 16, 24, 32, or 64. The physical medium used to connect a cluster controller port and a terminal or printer is usually a coaxial cable.

2.2.2 Different types

Depending upon the link-level protocol used, cluster controllers can be of two types—BSC or SNA/SDLC. The end-user terminal or printer is *always* the same whether it is connected to a BSC or an SNA/SDLC controller. The host software VTAM defines those two types in its tables using different sets of macros. It has to be ensured that the VTAM definition, the NCP definition, and the cluster controller type refer to the same protocol.

Any channel-attached cluster controller model which ends in D (e.g., IBM 3274 Model 41D) is a channel-attached non-SNA controller. If the model number ends in A (e.g., IBM 3274 Model 41A), it is a channel-attached SNA controller. A link-attached cluster controller model ends in C (e.g., IBM 3274 Model 41C). Any C-model controller can be either BSC or SNA/SDLC depending upon the software used to configure it. All the cluster controllers, channel-attached as well as link-attached, come with configuration disks to do port assignments and configure them for different types of printers or terminals.

How do you know whether the IBM 3278 terminal you have is attached to a BSC or an SNA/SDLC controller? Look at the lower left-hand corner of the terminal. If you have the character A, it is a BSC controller. Otherwise, the presence of the character B indicates an SNA/SDLC controller.

The devices that can be attached to a cluster controller are grouped into two categories, A and B. Type A devices are the new models of terminals and printers. They support extended attributes such as highlighting, reverse video, and color capability (e.g., in IBM 3279 display terminals). In addition, they also support user-defined charac-

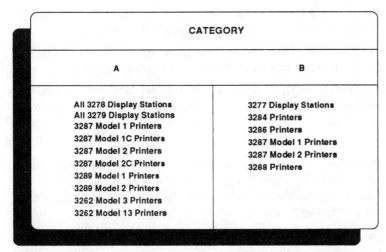

CATEGORY	
A	**B**
All 3278 Display Stations	3277 Display Stations
All 3279 Display Stations	3284 Printers
3287 Model 1 Printers	3286 Printers
3287 Model 1C Printers	3287 Model 1 Printers
3287 Model 2 Printers	3287 Model 2 Printers
3287 Model 2C Printers	3288 Printers
3289 Model 1 Printers	
3289 Model 2 Printers	
3262 Model 3 Printers	
3262 Model 13 Printers	

Figure 2.2 Category A and B terminals in the 3270 family of display and printer devices.

ter or graphic symbols. Category B terminals do not have such capabilities. You can mix category A and B terminals in an IBM 3274 cluster controller. A list of such devices is given in Figure 2.2.

An SNA/SDLC cluster controller is known as Physical Unit Type 2, or PU Type 2. A display device (e.g., IBM 3278) attached to a PU Type 2 is known as Logical Unit 2, or LU 2. A printer device attached to a PU Type 2, depending upon the device and the definition, can be LU 1 or LU 3. The meaning and significance of these acronyms will be discussed in a later chapter. However, we will start introducing the terminology now to enhance your understanding of them later on.

2.2.3 Models and their capacities

Cluster controllers can be broadly categorized into two groups, the *old* IBM 3274 family and the IBM 3174 family of controllers. IBM 3274 controllers come in various capacities and characteristics and still have a huge installed base in IBM shops. However, as the industry moves slowly from a dumb end-user terminal environment to intelligent workstations, these controllers will be replaced by IBM 3174s or similar devices.

IBM 3274 Cluster Controllers. Some of the widely used IBM 3274 controllers are explained as follows:

- IBM 3274 Model 41A is an SNA channel-attached controller and can support up to 32 category A and B devices.

- IBM 3274 Model 41D is a non-SNA channel-attached controller which can support up to 32 category A and B devices.

- IBM 3274 Model 41C can be a link-attached BSC or SNA/SDLC controller which can support 32 terminal devices.

- IBM 3274 Models 51C, 61C, and 31C can be BSC or SNA/SDLC controllers which can support 8, 16, and 24 terminal devices, respectively.

IBM 3174 Establishment Controller.   Notice that 3174 cluster controllers are called ECUs in IBM lingo. The capabilities of IBM 3174 ECUs are far beyond what is possible with the IBM 3274s. Some of their prominent features are listed below:

- Provides for customizing disks of remote sites at a central site location and maintaining them in a library.

- Provides for electronic distribution of microcode or customization parameters for remote IBM 3174s.

- Supports Intelligent Printer Data Stream (IPDS), which supports All Points Addressability (APA) to position text, bar codes, images, and vector graphics on a page.

- Supports up to two 20-megabyte (MB) fixed disk drives.

- Provides for IBM token-ring attachment and host communications either through a gateway feature or through an IBM 372X or 3745 communications controller's Token Interface Card (TIC) feature.

- Provides for network asset management by supporting retrieval of an ECU's hardware product information from a central host. Such product information can be machine serial number, model, machine type, etc.

- Provides for response time monitor (RTM).

- Allows for expansion of ECU's main storage to 3 MB.

- Supports one or more ASCII host access to its attached terminals.

Various models of IBM 3174s and their characteristics are explained as follows:

3174 Model 1L.   This model is a channel-attached ECU which can be SNA or non-SNA. As a general convention, all 3174 ECUs having the suffix L are local (channel-attached), while those having the suffix R are remote (link-attached) cluster controllers. You can attach up to 32 terminals to it. The basic configuration comes with 1 MB of main storage and a 1.2-MB high-density disk drive. Some of the optional features you can install are the IBM Token Ring Adapter and the Token Ring 3270 Gateway. It also support an Asynchronous Emulation Adapter, which can support up to 24 ASCII devices.

3174 Model 1R. This model is a link-attached ECU that can be SNA/SDLC or BSC. You can attach up to 32 terminal devices to it. The basic configuration comes with 1 MB of main storage and a 1.2-MB high-density disk drive. It supports ASCII terminals through the installation of an optional Asynchronous Emulation Adapter (AEA) feature. You may also install the Token Ring Adapter and the Token Ring 3270 Gateway feature. In fact, a 3174 Model 1R with the Token Ring attachment features is called "3174 Model 3R." The ECU supports a multitude of physical interfaces for DCE connectivity. Such interfaces are EIA-232-D, V.24, and V.35 and are discussed in Chapter 3.

3174 Model 2R. This model has the same characteristics and capabilities as the 1R except that the physical interface is an X.21 (CCITT V.11). It can also be upgraded (like Model 1R) to a Model 3R for token-ring connectivity.

3174 Model 3R. This model is in fact a Model 1R or 2R with Token Ring attachment. All other features and capabilities are the same as the other models. Model 3R can also be converted back to Model 1R or 2R by installing a Type 1 communications adapter for Model 1R conversion and a Type 2 communications adapter for Model 2R conversion. Remember, although Model 3R has token-ring connectivity, it *cannot* be a Token Ring 3270 Gateway.

3174 Model 51R. This model has the same capabilities as Model 1R except that it supports a maximum of 16 terminal devices instead of the 32 supported by 1R. This model can have a Token Ring attachment and can also be a Token Ring 3270 Gateway. Its physical interface with DCE can be EIA-232-D, V.24, and V.35.

3174 Model 52R. This model has the same capabilities as Model 2R except that it supports a maximum of 16 terminal devices instead of the 32 supported by 2R. This model can have a Token Ring attachment and can also be a Token Ring 3270 Gateway. It has an X.21 (CCITT V.11) interface.

3174 Model 53R. This model has the same capabilities as Model 3R except that it supports a maximum of 16 terminal devices instead of the 32 supported by 3R. This model can have a Token Ring attachment but *cannot* be a Token Ring 3270 Gateway.

3174 Model 81R. This model can attach to a maximum of eight terminal devices. Its physical interface with DCE can be an EIA-232-D, V.24, or V.35. It *cannot* have a Token Ring attachment.

3174 Model 82R.   This model can attach to a maximum of eight terminal devices. Its physical interface is X.21 (CCITT V.11). It *cannot* have a Token Ring attachment.

2.2.4   3174s and token-ring features

In the previous section we mentioned the Token Ring attachment and Token Ring Gateway features. Let's see what these features do to enhance connectivity.

Token-ring is a local area network (LAN) implementation for the IBM environment. A LAN enhances connectivity between multiple scattered devices (e.g., between the floors of a single building) and helps them share their resources, such as files, printers, programs, etc. LANs are normally used to connect microcomputers (e.g., IBM PCs and PS/2s) for sharing resources and for communicating with each other. IBM has gone one step further to provide connectivity between IBM 3174 ECUs over a token-ring-based LAN.

How do 3174s fit in a LAN? Under normal circumstances, a cluster controller is either directly attached to a host channel or attached to a communications controller's port over a communications link. You can also multidrop a number of cluster controllers over a single link. A token-ring LAN provides an alternative to the multidrop approach for cluster controllers housed in a single building location.

Devices such as IBM PCs, PS/2s, and 3174 ECUs can be attached to a token-ring LAN through a MultiSystem Access Unit (MAU). When we talk about upgrading a 3174 to support token-ring features, we are referring to the hardware, microcode, software, and physical interface needed to support token-ring protocols and physical connectivity to MAU. MAU looks like a wall-mounted plate with 10 physical interfaces. Two of the interfaces connect to other MAUs (i.e., MAU-in and MAU-out), and the other eight interfaces are used by the devices needing connectivity to the LAN.

Token-ring connectivity versus Token Ring 3270 Gateway.   The token-ring connectivity feature enables an IBM 3174 ECU to be connected to a token ring through an MAU port. The Token Ring 3270 Gateway feature enables a single IBM 3174 ECU to act as a gateway to the SNA network for *all* the IBM 3174 ECUs attached to the token-ring network through the token-ring connectivity feature.

While the connectivity feature provides for physical connectivity to the token-ring, the 3270 gateway feature proves the means to be connected to the outside world. A single 3270 gateway IBM 3174 can provide connectivity to a number of LAN-attached 3174s (see Figure 2.3).

We have a channel-attached IBM 3174 ECU with the Token Ring 3270 Gateway feature. It provides connectivity for itself and the other

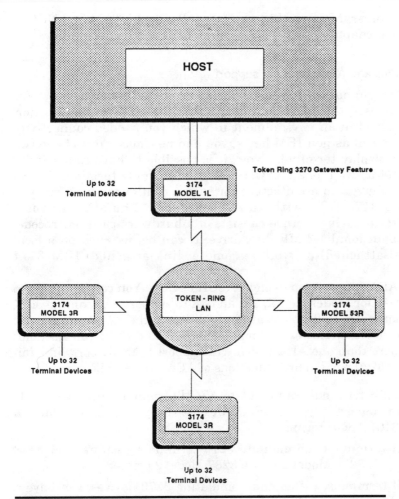

Figure 2.3 An example of a single IBM 3174 ECU with a Token Ring
3270 Gateway feature providing connectivity for itself and a number
of other IBM 3174 ECUs on a token-ring LAN.

three IBM 3174s through the token-ring connectivity feature. The
gateway ECU does not have to be channel-attached; it can be remote
(link-attached).

Only IBM 3174 Models 1L, 1R, 2R, 51R, and 52R support the 3270
Gateway feature. Models 3R and 53R can support connectivity to a
token-ring LAN but cannot act as gateways. Models 81R and 82R
support neither LAN connectivity nor the 3270 Gateway feature.

The channel-attached IBM 3174 Model 1L acting as a 3270
Gateway needs a single subchannel address to the host. A link-
attached IBM 3174 acting as a 3270 Gateway requires a single SDLC

station address. More on this when we talk about VTAM and NCP in subsequent chapters.

2.2.5   3174s and ASCII terminal support

The IBM terminal world consists primarily of 3270-like devices. For non-IBM hosts, the equivalent device is an ASCII terminal. Earlier, if you worked in an environment in which you needed connectivity to IBM as well as non-IBM hosts, you had no choice but to have two separate display terminals. Now, it is possible to have connectivity to an IBM as well as an ASCII host from a single terminal device. Figure 2.4 gives an architecture for such a configuration. What you need is a 3174 ECU with an Asynchronous Emulation Adapter (AEA) feature. This feature consists of a hardware card, microcode, and an additional 1.2-MB disk drive. It can be installed on a local (channel-attached) as well as remote (link-attached) IBM 3174 ECU.

Each AEA supports up to eight ASCII devices. You can install up to three AEAs for Model 1L, 1R, 2R, and 3R to support a maximum of 24 ASCII ports. These 24 ports are in addition to the 32 terminal ports for 3270 devices. However, for models 51R and 52R, you may not install more than one AEA which will support 8 ASCII ports in addition to 16 3270 ports. Three functions of AEAs are as follows:

- IBM 3278 terminals can emulate ASCII terminals such as the VT-100 for connectivity to a DEC/VAX host. They can also emulate an IBM 3101 ASCII terminal.

- ASCII terminals can emulate 3270 terminals such as IBM 3178 Model C2, 3279 Model 2A, or a 3287 Model 2 printer.

- ASCII terminals, rather than emulating 3270 devices, can have a straight pass-through to an ASCII host or a Public Data Network (PDN).

2.2.6   3174s and APPN support

The IBM 3174 Establishment Controller supports IBM's Advanced Peer-to-Peer Network (APPN) architecture. The 3174 can act as an APPN Network Node (NNs) and provide directory services for down-stream token-ring-attached APPN End Nodes (ENs). In theory one could implement an APPN network based on IBM 3174s with IBM OS/2 servers providing the applications as APPN ENs. Microcode-level support C is needed to implement APPN NN support on the IBM 3174. Figure 2.5 illustrates a possible APPN configuration using IBM 3174.

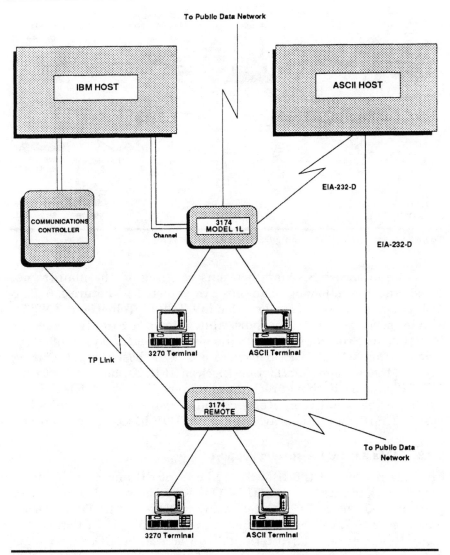

Figure 2.4 Interconnectivity between the IBM 3270 and ASCII terminals to the IBM and ASCII hosts.

2.2.7 3174s and TCP/IP support

Ethernet and token-ring support along with Microcode-level support C enable the IBM 3174 to act as a gateway to TCP/IP hosts off an Ethernet or token-ring network. Downstream token-ring devices can use the 3174 TCP/IP support to access TCP/IP hosts located on another or the same token-ring segment. Likewise, Ethernet-attached

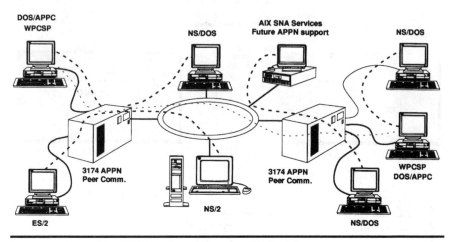

Figure 2.5 A possible APPN configuration using IBM 3174.

devices can access SNA applications residing on the mainframe. These and the following examples are depicted in Figure 2.6. The TCP/IP support provided allows for full TCP/IP Telnet and TN3270 sessions. The TCP/IP support also allows a Simple Network Managment Protocol (SNMP) manager, such as NetView/6000, to manage the IBM 3174 as if it were a TCP/IP host computer using SNMP Management Information Block MIB-II. An advantage to the TCP/IP support is that direct-attached terminals (e.g., CUT, DFT, ASCII, or personal computers) can also connect to the character-based TCP/IP applications available on TCP/IP-based host computers.

2.2.8 3174s and PC file-sharing support

The AEA feature of the IBM 3174 allows ASCII and asynchronous-type devices to attach to the 3174. One function, provided by a third party named The Software Lifeline, gives CUT- and DFT-type terminals access to LAN character-based applications. As in Figure 2.7, using this feature, a 3270 terminal can access the DOS versions of Lotus cc:Mail, Microsoft WORD, and Lotus 1-2-3 through the OS/2 LAN gateway provided by The Software Lifeline product named Remote-OS. This capability allows network designers to eliminate costs, and provide for an evolutionary migration path to a LAN environment while protecting the companies' investments.

2.2.9 3174s and multihost SNA/3270 access

Enhanced connectivity capabilities are now available on the IBM 3174. Using a concurrent communications adapter and configuration

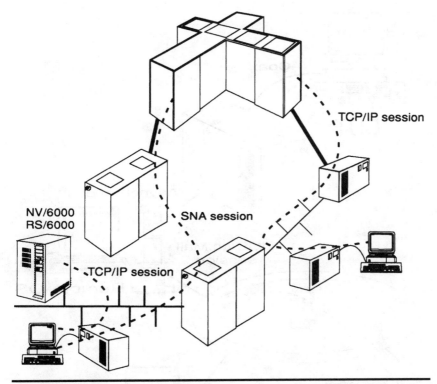

TCP/IP session

SNA session

NV/6000
RS/6000

TCP/IP session

Figure 2.6 Ethernet-attached devices can access SNA applications residing on the mainframe. The TCP/IP support provided allows for full TCP/IP Telnet and TN3270 sessions.

support C microcode allows the IBM 3174 to be active to multiple SNA hosts concurrently. As shown in Figure 2.8, the 3174 can be attached to a token-ring, SDLC line, BSC line, or X.25 network to support multiple hosts. This configuration is called single-link multi-host. In addition, a single 3174 can act as a gateway to several down-stream Ethernet or token-ring devices to multiple SNA hosts. This configuration is called a multihost SNA LAN gateway.

2.2.10 3174s and frame relay support

In Figure 2.9 we can see how the IBM 3174 can be used in a frame relay network. As shown, the 3174 with token-ring support can provide APPN, SNA, and TCP/IP protocol support, allowing multiprotocol network transport to occur. Along with the frame relay support, the IBM 3174 can act as a Remote Source Route Bridge (RSRB) between token-ring segments. The illustration in Figure 2.9 shows a remote token-ring attached to a 3174 with frame relay support. The worksta-

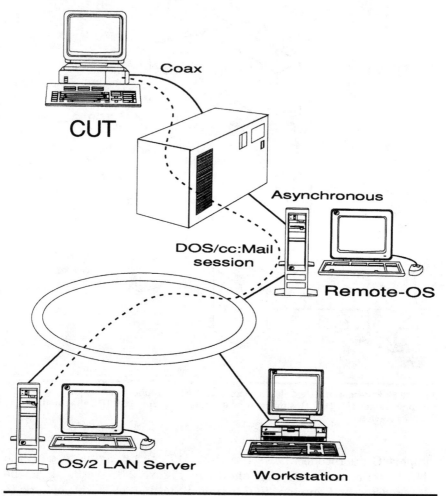

Figure 2.7 A 3270 terminal can access the DOS versions of Lotus cc:Mail, Microsoft WORD, and Lotus 1-2-3 through the OS/2 LAN gateway provided by The Software Lifeline product named Remote-OS.

tion on this ring can communicate with a server on the remote ring using the 3174s as an RSRB bridge through the frame relay network.

2.2.11 AS/400 and APPN support

The AS/400 and its predecessor, the System/36 minicomputers, were the proving ground for APPN. Large APPN networks have been implemented worldwide using AS/400s. APPN on the AS/400 can be supported using communications line and token-ring connections. The AS/400 has the largest installed base of any IBM business com-

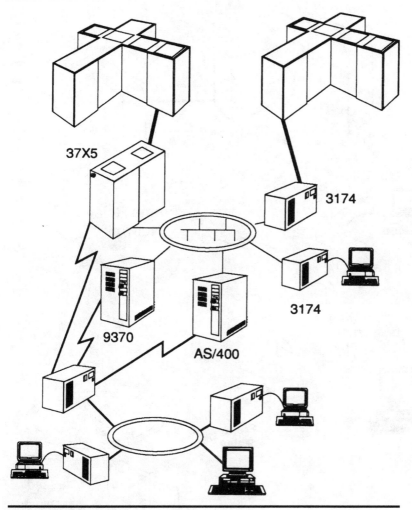

Figure 2.8 The 3174 can be attached to a token-ring, SDLC line, BSC line, or X.25 network to support multiple hosts.

puting platform to date. Figure 2.10 shows an APPN network of AS/400s as APPN NNs, the mainframe as an APPN EN, and an OS/2 workstation and RS/6000 workstation as APPN ENs. This picture alone should tell you that the hierarchical world of SNA has been flattened. In a network configuration like this, the mainframe requests session setup services from the AS/400.

2.2.12 Additional considerations

Different ports on a cluster controller can be configured from a 3270 terminal attached to port 00 (first port). Later on, the same port can

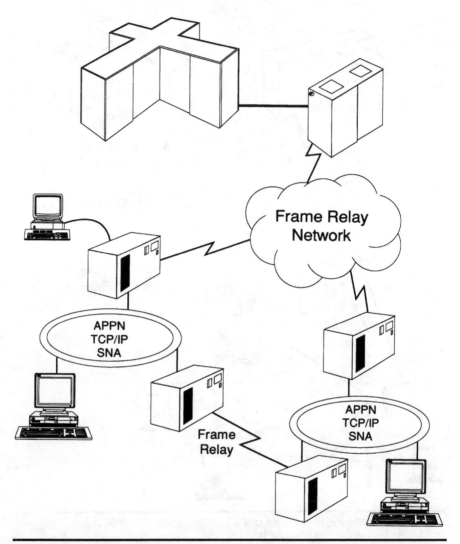

Figure 2.9 The IBM 3174 can be used in a frame relay network. The 3174 with token-ring support can provide APPN, SNA, and TCP/IP protocol support, allowing multiprotocol network transport to occur.

be used for a regular host session. The configuration considerations depend on whether a port is to be used for a printer, a monochrome display device, or a colored display device and the model of the termi-nal. The port configurations must be in synchronization with the VTAM and NCP definition for that port address.

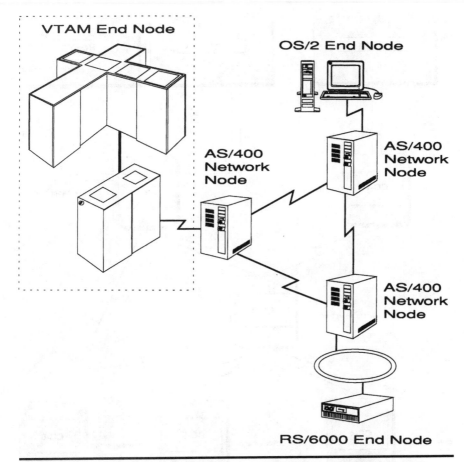

Figure 2.10 An APPN network of AS/400s as APPN NNs, the mainframe as an APPN EN, and an OS/2 workstation and RS/6000 workstation as APPN ENs.

2.3 Communications Controllers

2.3.1 Function

The communications controllers are intelligent systems which are dedicated to the control of communications lines and devices. Any device which cannot be channel-attached because of the distances involved has to be connected through a communications controller, which can be split into the following three functional categories:

1. *Front-End Processor* (*FEP*).   It is attached to the channels of one or more host systems, and it accepts data from its host processors for subsequent transmission to the appropriate devices. It also accepts

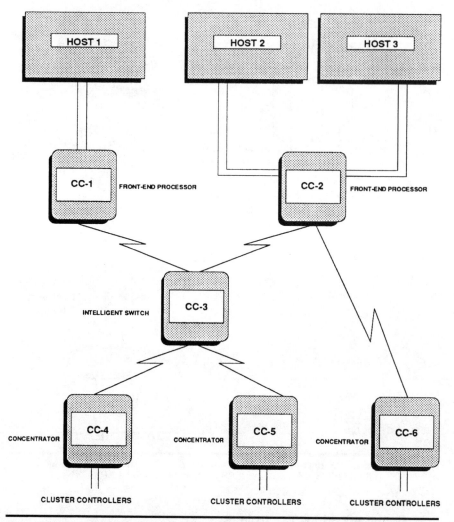

Figure 2.11 A sample network in which different communications controllers perform the functions of a front-end processor, a concentrator, and an intelligent switch.

data from the terminal devices or other communications controllers for routing to the proper host system. In Figure 2.11, CC-1 and CC-2 are the FEPs. In this case, CC-2 is attached to two different hosts.

2. *Concentrator.*   A communications controller acting as a concentrator controls a number of cluster controllers, terminals, RJE/RJP devices, and other communications devices. In Figure 2.11, CC-4, CC-5, and CC-6 provide the concentrator function in a communications controller.

3. *Intelligent switch.* An intelligent switch is an intermediary node which provides switching and routing functions between other communications controllers. CC-3 in Figure 2.11 is an intelligent switch.

Remember that we are classifying the communications controllers by the functions they perform. Such functions can be overlapping. For example, CC-3 in our example can be an intelligent switch and a concentrator at the same time. The communications controllers run under the control of the Network Control Program, or NCP.

Channel-attached and link-attached. A communications controller can either be attached to the host channel or link-attached to another communications controller. In Figure 2.11, CC-1 and CC-2 are channel-attached, while CC-3, CC-4, CC-5, and CC-6 are link-attached. Some of the top-of-the-line communications controllers (e.g., IBM 3745 Model 410) have the capability of having up to 16 channel attachments.

Major functions. A communications controller offloads communications-related functions from the host, thus leaving it free to perform what it was meant for—running application systems. Without communications controllers, a host would be interrupted every time data is transmitted or received from another source. Major functions of a communications controller are as follows:

1. It provides data buffers for temporary storage of data coming from the communications lines or host for subsequent transmission to the target location.

2. It controls and corrects transmission errors and ensures that data is retransmitted if it is erroneous.

3. It controls the physical transmission and receiving of data on various communications lines.

4. It controls message pacing to ensure that it regulates the flow of data to devices depending upon their capacity.

5. It selects appropriate routes for the data depending upon its destination.

6. It sequences the messages to ensure that the receiving device can reconstruct them from the sequence numbers assigned to message segments.

7. It frees the host applications from the burden of polling the devices having a session with that application.

8. It establishes sessions with hosts and other communications controllers to provide logical connectivity between various nodes of the network.

9. It logs and transmits transmission error data to the host.

10. It provides concentrator and intelligent switching functions.

2.3.2 Architecture

Although different communications controllers are architected in different ways, they basically consist of three subsystems:

1 Control

2. Transmissions

3. Maintenance and operator

Figure 2.12 gives a very simple architectural view of such a communications controller.

The control subsystem is the brain of the system. Its channel adapter interface receives and drives channel signals in a channel-attached FEP. The memory component provides temporary buffers for data coming from or going to the channels or the transmission subsystem. The CPU provides the necessary cycles to perform these functions.

The transmission subsystem is the primary interface with the communications lines. It is a three-tiered architecture. Scanners, which are powerful microprocessors, provide for the control of a number of line interface couplers. A line interface coupler controls a line set consisting of a number of transmission lines. Line speeds of the communications lines determine how many of them can be used in a coupler. Finally, the speed of a scanner determines the ultimate transmission load the scanner can handle. The underlying principle is that every time a bit of data comes on the line, the scanner should be ready to grab the bit and pass it over the bus to its memory buffers. Load balancing of communications lines over multiple scanners is a design issue and a subject in itself. Readers are advised to consult appropriate IBM manuals to become more familiar with it.

The Maintenance and Operator Subsystem (MOSS) provides the maintenance functions for the controller. It provides an operator interface through a console and a keyboard. The MOSS feature was not provided in old communications controllers such as the IBM 3705. When used, it helps in on-line diagnostics, Initial Program Load (IPL) of the controller, Initial Microcode Load (IML) of a scanner, generating alarms and alerts, and isolating failures to a specific component.

The communications controllers family consists of four different models—IBM 3705, IBM 3725, IBM 3720, and IBM 3745. They are discussed in the next sections.

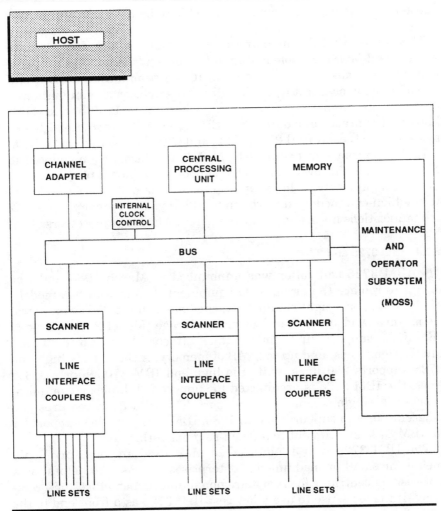

Figure 2.12 A simplified version of the architecture of a communications controller.

2.3.3. IBM 3705

The IBM 3705 is the largest-selling communications controller ever. Although its successor models such as the IBM 3725, 3720, and 3745 are advanced controllers, the IBM 3705 still has the largest user base. It is estimated that there are over 48,000 units installed throughout the world.

IBM announced this model in November 1975 and started shipping it in the middle of 1976. After dominating the communications scene for over a decade, it is at the end of its life cycle, and IBM does not market it anymore. It is being discussed here because of its large user base.

This is the *only* IBM communications controller which does not have an attached MOSS console and therefore depends upon a host for control and diagnostics. In a network, it can function as a front-end processor, a concentrator, an intelligent switch, or a combination of these. It can support line speeds anywhere from 1200 bps to 230.4 kbps and can accommodate BSC, SDLC, or ASCII protocols. It can also emulate the old IBM 2701, 2702, and 2703 hard-wired controllers.

IBM 3705 comes in two models. Model 80, which was announced in March 1981, is the low-end model; it supports 256 kbytes of memory, up to 128 communications lines, and up to two channel interfaces. Model II, the high-end model, supports up to 512 kbytes of memory, up to 352 communications lines, and eight channel-attached host processors.

2.3.4   IBM 3725 and 3720

The IBM 3725 controller was announced in March 1983 and was meant to replace the aging 3705 equipment. It comes in two models. Model 1 supports eight channel-attached hosts, up to 256 communications lines, and up to 3 MB of main storage. Model 2, the low-end IBM 3725, supports up to four channel-attached hosts, up to 80 communications lines, and up to 2 MB of memory. Although the high-end 3725 supports 256 lines while the high-end IBM 3705 supports 352 lines, the IBM 3725 is architected to support full-duplex communications on a single communications port. The IBM 3705 requires two adjacent ports to support this. In 1986, IBM also provided support for an IBM 3725 attachment to a token-ring network.

The IBM 3720 was announced in May 1986 and was primarily meant for small or medium-size enterprises. It was also helpful as a low-cost concentrator for remote regional and branch offices of a company. It makes sense to use a low-cost IBM 3720 as a front end to the 9370 family of processors. Otherwise, an IBM 3725 could cost more than the IBM 9370 host itself. The IBM 3720 comes in four models. Model 1 is for channel attachment and can support a maximum of 28 lines. Model 2 is the remote attachment model and can also support 28 lines. A 3721 expansion unit can enhance Model 1 or 2 and provide for 32 additional lines. Models 11 and 12 are channel-attached and link-attached models, respectively, which support two IBM Token Ring attachments also. These Token Ring Interface Couplers (TICs) reduce the number of lines supported from 28 to 16. Models 1 and 2 can be upgraded to Models 11 and 12, respectively.

Now that we are familiar with some of the capabilities of the IBM 3725 and IBM 3720, let's look at some of the terminology that is used to refer to their various components.

2.3.5   Line Attachment Base (LAB) and Line Interface Coupler (LIC)

A LAB provides the interface between a communications controller's CPU and its scanners. Depending upon its type, a LAB can interface with one or more scanners. On the other end, a LAB interfaces with multiple LICs. LIC ports house the direct physical interface with the communications lines. Although you are aware of the LABs and LICs, scanners are basically invisible entities. Let's look at these elements from a hierarchical perspective.

- A communications controller has a number of LABs. Each LAB can be of a different type.

- Each LAB can control a number of LICs. There are various types of LICs too. Some LICs can be attached only to certain types of LABs.

- Communications lines are attached to various ports of an LIC.

- Physical ports on an LIC signify different physical interfaces and determine the LIC type. Examples of a physical interface are RS-232-C, V.35, etc.

Figure 2.13 gives the characteristics of various types of LABs.

CLAB stands for Channel and Line Attachment Base. Each CLAB comes with one channel adapter and up to 32 line attachments (via LICs, of course). In the case of an IBM 3725 Model 1, two CLABs come with the box. LAB A is generally meant for low-speed lines with a maximum of 19.2 kbps. LAB B is used for high-speed lines and can support up to 256 kbps. LAB C is used for token-ring attachment. A

TYPE	NUMBER OF SCANNERS USED	TYPE OF LICS SUPPORTED	MAXIMUM LINE ATTACHMENTS
CLAB	1	1 or 4A	32
LAB A	1	1 or 4A	32
LAB B	2	1, 2, 3, 4A, or 4B	32
LAB C	1		4 Token Interface Couplers (TICs)

Figure 2.13   Characteristics of various types of LABs for the IBM 3725.

3725 Model 1 can have a maximum of eight LABs, while the Model 2 can have only three.

Communications Scanner Processors (CSPs). A CSP is actually a microprocessor. While a LAB provides the interface between a communications controller's CPU and the CSP, the CSP provides interface between the LAB and the LICs. There is only one type of CSP. An IBM 3725 Model 1 can have a maximum of 14 CSPs, while for Model 2 the limit is 4. CSP also provides one integrated clock controller (ICC) per LAB. ICC is required for direct attachment to non-clocking modems, e.g., asynchronous modems. In other words, an ICC when used with a special cable acts as a modem eliminator.

2.3.6 Types of line interface couplers

While a single LAB supports multiple LICs, a single LIC may support one or more communications lines. Various types of LICs provide different capabilities and features. Figure 2.14 gives a summary of various LIC types and their characteristics.

Although lines at T1 speeds (1.544 Mbps) are not supported, a Request for Price Quotation (RPQ) has been available since June 1987 to support T1 lines on an IBM 3725. Such an alteration can use six scanners. However, one might like to consider an IBM 3745, which supports such line speeds more efficiently.

2.3.7 LIC weight for IBM 3725

You cannot connect communications lines to the IBM 3725 ports at random. Connection is strictly controlled by considerations of LIC weight for the scanner that controls that line. The basic principle is as follows: No scanner can have an LIC weight of more than 100 units. If an LAB is a single-scanner LAB (i.e., CLAB, LAB A, and LAB C), that LAB cannot have an LIC weight of more than 100 units either. If the LAB is a two-scanner LAB (i.e., LAB B), it can have a weight of up to 200 units. Now how do you determine the LIC weight? IBM supplies a table (Figure 2.15) which gives the LIC weight based on LIC type, line speed, and link-level protocol used.

Now make a chart of the LICs within an LAB and the physical lines within an LIC. Put the weight of each line from Figure 2.15 into that table. Determine the line with *maximum* value within each LIC. Determine such maximum values for each LIC in the LAB. Add them up. The sum should not be more than 100 for a single-scanner LAB and 200 for a double-scanner LAB. Figure 2.16 gives an example; it is an LAB A with eight LICs. Each LIC could have a maximum of four physical ports. LIC-1 has four lines each with an LIC weight of 12.

LIC TYPE	INTERFACE	PROTOCOL	LINE SPEED (BPS)	LIC WEIGHT
LIC 1	RS-232-C	BSC EBCDIC or SDLC (HDX)	9600	12
	V.24	BSC ASCII or SDLC (FDX)	4800	12
	X.21 BIS	BSC ASCII or SDLC (FDX)	9600	25
		BSC EBSDIC or SDLC (HDX)	19200	25
		BSC ASCII or SDLC (FDX)	19200	50
	RS366 or V.25	AUTOCALL		12
		Burst Mode ASYNC (S/S)	1200	12
		Burst Mode ASYNC (S/S)	2400	25
		Burst Mode ASYNC (S/S)	4800	50
		Burst Mode ASYNC (S/S)	9600	100
		Burst Mode ASYNC (S/S)	19200	100*
		Character Mode ASYNC (S/S)	300	12
		Character Mode ASYNC (S/S)	600	18
		Character Mode ASYNC (S/S)	1200	37
		BSC Tributory	1200	42
		BSC EBCDIC or SDLC (HDX)	14400	20
LIC 2	Wideband 8751, 8801, or 8803	BSC EBCDIC or SDLC (HDX)	64000	25
		BSC ASCII or SDLC (FDX)	32000	25
		BSC ASCII or SDLC (FDX)	64000	25
		SDLC (FDX/HDX)	128000	50
		SDLC (FDX/HDX)	230000	100
LIC 3	V.35	BSC EBCDIC or SDLC (HDX)	64000	25
		BSC ASCII or SDLC (FDX)	32000	25
		BSC ASCII or SDLC (FDX)	64000	25
		SDLC (FDX/HDX)	256000	100
		SDLC (FDX/HDX)	128000	50
LIC 4A	X.21	SDLC (HDX)	9600	12
		SDLC (FDX)	4800	12
		SDLC (FDX)	9600	25
LIC 4B	X.21	SDLC (HDX)	64000	25
		SDLC (FDX)	64000	25

HDX = Half Duplex   FDX = Full Duplex
* Only Two Points Usable per LIC

Figure 2.14 Characteristics of various types of LICs for the IBM 3725.

Thus the LIC weight for LIC-1 is 12. *Remember that LIC weight is the maximum value within an LIC and not a sum of all the values.* LIC-2 has two ports of weight of 12 each and a third of 25. Thus the maximum value is 25. Notice that the fourth port in LIC-2 is not used. If we were to add another full-duplex line at 9600 bps (weight 25), LIC-2's empty port would be the right choice because it would not affect the LIC weight at all. Similarly, in LIC-3 and LIC-5 the maximum value is 12 each. In LIC-4, the maximum value is 37. The sum total of

LIC TYPE	PHYSICAL INTERFACE	MAXIMUM LINE SPEED (KBPS)	NUMBER OF PORTS
LIC 1	RS 232 / V.24	19.2	4
	RS 366 / V.25		
	X.21 BIS		
	Direct	19.2	
LIC 2	Wideband for 8801, 8803, or 8751 Service	64 (BSC) 230.4 (SDLC)	1
LIC 3	V.35	256	1
	Direct	240	
LIC 4A	X.21	9.6	4
	Direct	9.6	
LIC 4B	X.21	64	1
	Direct	56	

Note: X.21 is not available in U.S.A.

Figure 2.15 Table showing relationship between LIC weight and LIC type, line speed, and link-level protocols for an IBM 3725.

all such values up to LIC-5 is already 98. Therefore we cannot use LIC-6, LIC-7, and LIC-8 because LAB A, being a single-scanner LAB, has an upper-limit LIC weight of 100.

Configuration rules. LIC weight considerations and other configuration rules are summed up as follows:

1. Total LIC weight on a scanner cannot exceed 100.

LAB A

PORT	SCANNER							
	LIC 1	LIC 2	LIC 3	LIC 4	LIC 5	LIC 6	LIC 7	LIC 8
1	12	25	12	18				
2	12	12	12	12				
3	12	12		37	12			
4	12							
Maximum Value	12	25	12	37	12			

Total LIC Weight = 12 + 25 + 12 + 37 + 12 = 98

Figure 2.16   An example to determine LIC weight for a scanner.

2. Given the total scanner capacity of 307,200 kbps, calculate the LIC capacity as follows:

$$\text{Total LIC capacity} = \frac{\text{total scanner capacity}}{\text{number of last LIC position physically installed on the scanner}}$$

3. After finding out the LIC capacity from 2 above, the maximum line speed for an LIC is calculated as follows:

$$\text{Maximum line speed for an LIC} = \frac{\text{total LIC capacity}}{\text{number of ports per LIC}}$$

The above rules must be strictly followed for the communications controller to function properly. If the rules are not followed properly, the Physical Unit (PU) activation fails after NCP load, and corrupted data is transmitted over the line.

Line weight for 3720.   The IBM 3720 communications controller's configuration is based not upon LIC weight but rather on line weight. Since the rules for an IBM 3745 are also based on line weight, refer to the discussion in Section 2.4.5 of this chapter.

2.4   IBM 3745 Communications Controller

The IBM 3745 Communications Controller has spearheaded IBM's direction into internetwork support for established enterprise-wide networking. Since its introduction in January 1988, the IBM 3745 has met end-user demands for enhanced connectivity, performance, and availability. There are seven different models in the IBM 3745 family. Each model provides a different solution for various end-user requirements. The ability to provide enhanced networking connectivity for token-ring connections, Ethernet connections, high- and low-speed serial connections, Enterprise System Connection (ESCON), and channel connections enables established SNA backbone networks to evolve and grow with the needs of the corporation's information requirements.

2.4.1   IBM 3745 Models 130/150/160/170

The IBM 3745 Models 130/150/160/170 are the low end of the IBM 3745 family product offerings. These communications controllers, however, are powerful and fulfill specific connectivity requirements. The Model 130 provides for high-speed T1 connectivity and token-ring connectivity or Ethernet connectivity and channel attachment. The Model 130 can have up to four channel adapters, two high-speed scanners, two token-ring adapters, and two Ethernet LAN adapters. The Central Control Unit (CCU) on the Model 130 can access main storage of up to 8 MB and provide for an NCP load module of up to 6 MB. The Model 130 does not support low-speed lines. It can be field upgraded to a Model 170. Figure 2.17 illustrates a typical use for the Model 130 as a concentrator for LAN access to the mainframe.

Models 150 and 160 are used as remote low-end communications controllers. These models provide for low-speed connectivity but do not support channel attachment. This is in contrast to the Model 130, which supports channel attachment but does not provide for low-speed line connectivity. Like the Model 130, Models 150 and 160 support a single CCU with a 32-KB cache memory, 8 MB of storage, and an NCP load module size of up to 6 MB. Both models support up to 32 low-speed lines and one token-ring adapter. The Model 150 can also support a high-speed scanner and one Ethernet LAN adapter. The Model 160 supports a second high-speed scanner and a second Ethernet LAN adapter. Figure 2.18 diagrams the use of Models 150 and 160 as remote communication controllers attached to a channel-attached Model 130.

The Model 170 is also a channel-attached communications controller. Like the Model 130, the Model 170 has a total of four channel adapters and two high-speed scanners. The Model 170 can support up to 112 low-speed lines through a combination of 96 externally con-

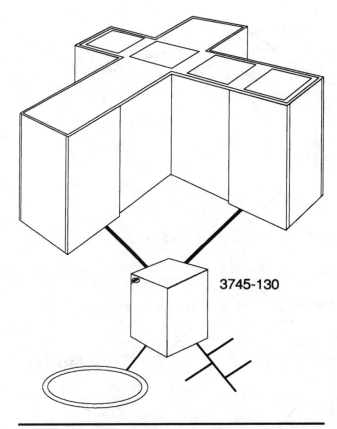

Figure 2.17 A typical use for the Model 130 as a concentrator for LAN access to the mainframe.

nected modems and 16 internally connected modems. It has a single CCU with 8 MB of memory and a 32-KB cache. It can also support one token-ring adapter and two Ethernet LAN adapters. The table in Figure 2.19 summarizes the maximum number of features available for the IBM 3745 1XX models.

2.4.2 IBM 3745 Models 210/410

In March 1988 IBM made available the IBM 3745 Model 210 Communication Controller, and in September 1988 the IBM 3745 Model 410 Communication Controller. The basic difference between the two models is the support for two CCUs in the Model 410. Each controller can handle up to 8 MB of CCU memory with 16 KB of cache. Each controller can have up to 16 channel adapters and 896 low-speed lines, four high-speed scanners, four token-ring adapters, and up to eight Ethernet LAN adapters.

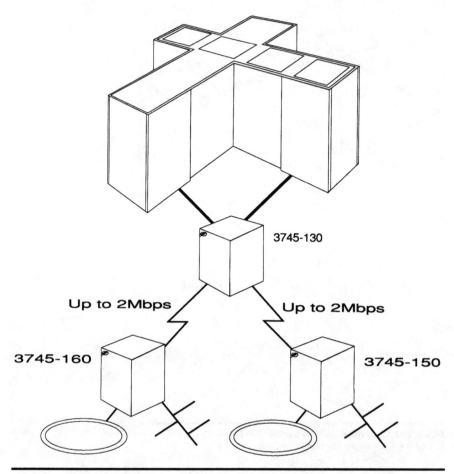

Figure 2.18 The use of Models 150 and 160 as remote communication controllers attached to a channel-attached Model 130.

The Model 210 is field upgradable to the Model 410. The Model 210 is twice as powerful as the IBM 3725, and the Model 410 operates at four times the capacity of the IBM 3725. The Model 410, with two CCUs, can be configured to run in three different modes of operation:

- *Twin dual mode.* Each CCU acts as a separate communications controller. Therefore, you can run two independent NCPs. It is just like having two separate Model 210 controllers standing side by side. It is important to note that the two NCPs in Model 410 have no internal link. An external communications facility, such as a channel, token-ring, or communication link, must be established.

- *Twin standby mode.* In this mode there is one active NCP while the second one is in a standby mode. This allows for quick recovery

Maximum	3745			
	130	150	160	170
CCUs	1	1	1	1
Storage (MB)	8	8	8	8
Cache (KB)	32	32	32	32
CAs	4	0	0	4
BCCA	Optional	Optional	Optional	Optional
ESCON ports	-	-	-	-
LSS	0	1	1	6
Lines (up to 256 Kbps)	0	32	32	112
HSS	2	1	1	2
Lines (up to 2 Mbps)	4	2	2	4
Token-ring ports	4	2	2	2
Ethernet ports	4	2	2	4
HDD (MB)	45	45	45	45
Local console	Mnd	Mnd	Mnd	Mnd
Remote/alternate console	Optional	Optional	Optional	Optional
Remote support facility (RSF)	Optional	Optional	Optional	Optional

Figure 2.19 Summary of the maximum number of features available for the IBM 3745 1XX models.

from storage-related failures and from CCU hardware checks. The idle CCU is totally inactive unless a failure is detected.

- *Twin backup mode.* In this mode, each CCU runs approximately half the network, hence each CCU is half full. It is like having two controllers, each running one-half of the network load. If one CCU fails, the other CCU can assume the load of the failed CCU. This also allows for easy recovery from storage-related failures and CCU hardware checks.

Figure 2.20 illustrates the three different modes of operation for the twin CCU model of the IBM 3745.

2.4.3 IBM 3745 Models 310/610

The most recent IBM 3745 models, the 310 and 610, match the physical capacity of the 210 and 410 as well as the features of each, respectively. The Model 310 has a single CCU with the capacity for up to 16 MB of main memory and a 64-KB cache. The Model 610 has a twin CCU with up to 16 MB of CCU memory per CCU and a 64-KB cache per CCU. Both of these controllers offer increased performance over the 210 and 410 by 33 percent. The cache cycle time in the 310 and 610 matches the CCU machine cycle time. This improves the average instruction time in the controllers and thereby, with the larger cache size, increases the

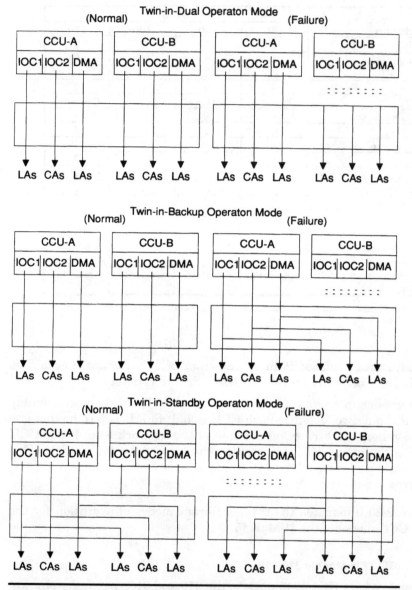

Figure 2.20 The three different modes of operation for the twin CCU model of the IBM 3745.

probability of finding data in cache by a factor of four. The improved architecture also eliminates the need for CCU cycles to transfer data on the internal busses of the 310 and 610. This provides for increased throughput while conserving CCU power. Figure 2.21 gives a table of the Model 210/310/410/610 maximum configurations.

Maximum	3745			
	210	310	410	610
CCUs	1	1	2	2
Storage (MB)	8	8	8	8
Cache (KB)	16	64	16	64
CAs	16	16	16	16
BCCA	Optional	Optional	Optional	Optional
ESCON ports	-	-	-	-
LSS	32	32	32	32
Lines (up to 256 Kbps)	896	896	896	896
HSS	8	8	8	8
Lines (up to 2 Mbps)	16	16	16	16
Token-ring ports	8	8	8	8
Ethernet ports	16	16	16	16
HDD (MB)	45/72	45	45/72	45
Local console	Mnd	Mnd	Mnd	Mnd
Remote/alternate console	Optional	Optional	Optional	Optional
Remote support facility (RSF)	Optional	Optional	Optional	Optional

Figure 2.21   A table of the Model 210/310/410/610 maximum configurations.

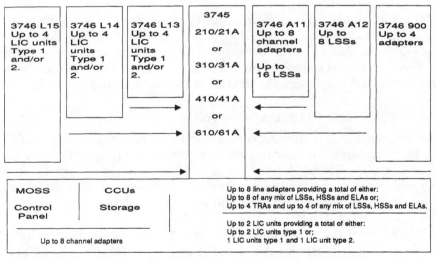

Figure 2.22   The expansion unit's configuration and feature support.

2.4.4   IBM 3746 expansion units

There are five expansion units available for the IBM 3745 210/310/410/610 models. The low-end models, 130/150/160/170, do not support expansion; however, the 130, 150, and 160 can be field upgraded to a Model 170. Figure 2.22 illustrates the expansion unit's configuration and feature support. The names of the expansion units are A11/A12 and L13/L14/L15. The A or L prefix indicates the type of expansion unit. The A prefix indicates that the expansion unit supports additional channel or line adapters, and the L prefix indicates expansion for Line Interface Coupler (LIC) Type 1 and 2.

- The Model A11 provides up to eight additional channel adapters, giving a total of 16 channel adapters. It also provides two more LABs with up to eight low-speed scanners each.

- The Model A12 requires the A11 to be installed and adds an additional eight low-speed scanners

- The Model L13 provides additional LICs. It provides a maximum of four LIC units. Each LIC unit can have up to four LIC Type 1 or four LIC Type 2 units. A mixture of up to four LIC Type 1 and 2 units is also possible.

- The Model L14 adds up to a maximum of four more LIC units with the same configurations as are possible in the L13. The L14 expansion unit requires the installation of the L13.

- The Model L15 provides the same maximum configuration as the L13 and L14, but requires the installation of the L14.

Note that none of the expansion units provide for the expansion of high-speed scanners, token-ring interfaces, Ethernet interfaces, or ESCON interfaces. The Ethernet and high-speed scanner interfaces can reside only on the base unit. The token-ring and ESCON interfaces can be expanded by using the IBM 3745 Model 900 expansion unit.

2.4.5   IBM 3746 Model 900 expansion unit

Support for ESCON channel connection to the mainframe on the IBM 3745 communication controllers was provided through engineering changes to the 210/310/410/610 models. These controllers are denoted by replacing the zero in the number with the letter A. The IBM 3746 Model 900 provides for ESCON and token-ring adapter expansion for Models 21A/31A/41A/61A.

The Model 900 can be connected directly to any of these base units or to the A11 or A12 expansion units. The importance of the Model 900 is its ability to off-load CCU cycles down into the adapters themselves. For example, the token-ring adapter does all of the layer 1 and layer 2 functions without using CCU cycles. This adapter also has 64 KB of buffer space for token-ring frames. This basically eliminates queuing frames in the Model 900 and saves up to 70 percent of CCU cycles if the token-ring adapter on the 900 is used instead of the base unit. This all happens because the adapters installed on the Model 900 are based on Intel 80486 microprocessor technology. Hence, each adapter card has its own CPU to process the data that passes through the adapter.

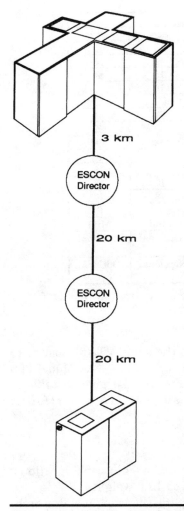

Figure 2.23 The ESCON channel adapters in both the base and Model 900 units provide for long-distance channel connectivity to the mainframe for the communication controllers and other channel-attached devices.

The ESCON channel adapters in both the base and Model 900 units provide for long-distance channel connectivity to the mainframe for the communication controllers and other channel-attached devices. For the IBM 3745 21A/31A/41A/61A, this fiber optic technology can allow the controller to be from 3 to 43 km away from the mainframe. Figure 2.23 diagrams this configuration. Each ESCON Channel Connector (ESCC) of an ESCON Channel Adapter (ESCA) can have

Maximum	3745			
	21A	**31A**	**41A**	**61A**
CCUs	1	1	2	2
Storage (MB)	8	16	8	16
Cache (KB)	16	64	16	64
CAs	16	16	16	16
BCCA	Optional	Optional	Optional	Optional
ESCON ports	64	64	64	64
LSS	32	32	32	32
Lines (up to 256 Kbps)	896	896	896	896
HSS	8	8	8	8
Lines (up to 2 Mbps)	16	16	16	16
Token-ring ports	17	17	17	17
Ethernet ports	16	16	16	16
HDD (MB)	45/72	45	45/72	45
Local console	Mnd	Mnd	Mnd	Mnd
Remote/alternate console	Optional	Optional	Optional	Optional
Remote support facility (RSF)	Optional	Optional	Optional	Optional

Figure 2.24 A table of the maximum connectivity provided by an IBM
21A/31A/41A/61A with all the attached expansion units.

16 logical connections. These can all be to the same VTAM host or to
different hosts or to other communication controllers. The Model 900
can support a maximum of four ESCAs. Figure 2.24 gives a table of
the maximum connectivity provided by an IBM 21A/31A/41A/61A
with all the attached expansion units.

2.4.6   Line weight for the IBM 3745 and 3720

While an IBM 3725 was configured using LIC weight considerations,
an IBM 3745 and 3720 are configured using line weights. While configuring an IBM 3725, we took the *maximum* LIC weight value from
each LIC and added them up for a specific LAB. We also made sure
that the total of all LIC weights did not exceed 100 for a single-scanner LAB and 200 for a dual-scanner LAB. While configuring an IBM
3745 and 3720, we take the line weight from each port of an LIC
within an LAB and add them together. Such line weights cannot
exceed 100 for each scanner.

Characteristics and types of different LABs and LICs are the same
as discussed before for an IBM 3725. You may refer to Figures 2.14
and 2.15 because the tables are applicable to IBM 3745/3720 as well.
Figure 2.25 gives line weights based upon line speed, link-level protocol, and scanner considerations. While configuring an IBM 3720 and
3745, line weights are considered to have a cumulative effect. Thus
you add up the line weights for each active port on an LIC and add up

PROTOCOL	LINE SPEED (BPS)	LINE WEIGHT
SDLC (FDX)	256000**	100
	64000**	25
	57000**	22.3
	19200**	7.5
	19200	10/12.5*
	14400	7.5/9.4*
	9600	5/6.2*
	4800	2.5/3.1*
SDLC (HDX) and BSC EBCDIC	256000**	60
	64000**	15
	57000**	13.5
	19200**	4.5
	19200	5.6/6.2*
	14400	4.2/4.7*
	9600	2.8/3.1*
	4800	1.4/1.6*
	2400	0.7/0.8*

HDX = Half Duplex
FDX = Full Duplex
* Lower Line Weights are for Double Scanner LABs and the Higher Line Weights are for Single Scanner LA
** One Port LICs

Figure 2.25 Table showing relationship between line weight, line speed, and link-level protocols used for IBM 3720 and 3745.

cumulative line weights of each LIC within an LAB. For example, Figure 2.26 shows an LAB with eight LICs. Each LIC could have a maximum of four physical ports. LIC1 has four physical ports, each supporting an SDLC (HDX) at 9600 bps. Thus each line has a line weight of 3.1. LIC2 has two ports, each having SDLC (FDX) lines at 19,200 bps with a line weight of 12.5 each. LIC3 uses a single port for SDLC (FDX) with a line speed of 14,400 bps, thus having a line weight of 9.4. LIC4 has four ports, each having a line weight of 0.8 because they are BSC EBCDIC with a line speed of 2400 bps. We add all the line weights up and it comes to a total of 50. We can install more lines on other LICs up to a maximum line weight of 100. Notice that we took higher line weights from the table in Figure 2.26 because LAB A is a single-scanner LAB.

Configuration rules. Line weight considerations and other configurations rules are summed up as follows:

LAB A

PORT	SCANNER							
	LIC 1	LIC 2	LIC 3	LIC 4	LIC 5	LIC 6	LIC 7	LIC 8
1	3.1	12.5	9.4	0.8				
2	3.1	12.5		0.8				
3	3.1	12.5		0.8				
4	3.1			0.8				

Total LIC Weight = 12.4 + 25.0 + 9.4 + 3.2 = 50

Figure 2.26 An example for determining the line weight for a scanner in an IBM 3720 and 3745.

1. Total line weight of *active lines* on a scanner cannot exceed 100. Notice that while in the case of an IBM 3725 we consider the *installed lines,* for an IBM 3745 we take only *active lines* into account. It increases your responsibility to ensure that you do not activate more lines than a scanner can accommodate. Since line speed in the case of an IBM 3745 can be changed from a MOSS console, you could inadvertently exceed the line weight limit. Be very careful.

2. Maximum line speed per line must meet the following criterion:

$$MLS = \frac{307,200}{NAL \times NAP}$$

where MLS = maximum line speed
      NAL = total number of active LICs
      NAP = number of ports on that LIC

2.4.7 Token-ring adapter and token-ring interface coupler

Token-ring LANs are supported by IBM controllers in several configurations. The IBM 3725/3720 and IBM 3745 models all support token-ring adapters. The IBM 3174, IBM AS/400, and IBM RS/6000 devices also have token-ring adapter cards that provide token-ring LAN connectivity. Several OEM suppliers make token-ring adapter

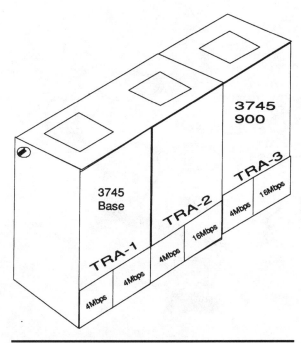

Figure 2.27   The use of TRAs on communications controllers.

cards for personal computers as well. Token-ring is widely used and has become a standard for many corporate LANs.

Token-ring adapters (TRAs) provided on the IBM communication controllers provide for two token-ring interfaces. The ports used by the token-ring adapters can only be port positions 1088 through 1095 on the IBM 3745. Figure 2.27 illustrates the use of TRAs on communications controllers. These token-ring interfaces are called token-ring interface couplers (TICs). The TRA comes in three flavors. TRA Type 1 operates at 4 Mbps. Each TIC on a TRA Type 1 must be dedicated to either Boundary Network Node (BNN) functions or Intermediate-Node (INN) functions. TRA Type 2 allows each TIC to operate at either 4 or 16 Mbps, and each TIC is capable of functioning as BNN and INN simultaneously or individually as BNN or INN. TRA Type 3 operates at 4 or 16 Mbps and can be installed only on the IBM 3746 Model 900 expansion unit. TRA Type 3 can be installed with only one or two TICs. Each TIC is installed using a cassette. The token-ring processor on the TRA itself is considered a cassette on the IBM 3746 Model 900 and performs the IEEE 802.2 Logical Link Control (LLC) and LAN Station Manager functions. The TIC cassette performs the Media Access Control (MAC) IEEE 802.5 functions.

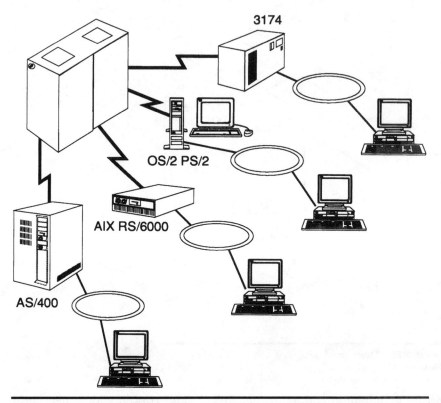

Figure 2.28 The token-ring adapters on the IBM 3174, AS/400, and RS/6000 as well as personal computers provide for token-ring LAN connectivity.

Thus, the TRA Type 3 does not require the NCP Token-Ring Interface (NTRI) functions of NCP, as do the TRA Types 1 and 2. This saves up to 70 percent of valuable CCU cycles.

The token-ring adapters on the IBM 3174, AS/400, and RS/6000 as well as personal computers provide for token-ring LAN connectivity. Figure 2.28 diagrams a configuration use for these devices. The IBM 3714 Establishment Controller can act as a gateway for downstream token-ring devices that need access to VTAM applications. The AS/400, with its token-ring adapter, can provide the same type of service as the RS/6000 and personal computers. Personal computers can act as "cluster controllers" themselves and access SNA applications without the aid of a gateway device as well as access LAN Network Operating System (NOS) functions.

The IBM 3172 Interconnect Controller, with its token-ring adapter and channel attachment for the VTAM mainframe, can provide a channel-attachment gateway function to the mainframe, thus reliev-

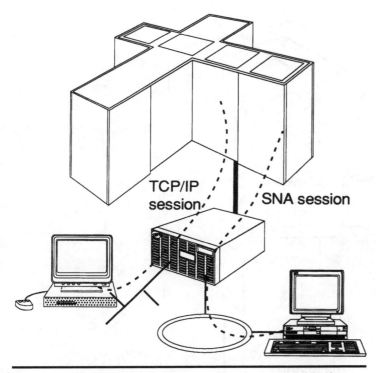

Figure 2.29 An IBM 3172 acting as a token-ring gateway to the mainframe for both SNA and TCP/IP.

ing the communications controllers of such duties. However, cycles are used up in the mainframe to service the SNA sessions through the IBM 3172. The IBM 3172 can also provide channel attachment to token-ring LANs that do not require SNA access on the mainframe but rather require TCP/IP access to the mainframe. Figure 2.29 depicts an IBM 3172 acting as a token-ring gateway to the mainframe for both SNA and TCP/IP.

2.4.8 Ethernet LAN adapter

The Ethernet LAN Adapter (ELA), like the TRA, has two physical LAN interfaces. Each interface can be physically connected and active. The ELA interfaces can be connected to the same Ethernet segment or different segments. Ethernet capability and support is valid only with the IBM 3745 communications controllers, and the NCP must be at Version 6 Release 1. These adapters support the IEEE 802.2 and CSMA/CD LAN protocols. The ELA ports are valid in port positions 1024 through 1039 on the IBM 3745 when TRAs are

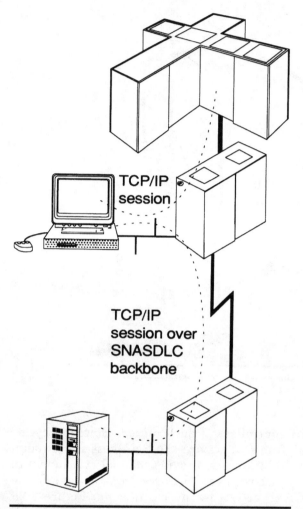

Figure 2.30   The functionality provided by the installation of ELAs.

not installed. When TRAs are installed, the ELA port address range is from 1028 through 1031.

The support for Ethernet allows IBM communications controllers to act as TCP/IP routers supporting TCP/IP sessions across the SNA backbone between two Ethernet workstations. Figure 2.30 illustrates the functionality provided by the installation of ELAs. Ethernet workstations can also access data on an IBM mainframe executing TCP/IP for MVS utilizing an interconnect controller like the IBM 3172. The ELA can also be used as an indirect route between communication controllers for TCP/IP traffic only. The Ethernet LAN adapter presently cannot be used to load or activate an NCP on the communications controller.

Like the TRA, ELAs are available on IBM 3174s, AS/400s, personal computers, and all UNIX-based workstations. The IBM 3174 provides SNA host access for the Ethernet devices, acting as a gateway for them. The AS/400 can also provide the same functionality as the IBM 3174.

2.5 Summary

In this chapter we discussed various communications hardware equipment, such as the cluster controllers and communications controllers. We also provided a global view of other communications hardware, such as terminals, network controllers, network conversion units, physical interfaces, modems, patch panels, multiplexers, etc. We also discussed the most recent enhancements to the IBM 3174 Establishment Controller and its role in preserving a corporation's investment in 3270 technology through the support of token-ring, Ethernet, APPN, and TCP/IP as well as support for frame relay and remote bridge functionality. The maximum configurations discussed for the IBM 3745 do not necessarily mean the maximum configuration for the model as a whole. The only way to determine the proper configuration for your network is to run the IBM-provided configuration program named CF3745. You can obtain this program through IBM's IBMlink or through your IBM customer representative. For all the IBM 3745 family, a high-speed scanner has the ability to have two high-speed lines connected, but only one can be active at any given time; each token-ring adapter has two token-ring interfaces that can be active simultaneously. Each Ethernet LAN adapter has two Ethernet interfaces that can be active simultaneously; however, for each Ethernet LAN adapter installed, a high-speed scanner interface becomes inoperable. In addition, each controller can come with a hard drive installed to allow for direct loading of the NCP from the IBM 3745 hard drive and for NCP load module dumping.

Communications Hardware— Miscellaneous

We learned about cluster controllers and communications controllers in the previous chapter. Although that takes care of the major components of communications hardware equipment, there are many other components to properly understand VTAM/NCP-based network configurations.

Such components are described in this chapter.

3.1 Protocol Converters and Concentrators

The IBM 3270 family of terminals has the largest installed base in the mainframe environment. In the non-IBM environment, the most popular terminals are the asynchronous ASCII devices. IBM PC, PS/2, and IBM 3101 are also asynchronous ASCII devices. ASCII terminals are generally far less expensive than the IBM 3270 family of terminals. Protocol converters provide for connecting such ASCII terminals to the SNA hosts and having them *emulate* IBM 3270 terminals.

3.1.1 IBM 3708 Network Conversion Unit

The IBM 3708 is a protocol converter that comes with 10 ports. One or two of these ports provide connectivity to one or two hosts through a communications controller (IBM 3725, 3720, or 3745). Other ports (the eight or nine remaining ones) provide direct connectivity to the asynchronous ASCII terminals. The protocol conversion software in the IBM 3708 makes the ASCII devices look like IBM 3270 terminals. Besides protocol conversion facilities, it also provides a pass-through mode in which the ASCII devices communicate directly with an

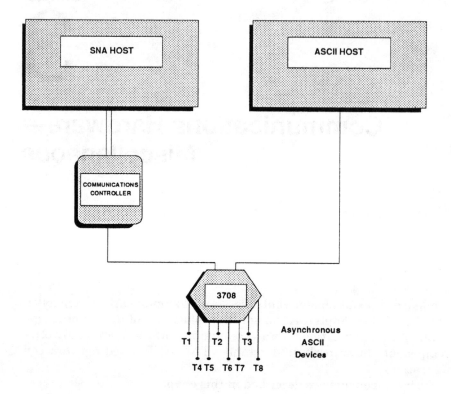

- ASCII Terminals T1, T2, and T3 are Communicating with the ASCII Host

- ASCII terminals T4, T5, T6, T7, and T8 are Emulating as 3270 Devices and Communicating with the SNA Host

Figure 3.1 A sample configuration using an IBM 3708 network conversion unit.

ASCII host. In this case, the IBM 3708 only redirects the data without any conversion. Figure 3.1 shows a configuration in which some of the terminals are emulating in the IBM 3270 mode, while others are accessing ASCII hosts.

Three terminals, T1, T2, and T3, have direct sessions with the ASCII host. Here, the IBM 3708 is acting in the pass-through mode. Five terminals, T4 through T8, have sessions with the SNA host. Here, the IBM 3708 is working as a protocol converter.

An IBM 3708 is defined as a Physical Unit Type 2 (PU Type 2) to VTAM/NCP. An emulating terminal appears as 3278 Model 2, 3178, 3179, or 3279. An ASCII printer appears as a 3287 Model 1 or 2 and looks like Logical Unit 1 (LU1) or Logical Unit 3 (LU3) to

VTAM/NCP. An IBM 3708 supports line speeds up to 19,200 bps to the SNA host and the ASCII host.

3.1.2 IBM 3710 Network Controller

The IBM 3710 provides the functions of a line concentrator and a protocol converter. As a line concentrator, it can concentrate up to 31 lines onto a single SDLC link going to the host. As a protocol converter, it can attach up to seven 8-Port Communication Adapters (8PCAs). Each 8PCA can have up to eight ASCII terminals attached to it. Thus as a protocol converter it can support up to 56 ASCII terminals. In addition, 8PCAs provide for attachment to a maximum of five SNA hosts.

The IBM 3710 supports line speeds up to 19.2 Kbps on an RS-232-C and up to 64 Kbps using a V.35 interface. The IBM 3710 itself is defined as a PU Type 2 to VTAM/NCP and is directly attached to a communications controller.

3.1.3 IBM 7171 Protocol Converter

While the 3708 and 3710 provide 3270 emulation capabilities for remote sites, the IBM 7171 provides similar features for a channel-attached local environment. The IBM 3708 and 3710 need a communications controller for host connectivity. The IBM 7171 is directly attached to a block multiplexer channel of an SNA host. Up to 64 ASCII terminals can be attached to it in increments of eight. Figure 3.2 shows a configuration with an IBM 7171 in it.

An IBM 7171 with up to 32 ASCII devices looks like a channel-attached cluster controller 3274 Model 1D. An IBM 7171 with more than 32 devices (up to a maximum of 64) looks like two IBM 3274 Model 1D cluster controllers. The ASCII devices are attached to it using the RS-232-C physical interface in the United States and the V.24/V.28 interface elsewhere. Only full-duplex transmissions between ASCII devices and the IBM 7171 are supported. Some of the popular ASCII terminals supported are IBM 3101, IBM PC, IBM PS/2, DEC VT100, and Televideo 912/920/950.

Since an IBM 7171 is attached to a block multiplexer channel, there is the possibility of attaching other devices on the same channel as well. Be aware that tape devices may adversely affect the IBM 7171's operation and therefore should not be put on the same channel.

3.1.4 SDLC-to-LLC2 converters

The growth of LANs as a means of connecting to the mainframe has made historical devices like the IBM 3274 outdated. These older

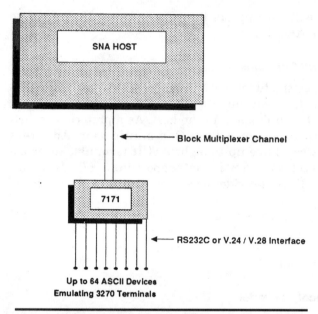

Figure 3.2 An SNA host configuration with the IBM
7171 protocol converter.

devices do not support token-ring adapters but do provide mainframe
connectivity. In many SNA networks where locations have been
upgraded with token-ring LANs, these antiquated devices still
require SDLC link connections to a communications controller.
Figure 3.3a illustrates the new network configuration dilemma that
has faced corporations investing in LANs alongside their SDLC networks. As you can see, we have not one but two networks to manage.
SDLC-to-LLC2 converters allow corporations to preserve their investment in older SNA/SDLC equipment by providing it with token-ring
connectivity to the communications controller for mainframe access.

These devices act as communications controllers to the SDLC
devices, providing all of the boundary network node functions that
the NCP provides in an IBM 37x5 communications controller. This is
shown in Figure 3.3b. SDLC-to-LLC2 converters receive and send
SDLC frames to the link-attached devices. The SDLC frames are converted to token-ring LLC2 frames before being sent out over the
token-ring network. As frames return from the mainframe, the LLC2
frames are converted to SDLC frames before they are sent out to the
SDLC device. These devices not only save reinvestment in 3270
access devices to support token-ring connectivity but also remove the
need for the two networks, SNA/SDLC and token-ring, to coexist.

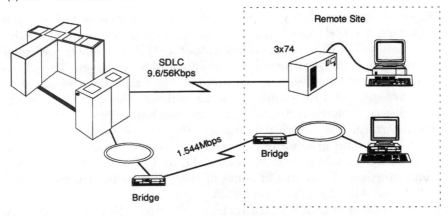

(a) Dual Communications Lines

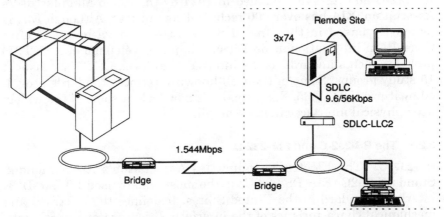

(b) Single Communications Line with SDLC-LLC2 Conversion

Figure 3.3 (a) Network configuration dilemma that has faced corporations investing in LANs alongside their SDLC networks. (b) Communications controllers to the SDLC devices.

SDLC-to-LLC2 conversion is provided by several vendors. Among them are Sync Research, NetLink, IBM, Cisco Systems, and WellFleet. Sync Research and NetLink have created devices whose sole purpose is to convert SDLC frames to LLC2 frames. They have each expanded on these functions to provide network management of the downstream SDLC devices from NetView residing on the mainframe as well as X.25, frame relay, and Ethernet support. The IBM, Cisco Systems, and WellFleet solutions are provided by each of the vendor's router offerings.

3.2   Physical Interface Standards

Physical interfaces provide the connectivity between Data Terminal Equipment (DTE) and Data Circuit-Terminating Equipment (DCE). As discussed in Chapter 1, an example of a DTE is a cluster controller or a communications controller. A DCE could be a modem. The physical interfaces between such equipment have long been established and are governed by standards organizations. The standards define the electrical characteristics of and the mechanical specifications for the interface. The electrical characteristics include voltage levels, current levels, and grounding requirements. The mechanical specifications include cable length (between DTE and DCE), design of the pin connector, etc. Examples of some of the physical interfaces are RS-232-C, RS-449, V.24, X.21, and V.35.

Different organizations are instrumental in setting up these standards. In the United States, the Electronic Industries Association (EIA) controls some of the most widely used standards, such as the RS-232-C and RS-449. Founded in 1924 by the Radio Manufacturers Association, EIA has over 200 technical committees. Although EIA is primarily based in the United States, the Geneva-based Comité Consultatif Internationale de Télégraphique et Téléphonique (CCITT) provides such standards at the international level. CCITT consists of 15 study groups. Some of the well-known interface standards provided by it are V.24, V.35, X.21, etc. Let's now look at some of the widely used physical interfaces in more detail.

3.2.1   The RS-232-C and EIA-232-D

RS-232-C.   The Electronic Industries Association's Recommended Standard 232C (EIA RS-232-C) is the most widely used DTE-to-DCE interface standard in the United States. It defines the electrical and mechanical characteristics of the interface. It is functionally compatible with the CCITT's recommended standard V.24. Some salient features of RS-232-C are:

- It supports asynchronous (start/stop) as well as synchronous (e.g., BSC or SDLC) transmissions.

- It supports data speeds up to 19,200 bps.

- It is applicable to point-to-point or multipoint configurations.

- It supports private as well as common carrier lines. In common carrier circuits, it supports switched as well as dial-up lines. Lines can be two-wire or four-wire.

- Transmission can be half duplex or full duplex.

- The mechanical interface consists of a 25-pin connector.

■ The recommended cable length between the DTE and the DCE is 50 feet. However, larger cable lengths are supported in case of low data rates or low-capacitance cables.

The standard defines the 25 pin assignments. Each pin has a specific function. The connector has 13 pins in the top row and 12 pins in the bottom row to make it difficult to make a wrong connection. Figure 3.4a shows the pin assignments, and Figure 3.4b gives the function of each pin.

Each pin has a specific function, and one has to follow the rules to

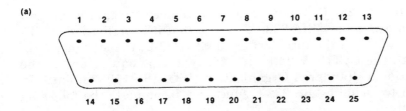

(a)

PIN NUMBER	PIN FUNCTION	V.24 EQUIVALENT
1	Protective Ground	101
2	Transmitted Data	103
3	Received Data	104
4	Request to Send	105
5	Clear to Send	106
6	Data Set Ready	107
7	Signal Ground / Common Return	102
8	Received Line Signal Detector	109
9	Reserved for Testing	-
10	Reserved for Testing	-
11	Not Assigned	-
12	Secondary Received Line Signal Detector	122
13	Secondary Clear to Send	121
14	Secondary Transmitted Data	118
15	Transmitter Signal Element Timing	114
16	Secondary Received Data	119
17	Receiver Signal Element Timing	115
18	Not Assigned	-
19	Secondary Request to Send	120
20	Data Terminal Ready	108.2
21	Signal Quality Detector	110
22	Ring Indicator	125
23	Data Signal Rate Selector	111 / 112
24	Transmitter Signal Element Timing	113
25	Secondary Clear to Send	-

(b)

Figure 3.4   (a) RS-232-C pin assignments; (b) RS-232-C pin functions.

be universally compatible. For example, pin 22 always transmits a *ring indicator* signal. In Figure 3.4b, you will also see the V.24 equivalent for the RS-232-C pin functions. V.24 is the CCITT equivalent for RS-232-C.

Since RS-232-C has limitations on the transmission speed and cable length, EIA came up with a new standard in 1977 called RS-449. Although RS-449 will be covered in a later section, it is important to know that you can use an "interface converter" to make RS-232-C-compliant equipment compatible with RS-449-compliant equipment.

EIA-232-D. RS-232-C was revised in November 1986, and the new standard was called "EIA-232-D." Notice that it is prefixed with EIA and not RS. With the introduction of EIA-232-D, the standard is in line with the CCITT V.24 and V.28. It is also equivalent to the International Standards Organization (ISO) IS 2110 standard. In addition to including RS-232-C functions, it supports the following new features:

- Local loopback
- Remote loopback
- Test mode

3.2.2 RS-449

Since RS-232-C could support only data rates up to 19,200 bps and also had cable length limitations, EIA came up with the RS-449 standard in 1977. This standard can support data rates up to 2 million bps, which is 100 times the maximum data rate for RS-232-C. RS-449 was developed in cooperation with ISO and CCITT; therefore, it is compatible with CCITT's V.24 and ISO's 4902 standards.

RS-449 comes with a 37-pin connector. The electrical characteristics are specified with two additional standards—RS-422-A for balanced circuits and RS-423-A for unbalanced ones. In 1980 the RS-449-1 standard was issued, which differed from the previous one in specifying changes to pin 34.

RS-449 is compatible with RS-232-C if an interface converter is used. However, you must not exceed the limitations imposed by RS-232-C in that case; i.e., the data rate must not exceed 19,200 bps, and the cable length between DTE and DCE must not exceed 50 feet.

RS-449 did not become a very popular standard. For data speeds below 19,200 bps, vendors still use RS-232-C. For data speeds above 19,200 bps, V.35 is the more commonly used interface.

3.2.3 V.35 interface

The explosion of LANs has required higher bandwidth on the network. Most corporate networks now have T1 backbones. The V.35 interface with its 34-pin assignment was developed by the CCITT to support this higher bandwidth. The highlights of V.35 are:

- It supports synchronous and isochronous transmission.

- Data speeds can be up to 1.544 Mbps.

- It is used in point-to-point configurations only.

- It supports common carrier lines using four-wire circuits in full-duplex transmission.

- The mechanical interface is a 34-pin connector.

- Typical cable length is up to 50 feet.

The V.35 standard defines its 34-pin assignments with specific functions. Instead of numbers, the CCITT standard uses letters to identify the pins. The interface has four alternating rows of 8 and 9 pins. Figure 3.5 illustrates the V.35 interface and its pin assignments.

3.2.4 Additional standards

V series interfaces. Besides RS-232-C, EIA-232-D, and RS-449, there are other standards that define the physical interface between DTE and DCE. Figure 3.6 gives the salient features of V series interfaces

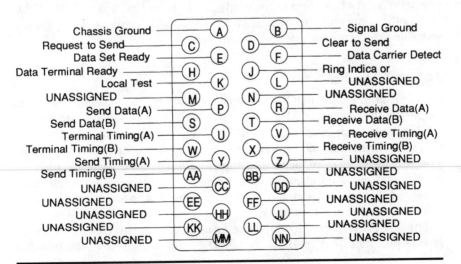

Figure 3.5 The V.35 interface and its pin assignments.

NUMBER	LINE SPEED	FDX / HDX	ASYNCHRONOUS / SYNCHRONOUS	SWITCHED LINE SUPPORT
V.21	300	FDX	Both	Yes
V.22	1200	FDX	Both	Yes
V.22 BIS	2400	FDX	Both	Yes
V.23	1200	HDX	Both	Yes
V.26	2400	FDX	Synchronous	No
V.26 BIS	1200	HDX	Synchronous	Yes
V.26 TER	2400	Both	Both	Yes
V.27	4800	Both	Synchronous	No
V.27 TER	4800	HDX	Synchronous	Yes
V.29	9600	Both	Synchronous	No
V.32	9600	FDX	Synchronous	Yes
V.35	48000	HDX	Synchronous	No

Figure 3.6 CCITT's V series interfaces for DTE-to-DCE connectivity.

developed by CCITT. Currently they are the most widely used standards in the world. They were last revised in 1984.

EIA-530. The EIA-530 uses the 25-pin mechanical connector of EIA-232-D and is intended to replace the RS-449 standard. It was approved in 1987 by EIA. It is used for data rates from 19,200 bps to 2 Mbps.

EIA-366-A. The EIA-366-A defines the interface for a DTE, DCE (modem), and Automatic Call Unit (ACU). It is a dial and answer system which gets the phone number from DTE and makes sure that the DCE is in operable condition. It also scans the call on a continual basis and determines when to disconnect.

3.2.5 X.21 interface

In traditional DTE-to-DCE interfaces, each pin has a specific function to perform. Therefore they are called pin-per-function interfaces. Examples of such interfaces are RS-232-C, CCITT's V.24, and RS-449. CCITT's X.21 interface uses coded character strings for each function. The pin-per-function technique puts a limitation on the number of functions supported because of the size of the mechanical connectors involved. However, a coded character string approach can help in assigning unlimited functions.

The X.21 interface was developed and approved by CCITT in 1972. It was last modified in 1980. The X.21 is the designated interface for

X.25-based packet switching networks. Examples of packet switching networks in North America are Tymnet, Telenet, and Datapac. The X.21 can also be used in environments which are not related to packet switching. Because of the extreme popularity of RS-232-C in the United States, X.21 has not really gained wide acceptance. Another reason for its lack of acceptance is the cost factor. It needs an intelligent device to interpret the character strings.

X.21 bis is the interim solution for X.25 packet switching networks. X.21 bis is compatible with RS-232-C and V.24 interfaces. Instead of pins, it defines 25 interchange circuits between DTE and DCE.

3.3 Terminals

Terminals provide the visual interface between human beings and the computer. Information requested is transmitted from the host to the terminal via the communications controller, modems, communications lines, and cluster controller. Display or printer terminals are attached to the cluster controller via a coaxial cable.

For the mainframe environment, IBM terminals can be classified in two categories:

1. The IBM 3270 family of terminals

2. ASCII terminal devices

3.3.1 The IBM 3270 family of terminals

The IBM 3270 family of terminals is the most widely used set of terminals in the SNA environment. If offers the user many choices and models to select from, each having different shapes, sizes, screen display capacity, color features, and other characteristics.

In an SNA environment, a terminal uses the IBM 3270 Data Stream protocol. An SNA/SDLC display terminal looks to VTAM like a Logical Unit 2 (LU2), while a BSC terminal looks like a 3270 device. A printer terminal in an SNA/SDLC environment looks like LU1 or LU3, and in a BSC environment, like a 3286. In fact, a 3270 display terminal or printer does not know about SNA/SDLC or BSC protocols. It is the cluster controller that makes it look like one. You may disconnect a 3278 terminal from a 3274 Model 41D (channel-attached non-SNA) and connect it to a 3274 Model 41A (channel-attached SNA/SDLC device) without any hardware consideration. Thus a terminal device is independent of the link-level protocols; all the intelligence to interpret them lies in the cluster controller. The following are some of the widely used IBM 3270 terminals:

IBM 3270 display station. This comes in five models. Model 1 displays 12 lines by 80 columns. Model 2 is the most widely used display terminal and comes in a 24-row by 80-column display. Model 3 displays 32 lines, while Model 4 displays 43 lines, both on 80-column screens. Model 5 gives a 27-row by 132-character display.

IBM 3178 display station. This is functionally equivalent to the 3278 Model 2 and provides a 24-row by 80-column display. Models C1, C2, C3, and C4 differ only in keyboard capabilities. The 3178 is less expensive than the 3278.

IBM 3180 display station. This is functionally equivalent to the 3278 Models 2 through 5. It comes in two models. Model 1 has four operator or program selectable screen formats. They provide the same capabilities as 3278 Models 2, 3, 4, and 5, respectively. Model 2 is used for connectivity to the System/36 and System/38.

IBM 3279 color display terminal. This comes in three standard models. Model S2A has a 24-line by 80-column screen and supports base color mode. Model S2B also supports extended color mode and extended highlighting. Model S3G is the top of the line and supports 32 lines by 80 columns, extended highlighting, extended color mode, and programmed symbols. Base colors are red, green, blue, and white. The extended colors are yellow, pink, and turquoise. Extended highlighting includes reverse video, blinking, and underscoring.

IBM 3179 color display station. This comes in four models. Model 1 can function in the 3279 Model S2A and S2B mode. The configuration depends upon how the cluster controller (e.g., the 3274) is configured. Model 2 works with System/36 and System/38. Models G1 and G2 have Model 1 characteristics and also support All-Points-Addressable (APA) graphics. In the graphics mode an eighth color, black, is also supported.

IBM 3191 display station. This comes in four models. Models A10 and B10 display green phosphor characters, while Models A20 and B20 provide amber-gold phosphor characters. All models display 24 lines by 80 columns.

IBM 3193 display station. This comes in two models. Both models support the equivalent of 3278 Model 2 (24X80), Model 3 (32X80), and Model 4 (43x80). In addition, they also support a display of 48 lines by 80 columns. A 3193 can also act as two logical terminals. It supports APA capability.

IBM 3194 color display station. This comes in two models. Both models support seven colors with a 24-row by 80-column display. It supports up to four host sessions and two local notepad sessions. Various sessions can be displayed simultaneously by manipulating the windows.

IBM 3270 personal computer. This can run up to four host sessions, one PC-DOS session, and two notepad sessions. Multiple sessions can be displayed simultaneously by using the windows. The 3270 PC is defined as a Distributed Function Terminal (DFT) when required to create multiple host sessions.

IBM 3290 information panel. This is a flat gas plasma display panel which has the capacity to display up to four host sessions in four different quadrants of the panel. Each quadrant can display a 24-row by 80-column screen. Although a 3270 PC can also display four host sessions simultaneously, they overlap each other as far as the display is concerned. In a 3290, you can have concurrent viewing of up to four 3278 Model 2 screens or two 3278 Model 3 screens or two 3278 Model 4 screens or two 3278 Model 5 screens. It can also display 62-row by 132-column full-resolution graphics. A 3290 is defined as a DFT terminal (four session maximum) rather than a Control Unit Terminal (CUT) to the VTAM. A CUT-mode terminal can have only one host session. Figure 3.7 summarizes the characteristics of 3270 display terminals.

3.3.2   The IBM 3270 family of printers

The 3270 family of printers communicates with the host via a cluster controller. They are defined to VTAM as LU1 or LU3 and are called VTAM printers. Their physical interface to the cluster controller is through a coaxial cable. They are different from the high-speed JES printers, which are completely under the control of a spooling software such as JES2 or JES3. In comparison, JES printers are either channel-attached to a host or connected via an RJE/RJP device through a communications controller to the host. Printouts to a VTAM printer are sent by a VTAM application such as CICS or IMS/DC. Printouts to a JES printer are sent by JES2 or JES3 and are controlled through MVS JCL statements. The following are some of the commonly used VTAM printers.

IBM 3268 matrix printer. This prints at a speed of 340 characters per second (cps) and can support six-part paper. It can look like an LU1 or LU3 to VTAM. Maximum line length is 132 characters. IBM 3268 Model 2C is a color printer which supports base colors (black, red,

TERMINAL	COLOR	SCREEN SIZE ROWS x COLUMNS	GRAPHICS SUPPORT	DFT / CUT *	COMMENTS
3278	Green	12 x 80, 24 x 80, 32 x 80, 43 x 80, 27 x 132	No	CUT	• Large User Base • Comes in 5 Models
3178	Green	24 x 80	No	CUT	• Functionally Equivalent to 3278 Models 2 thru 5
3180	Green	24 x 80, 32 x 80, 43 x 80, 27 x 132	No	CUT	• Equivalent to 3278 Model 2
3279	Multicolor	24 x 80 32 x 80	Optional	CUT	-
3179	Multicolor	24 x 80	No	CUT	-
3191	Green or Amber	24 x 80	No	CUT	-
3193	Black / White	24 x 80, 32 x 80, 43 x 80, 48 x 80	Images	CUT	• Can Be Two Logical Terminals
3194	Multicolor	24 x 80	No	CUT	• Can Also Be Monochrome
3270 PC	Multicolor	24 x 80	Yes	DFT	• One PC-DOS Session • Two Notepad Sessions
3290	Amber	24 x 80, 32 x 80, 43 x 80, 27 x 132	Yes	DFT	• Flat Gas Plasma Display • Each Session is in a Different Quadrant

* CUT = Control Unit Terminal; Single Host Session
 DFT = Distributed Function Terminal; Up to Four Host Sessions

Figure 3.7   Characteristics of IBM 3270 display terminals.

blue, and green) and extended colors (pink, yellow, and turquoise). The base colors are selected at the field level, which may consist of many characters. Extended colors can be selected at the individual character level.

IBM 3262 line printer.   These are relatively fast printers, and their speed can range from 12 to 650 lines per minute (lpm). They do not come in color models.

IBM 3287 matrix printers.   These are the desktop models and are relatively low-speed printers. Model 1 operates at 80 cps, while Model 2 supports a 120-cps print speed. Both can print up to 132 characters on a line. Models 1C and 2C are the color models, but they support colors from position 1 through position 120 only. Positions 121 through 132 can be printed only in black.

IBM 4250 APA printer.   This is a very high-resolution APA printer with a resolution of 600 dots per inch (dpi). It supports multiple fonts, typefaces, and graphics objects. Figure 3.8 summarizes the characteristics of the IBM 3270 family of printers.

3.3.3   ASCII terminals

While the most widely used terminals in the IBM environment are the 3270-family terminals, for other vendors they are the ASCII terminals. While the 3270 terminals are connected to a cluster controller via a BNC connector and a coaxial cable, the ASCII terminals are usually connected through an RS-232-C interface. IBM entered the ASCII terminal market in 1979 with the introduction of the 3101 terminal. Other popular vendors for ASCII terminals include TeleVideo, Wyse Technology, and Qume. Characteristics of ASCII terminals are as follows:

IBM 3101 terminal.   This is a TTY-compatible terminal which comes in two models. Model 13 supports character mode transmission, while Model 23 supports both character and block mode. In the latter case, either mode is switch selectable. It supports both the RS-232-C and RS-422-A communications interfaces, and is also switch selectable.

PRINTER	COLOR / MONOCHROME	WIDTH IN CHARACTERS	SPEED MAXIMUM	COMMENTS
3268	Both Models	132	340 CPS	-
3262	Monochrome	132	650 LPM	-
3287	Both Models	132	120 CPS	-
4250	Monochrome	-	-	All Points Addressable with 600 dots per inch resolution

Figure 3.8   Characteristics of IBM 3270 printers.

With the introduction of the 316X family of ASCII terminals in 1985, it looks like IBM is phasing out the 3101 models.

IBM 316X family of terminals. This series consists of four models. The IBM 3151 supports 24-row by 80-column display and comes in green or amber character display. The 3162 supports 24-row by 80-column or 80-row by 132-column display and also comes in green or amber character display models. The 3163 supports 24-row by 80-column display and has memory for 7680 characters. The 3164 is a color model. All models support RS-232-C and RS-422-A physical interfaces. They emulate certain well-known and established ASCII terminals such as DEC's VT52, VT100, and VT220, Lear Siegler, ADDS Viewpoint, Hazeltine 1500, TeleVideo 910, WYSE, and a few others.

Figure 3.9 summarizes the characteristics of IBM ASCII terminals.

3.3.4   IBM 3299—the terminal multiplexer

To provide physical connectivity between a terminal and a cluster controller, you need a coaxial cable. If the cluster controller can provide attachments for 32 devices, you need to run up to 32 individual coaxial cables between the controller and the devices. The cost of running cables under these circumstances can be fairly high. The problem is more severe if the controller and the devices are on different floors. Old buildings in metropolitan areas do not have risers large

TERMINAL	COLOR	SCREEN SIZE	GRAPHICS SUPPORT	EMULATION
3101	Green	24 x 80	No	TTY
3161	Green / Amber	24 x 80	Line Drawing Set	Televideo 910, Hazeltine 1500, Leer Siegler, ADDS Viewpoint, ADM3A
3162	Green / Amber	24 x 80 80 x 132	Line Drawing Set	WYSE, Hazeltine 1500, ADDS, Televideo, VT52, VT100, VT220, Lear Siegler
3163	Green / Amber	24 x 80	Line Drawing Set	VT52, VT100
3164	Multicolor	24 x 80	Line Drawing Set	VT52, VT100

Figure 3.9   Characteristics of IBM ASCII terminals.

enough to accommodate the cable requirements for increased use of terminals. Under these circumstance, you need a device which can concentrate data from multiple terminals in one physical location of a floor and provide connectivity to the cluster controller on a single cable. The IBM 3299 terminal multiplexer provides such capabilities. It can concentrate data from up to eight category A devices and transmit it over a single cable to the controller (Figure 3.10).

As shown in Figure 3.10, there are up to eight cables going from the terminals to the 3299, but only a single cable going from the terminal multiplexer to the cluster controller. Thus you can run only four cables from four different terminal multiplexers to provide connectivity for up to 32 terminals. The IBM 3299 comes in two models, 2 and 3. Model 2 can use a coaxial cable as well as the IBM cabling system. Model 3 can be used to attach terminals using twisted-pair wire. Since the terminals have a coaxial connector at their end, you need a coax-to-twisted-pair adapter for terminal-to-3299 connectivity. An IBM 3274 Model 51C is *not* supported by the terminal multiplexers.

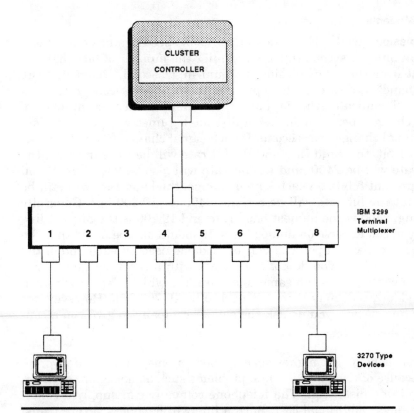

Figure 3.10   Use of an IBM 3299 terminal multiplexer.

3.4   Modems and Multiplexers

For computer equipment to communicate with each other, there must be a physical medium over which the data can travel. We learned in Chapter 1 that such computer equipment is Data Terminal Equipment, or DTE for short. The physical medium can be copper wire, optical fiber, microwave, or any other possible medium. The medium can be owned by the enterprise, or the services for such a medium can be provided by a common carrier such as AT&T, MCI, Sprint, and the Bell holding companies. In either case, the transmission technology can be based on analog or digital transmission. The equipment that provides the interface between the DTE and the communications media is Data Circuit-Terminating Equipment, or DCE for short. In the case of analog transmission lines, the DCE is a modem. For transmission media that provide digital facilities, the DCE is DSU/CSU. Figure 3.11 gives a pictorial representation of this.

3.4.1   Modems

Transmission speed.   The data transmission speed of a modem is measured in bits per second (bps). It signifies the number of bits that can be sent over the transmission medium in 1 second. Generally, the term "baud" is also used to represent the bits per second rate for a modem. Technically, this is inaccurate. Baud denotes the number of signal changes per second. Generally, analog lines can accommodate 2400 signal changes per second. If each signal change can be translated into 1 bit, the baud rate and the bit rate will be the same; i.e., the baud rate will be 2400, and the bit rate will also be 2400 bps. If you can represent 8 bits on each signal change, the baud rate will still be 2400, but the bit rate will be 8 times 2400, or 19,200 bps. Generally speaking, 2400 is the highest baud rate and 19,200 is the highest line speed you can achieve on analog lines. It should also be noted that the line speed depends upon the pair of modems used on each end of the transmission line. You do *not* order lines with a certain speed from a common carrier. It is the same four-wire line which, depending upon the modem used, may transmit at 110 or 19,200 bps. However, for higher line speeds, you do ask the carrier company to give you a conditioned line.

Line conditioning.   At higher speeds, modems may not function properly because of transmission line problems such as noise, phase jitter, or amplitude distortion. The telephone company can supply a conditioned line at additional cost. Such a line will have less severe trans-

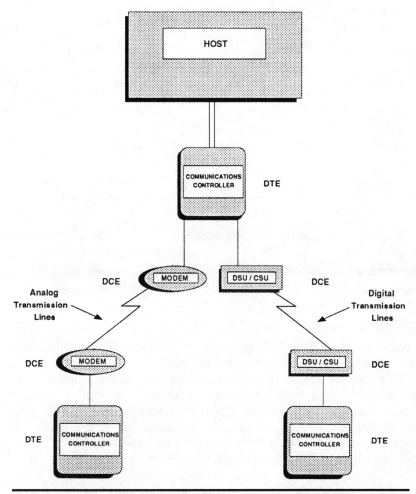

Figure 3.11 A diagram showing the relationship between DTEs, DCEs, modems, and DSU/CSUs.

mission problems. A conditioned line generally involves the use of amplifiers and attenuation equalizers.

Modem standards.   In the predivestiture era, AT&T's standards were the most widely used modem standards in the United States. Although this is changing very rapidly, they are still being used quite extensively. Conforming to such standards ensured that two different modems from two different vendors would work with each other without any problems. AT&T's modem specifications define standards for

NAME	TRANSMISSION RATE (BPS)	ASYNCHRONOUS / SYNCHRONOUS	OTHER CHARACTERISTICS
103	300	ASYNCHRONOUS	POINT-TO-POINT
212	1200	BOTH	POINT-TO-POINT
201B,C	2400	SYNCHRONOUS	POINT-TO-POINT
208A,B	4800	SYNCHRONOUS	POINT-TO-POINT MULTIPOINT (for 208A)
209	9600	SYNCHRONOUS	POINT-TO-POINT MULITPOINT CONDITIONING REQUIRED

Figure 3.12 Some of the more commonly used AT&T standards for modem specifications.

data rates varying from 300 to 9600 bps. Figure 3.12 outlines the characteristics of such modem specifications.

CCITT's modem standards, though biased toward European telephone facilities, are more international in nature. Generally known as V standards for modem specifications, they specify line speeds from 200 to 14,400 bps. Figure 3.13 outlines some of the commonly used specifications and their characteristics. The suffixes "bis" or "ter" signify the second or third iteration of the standard, respectively.

Limited Distance Modems (LDMs).   Also called short-haul modems, they are used to transit data over short distances. Both the DTEs for such transmissions are usually in the same location. LDMs cost less than regular modems and are used to bypass the telephone company's transmission facilities for short-distance communications.

Modem eliminators.   When distances between DTEs are less than 1000 feet, you may replace two LDMs with a single modem eliminator. A modem eliminator sits midway between the DTEs. By using low-capacitance cable, you may increase the distance between DTEs to more than a thousand feet.

Line drivers.   They act as repeaters and can drive a signal from a few hundred feet to several miles. Generally, they are used to connect DTEs within the same premises.

NAME	TRANSMISSION RATE (BPS)	ASYNCHRONOUS / SYNCHRONOUS	OTHER CHARACTERISTICS
V.22	300	BOTH	POINT-TO-POINT
V.22 BIS	1200	BOTH	POINT-TO-POINT
V.26	2400	SYNCHRONOUS	POINT-TO-POINT MULTIPOINT (for 208A)
V.26 BIS	2400	SYNCHRONOUS	POINT-TO-POINT
V.27, V.27 BIS	4800	SYNCHRONOUS	POINT-TO-POINT MULTIPOINT (for 208A)
V.27 TER	4800	SYNCHRONOUS	POINT-TO-POINT
V.29	9600	SYNCHRONOUS	POINT-TO-POINT MULTIPOINT (for 208A)
V.32	9600	SYNCHRONOUS	POINT-TO-POINT
V.33	14400	SYNCHRONOUS	POINT-TO-POINT

Figure 3.13   Some of the commonly used CCITT standards for modem specifications.

3.4.2   Data Service Unit/Channel Service Unit (DSU/CSU)

DSU/CSUs are to digital lines what modems are to analog lines. They are required for AT&T's Dataphone Digital Service (DDS) and Accunet T1.5 Service. DDS is available at line speeds of up to 56 Kbps. Accunet T1.5 supports digital signals at 1.544 million bps. Such service is also commonly known as T1 lines.

For digital services, modems are not required at all. All the necessary interface functions such as timing, control signals, signal translation, and data regeneration are supplied by the DSU/CSU. The physical interface between the DTEs and DSU/CSUs for 56-Kbps DDS services is the CCITT's V.35. Recall from Chapter 2 that the V.35 interface exists on LIC Type 3, which is supported by LAB Type B.

3.4.3 Multiplexers

Corporations have been searching for ways to reduce communication line costs to remote locations. The drop in T1 leased-line costs has made it possible for corporations to deploy T1 service to many of their larger sites. The use of a multiplexer can divide the total T1 bandwidth into smaller bandwidths through time-division multiplexing. The advantage can be seen in Figure 3.14. In Figure 3.14a, a cor-

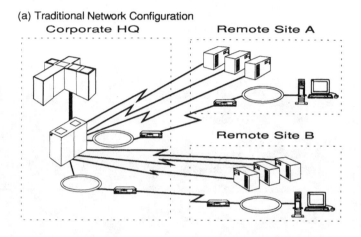

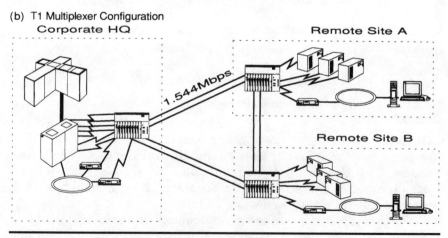

Figure 3.14   (a) The need for several leased lines to support remote devices in remote locations. (b) The new configuration with T1 service and multiplexers.

porate network has three locations. The corporate office is the location of the computing center, and the remote sites are large corporate offices. Figure 3.14a depicts the need for several leased lines to support the various remote devices in each remote location. These devices can be LAN bridges and routers, establishment controllers, communications controllers, or minicomputers. Using a T1 connection to each location in combination with a multiplexer will provide the same logical connections to the remote site, but over a single leased-line connection.

As shown in Figure 3.14b, the new configuration with T1 service and multiplexers provides the same leased-line connectivity for the remote site location, but reduces the number of leased lines. The multiplexer can provide up to twenty-four 64-Kbps lines over the T1. By using software provided by the multiplexer vendor, any combination of line bandwidths can be accommodated over the T1. As seen in the figure, several different bandwidths are provided between the remote sites, with extra bandwidth left over for future expansion to cover possible voice or image transmission.

As an added feature, many multiplexer vendors provide for alternative routing. For example, in Figure 3.14b, if the link from corporate headquarters to remote site B were to fail, communication to corporate headquarters could still be provided by the T1 link from site B to site A to corporate headquarters. In a design such as this, recovery for link outages is built into the network.

For many corporations, a full T1 may not be necessary. Many long-distance carriers provide an alternative to T1 service called fractional T1. Fractional T1 is provided to reduce high-speed line costs but still give the ability to multiplex the fractional T1.

3.5 IBM 3172 Interconnect Controller

The IBM 3172 Interconnect Controller connects the IBM SNA host to non-SNA hosts or workstations. The IBM 3172 uses the PS/2 Micro Channel Architecture–based, Intel 80386 or 80486 microprocessor–based computers. The Model 001 version comes with four LAN adapter slots and up to two channel adapter interfaces. The Model 002 increases the LAN adapter capabilities by one. The Model 003 is similar to the Model 001, but adds another LAN adapter slot and off-loads TCP/IP processing from the mainframe. The conceivable SNA physical unit support on an IBM 3172 is 1020. The IBM 3172 is called an interconnect controller because it connects dissimilar networks, as opposed to connecting similar networks as bridges do.

The IBM 3172 can be used at token-ring data rates of 4 or 16 Mbps as well as providing Ethernet 10-Mbps support, and IBM has stated

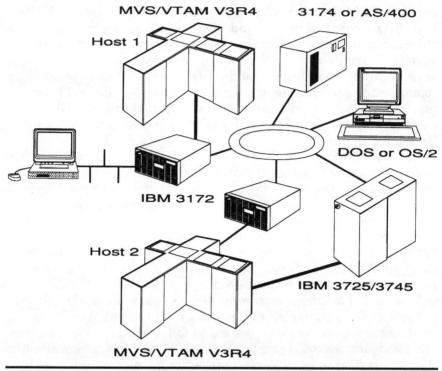

MVS/VTAM V3R4   3174 or AS/400

Host 1

IBM 3172

DOS or OS/2

Host 2

IBM 3725/3745

MVS/VTAM V3R4

Figure 3.15   Some of the possible configurations for the IBM 3172 Interconnect Controller.

an intention of supporting FDDI data rates of 100 Mbps. Additionally, the IBM 3172 can support SNA VTAM channel-to-channel connections using remote telecommunications lines as well as provide SNA Intermediate Network Node (INN) traffic through the token-ring adapter. Figure 3.15 diagrams some of the possible configurations for the IBM 3172 Interconnect Controller.

The IBM 3172 can be remotely supported through its remote configuration support feature. Using the Remote Configuration Support (RCS) program on a token-ring-attached OS/2 Extended Edition V1.2 or higher, the IBM 3172 can be configured, upgrades to the Interconnect Controller Program can be installed, and the error log can be accessed by the OS/2 RCS workstation for analysis. A second feature provided by the IBM 3172 is the ability to send alarm information to a communications network management application residing on the SNA host. This information can be useful in determining problems with stations off of the IBM 3172 or the IBM 3172 itself.

3.6 IBM 8228 Multistation Access Unit

The star-wired topology of the IBM Token-Ring Network has as its foundation the IBM 8228 Multistation Access Unit (MAU). The MAU itself will form the ring through an internal relay wiring mechanism without external power and is sometimes called a Passive Wiring Concentrator (PWC). Each MAU has ten ports. Eight of these are used to connect LAN stations. The remaining two are used for connection between MAUs, repeaters, or converters to extend the size of the ring. One port is called Ring-Out (RO), and one is called Ring-In (RI). The RO is the transmit port of the MAU. The RI is the receive port for the MAU. Multiple MAUs can be connected by attaching the RO of one MAU to the RI of the next MAU. Up to 33 MAUs can be interconnected in this way, allowing up to 260 devices to be attached and operational.

Guidelines for using Type 3 media allow for up to nine IBM 8228s, for a maximum of 72 attached devices. When using Type 6 cable, a total of 12 IBM 8228s and up to 96 devices can be attached to the ring. Figure 3.16 illustrates a small token-ring network using interconnected multistation access units.

3.7 IBM 8230 Controlled Access Unit

Controlled Access Units (CAUs) have their own power and thus can have built-in intelligence. The IBM 8230 CAU has two components: the base unit and the Lobe Attachment Module (LAM). A maximum of four LAMs can be attached to the base unit. Each LAM can attach 20 ring

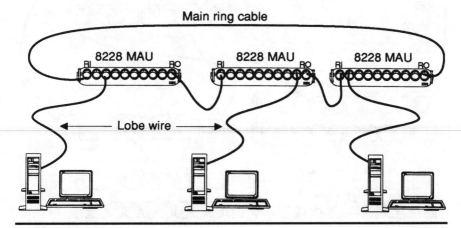

Figure 3.16 A small token-ring network using interconnected multistation access units.

stations. There are two types of LAMs that can be installed: LAMs with lobe interface ports supporting the IBM IEEE 802.5 connector ("ugly plug") or an RJ-45 connector to support Type 3 media. These LAM types can be mixed on the IBM 8230. LAMs can be installed or deinstalled nondisruptively to the stations on the ring while the CAU is operational. If you configure the IBM 8230 without the use of LAMs, it can be used as a repeater operating at either 4 or 16 Mbps. An IBM 8230 CAU can provide 80 ring stations with attachment to the ring.

The IBM 8230 can support either 4- or 16-Mbps data rates on the ring. However, if a Type 3 unshielded twisted-pair (UTP) medium is used for the lobes, the CAU will support only a 4-Mbps data rate with the use of the IBM 8230 4-Mbps media filter installed on the base unit. Both copper and fiber can be used for the RI and RO connections of the CAU. The connections can be both copper, both fiber, or one of each. Using these connections, the IBM 8230 can be installed with the IBM 8228. The IBM 8228 may be installed between connecting IBM 8230 CAUs, but cannot be connected to the CAU lobe ports. This configuration can be seen in Figure 3.17. The connection to the IBM repeaters and converters does not allow the IBM 8230 to be half of the repeater/converter pair.

3.8 Summary

In Chapter 2, we learned about the communications controllers and cluster controllers. In this chapter, we discussed data communica-

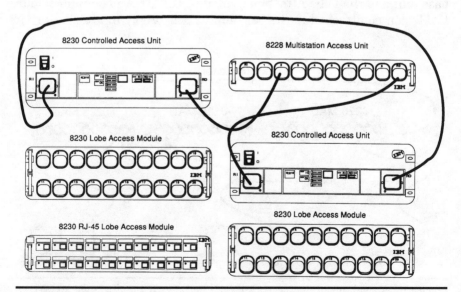

Figure 3.17 The IBM 8228 may be installed between connecting IBM 8230 CAUs, but cannot be connected to the CAU lobe ports.

tions, telecommunications, and local area network equipment. The IBM 3708 and 3710 are used for concentration and protocol conversion. We learned how SNA/SDLC devices can be attached to token-ring networks through the use of SDLC-to-LLC2 converter devices. Physical interfaces such as RS-232-C, RS-449, and V.35 provide interface between DTEs and DCEs. We also discussed various terminal types available in the 3270 and ASCII family. In the end we talked about modems for analog lines and DSU/CSUs for the digital lines. Modems and DSU/CSUs provide the DCE function in a network.

4

Workstations and SNA Connectivity

The advent of local area networking with the need to access data residing on the mainframe has fostered various ways for workstations to access the mainframe. Some have already been discussed, but here we will focus mainly on workstation access.

4.1 DOS and OS/2

Personal computers have become a staple for doing business in corporations. These desktop computers function as personal computers and cluster controller 3270 terminal devices. Using High-Level Language Application Program Interfaces (HLLAPI), Advanced Peer-to-Peer Communications (APPC), and Transmission Control Protocol/Internet Protocol (TCP/IP), these personal computers can provide access to data anywhere in the enterprise.

4.1.1 DOS workstations

The majority of personal computers in corporate networks have the Disk Operating System (DOS) installed, running Microsoft Windows. Workstations executing DOS can access SNA applications using any of the three functions mentioned previously. HLLAPI is provided by the SNA emulation software supplier. Some of the better-known providers of this type of emulation are IBM, Attachmate, Rumba, and DCA. The HLLAPI feature allows DOS applications to access SNA applications residing on the mainframe. This can be seen in Figure 4.1. It does this by providing 3270 emulation capabilities without the need to display

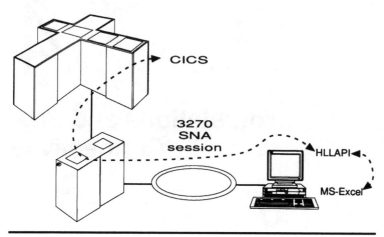

Figure 4.1 The HLLAPI feature providing SNA access for a DOS application.

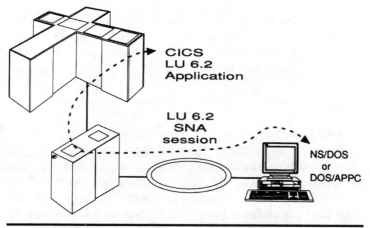

Figure 4.2 APPC under DOS providing LU6.2 session to the mainframe.

the actual screen provided by the application. HLLAPI will let a DOS application, for instance Microsoft Excel, access data from a CICS display for inclusion in an Excel spreadsheet.

APPC under DOS can provide LU6.2 session capability with an LU6.2 application residing on the mainframe. This configuration is depicted in Figure 4.2. Likewise, an LU6.2 application on the mainframe can access data that resides on the DOS workstation. As an example, the shipping accounting department keeps a daily record of accounts on a DOS workstation. The corporate accounting applica-

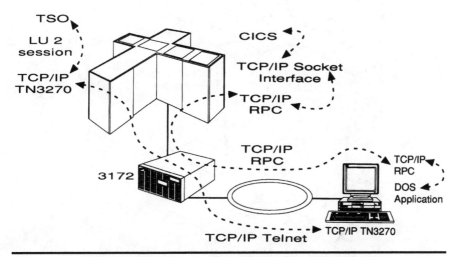

Figure 4.3 Example of a TCP/IP configuration from a PC to the mainframe.

tions residing on the mainframe can execute a CICS LU6.2 application that can access the data on the DOS data base for updates to the
corporate data base residing on the mainframe.

Instead of using SNA protocols like LU2 and LU6.2 to gain access to
SNA applications and their data, DOS workstations can use TCP/IP.
Figure 4.3 illustrates a TCP/IP configuration. TCP/IP provides a facility
called Remote Procedure Call (RPC) which allows a TCP/IP host (DOS
workstation) to execute a transaction on a remote TCP/IP host (CICS on
the mainframe). In the figure, a DOS workstation issues a CICS transaction using an RPC, and the result from the transaction is delivered to
the DOS workstation. TCP/IP providers for DOS also include a program
called TN3270. TN3270 allows full-screen 3270 emulation for SNA
applications without using SNA for the transport mechanism. Instead,
it uses TCP/IP architecture for the transport mechanism to gain full-
function 3270 data stream support. To provide this type of function,
TCP/IP must be installed on the mainframe, and SNA LUs for TCP/IP
to gain access to the SNA application must be defined.

4.1.2 OS/2 workstations

The latest release of OS/2 V2.2 includes full APPN functionality.
Therefore an OS/2 workstation can participate as an APPN Network
Node (NN) or as an APPN End Node (EN). As such, LU6.2 comes as
standard support for OS/2.

OS/2-based workstations have many of the same capabilities as
DOS workstations. However, an OS/2 workstation is a multitasking

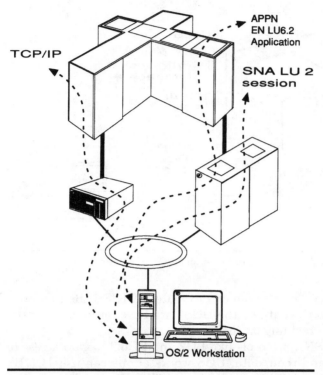

APPN
EN LU6.2
Application

TCP/IP

SNA LU 2
session

OS/2 Workstation

Figure 4.4 Multiple concurrent sessions from an OS/2 workstation.

operating system servicing multiple tasks asynchronously. As such, an OS/2 workstation can provide concurrent SNA and TCP/IP access to the mainframe. As shown in Figure 4.4, an OS/2 workstation can have multiple concurrent sessions, with an SNA application (CICS), an APPN application on an end node, and a TCP/IP application (FTP) all executing simultaneously on different mainframes.

4.2   UNIX Workstations

The surge in "right-sizing" mainframe applications has caused an explosion in the use of UNIX-based workstations to manipulate corporate-based data. There are several reasons why this has taken place. One reason is that the processing power provided by these workstations makes it possible to migrate applications that were once maintained on the mainframe. A second reason is the cost-effectiveness of these workstations. In many corporations the MIS department will create a chargeback system to recover the cost of operating a mainframe-based data center. A properly sized UNIX-based workstation can have a Return On Investment (ROI) of a few months if its cost is compared with the amount of money the department is charged for processing data on the mainframe. A third reason for the "right-sizing" effort is the quick responsiveness for new applications and the use of relational databases such as Sybase and Oracle. The cost-effectiveness, processing power, and responsive application generation of UNIX-based systems have fostered the need for interoperability between mainframe data and "right-sized" data.

Many vendors of UNIX-based workstations provide both SNA and TCP/IP capabilities. IBM provides the AIX UNIX-based system for its RS/6000 and mainframe-based offerings. Though AIX can execute on the mainframe, it is mostly found on the RS/6000 platform. As shown in Figure 4.5, IBM offers SNA and APPN support for its RS/6000. Similarly, Hewlett-Packard (HP) and Sun provide SNA-type connectivity but have yet to announce support for APPN.

All UNIX-based workstations come with the TCP/IP suite of applications. In Figure 4.5, an application on a UNIX workstation can issue a Structured Query Language (SQL) request both to a UNIX Sybase or Oracle server and to DB2 on the mainframe. The resulting informa-

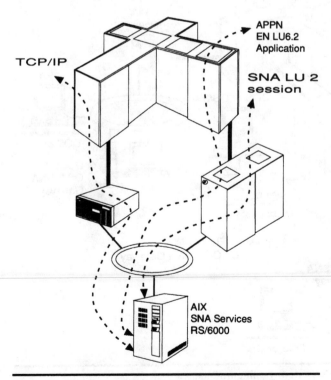

Figure 4.5 Example of a UNIX workstation with multiple sessions.

tion can be manipulated and presented to the end user as if only one request was made. This is an example of true cooperative processing. You will find that more and more you will become involved in providing SNA connectivity for non-SNA resources as the right-sizing effort and deployment of UNIX workstations moves forward.

4.3 SNA Gateways

Gateways provide a means for workstations to share resources and to access mainframe resources through nontraditional SNA network designs. IBM provides several different gateways for SNA connectivity. IBM-provided gateways are the OS/2 workstation, IBM 3174, AS/400, IBM 3172 Interconnect Controller, and IBM 37x5.

Figure 4.6 shows a configuration in which an OS/2-based workstation provides SNA connectivity for DOS-based and other OS/2-based workstations on a token-ring network. The downstream workstations

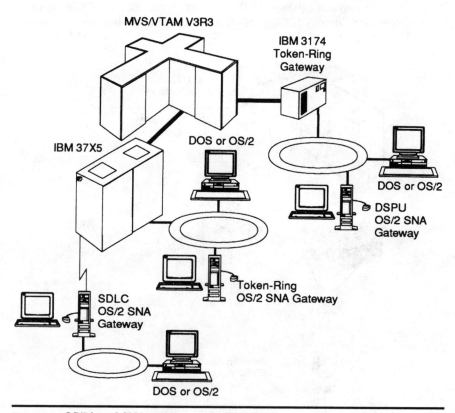

Figure 4.6   OS/2 based SNA gateway configurations.

to the OS/2 gateway can use standard LU2-based SNA protocols to the mainframe or dependent LU6.2 access to the mainframe. The use of an OS/2 workstation as the gateway for a new token-ring installation is more cost-effective than using an IBM 3174.

The IBM 3174 is quickly becoming more than just a cluster controller. With the most recent announcements, it is heading more toward being a communications controller. The IBM 3174, with token-ring or Ethernet capability, is well suited as an SNA gateway for an installation that has traditional coax-attached 3270-type terminals and is implementing token-ring. The new OS/2- or DOS-based workstations on the token ring can utilize the network access provided by the IBM 3174 to gain access to SNA applications on the mainframe. Figure 4.7 illustrates the use of the IBM 3174 as an SNA gateway for both OS/2 and DOS workstations. The IBM 3174 can be either channel-attached

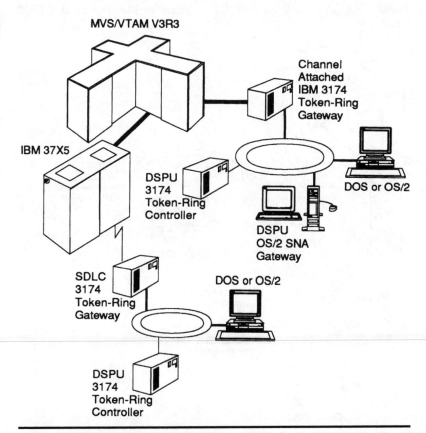

Figure 4.7 IBM 3174 SNA gateway configurations.

to the mainframe, SDLC link-attached to a communications controller, or token-ring-attached to a communications controller. In each configuration the IBM 3174 can provide up to 255 downstream PU Type 2.0 or SNA Node Type 2.1 connections. Note that the IBM 3174 can also provide SNA network access to other gateways that are downstream to the IBM 3174. These other gateways can be IBM 3174s, OS/2 gateways, or AS/400s. This ability to cascade gateways downstream to the IBM 3174 extends the reach of SNA to local area networks.

Installations that utilize an AS/400 can provide similar function for resources downstream to it on a token-ring network. Terminals that are usually used for connection to AS/400 applications can also reach SNA mainframe applications. This is shown in Figure 4.8.

For workstations on a token-ring network that have full SNA node Type 2.0 functionality, the IBM 3172 Interconnect Controller can provide a gateway function to the mainframe. The IBM 3172 can provide up to 1020 SNA Type 2.0 connections. Future releases of the IBM

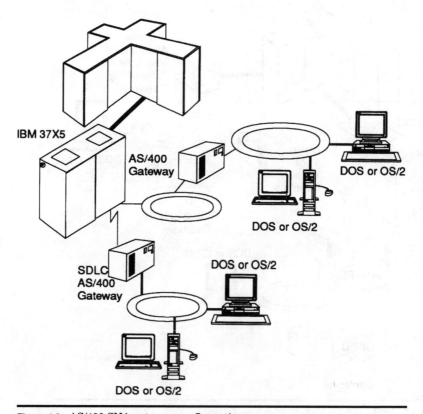

Figure 4.8 AS/400 SNA gateway configurations.

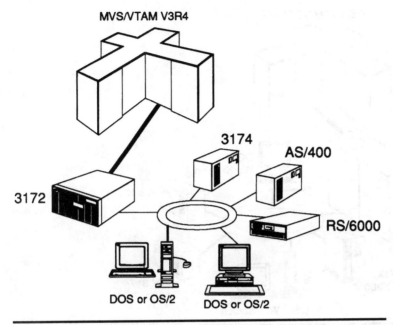

Figure 4.9 The IBM 3172 as an SNA gateway.

3172 Interconnect Controller Program (ICP) with OS/2 V2.2 will provide APPN NN capabilities, which will also allow it to support SNA Node Type 2.1 devices as well. As shown in Figure 4.9, the IBM 3172 allows a large number of SNA Type 2.0 devices to gain access to the mainframe without going through a communications controller. The advantage to using an IBM 3172 is that it is not as prone to downtime as a communications controller. A disadvantage to the IBM 3172 is that mainframe cycles are used to service the token-ring devices attached through the IBM 3172.

The IBM 37x5 communications controllers provide a wide array of connectivity options for network designers. These devices can provide gateway access, either channel-attached or remotely attached, for other gateways, as shown in Figure 4.10. IBM 37x5s can handle thousands of downstream devices that enter the network through the token-ring adapter on the IBM 37x5. As alluded to in the previous paragraph, the downside to using an IBM 37x5 is that the Network Control Program (NCP) must be loaded for changes to take permanent effect. The loading of the NCP makes the communications controller adapters inoperable, making connectivity to this communications controller impossible during the loading phase. However, the NCP still performs its main duty for token-ring- and Ethernet-

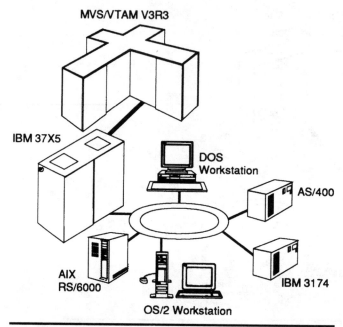

Figure 4.10   The IBM 37x5 Communication Controller configured as an SNA gateway.

attached devices, and that is to off-load processing cycles from the
mainframe for remote devices.

4.4 Summary

In this chapter we examined some of the ways workstations can be
networked to access SNA applications residing on the mainframe. We
read how OS/2 and DOS workstations can be used to access SNA
mainframe applications and how OS/2 workstations can be used as
SNA gateways for downstream token-ring-attached OS/2 and DOS
workstations that emulate IBM 3270 LU2 devices. We also explored
how UNIX-based workstations have created a right-sizing effort and
how these machines can be used in a cooperative processing environment with SNA-based applications. Finally, we reviewed the various
SNA gateways, their capabilities, and their use in providing SNA
access for LAN resources. Now that we have the necessary background in some aspects of the IBM SNA environment, we can learn
about SNA concepts and architecture.

SNA and Telecommunications Access Methods

5.1 Systems Network Architecture Concepts

The expanding thirst for quick and accurate retrieval of information by government, business, and the general public has necessitated rules to govern the interaction of components within the information network. A hierarchical structure composed of seven layers has become the cornerstone for handling the information demand. It was developed by IBM and is known as Systems Network Architecture (SNA).

5.1.1 Network components

A communications network comprises hardware and software components. As seen in Figure 5.1, these components include the processor, communications controller, cluster controller, workstation, distributed processor, and printer. These components make up the hardware portion of the network. The software components are found in the two main components of the hardware. The cornerstone of all communication software executes on the main processor and is known as the telecommunications access method. It is this software that correlates and manages the entire network. To allow the expansion of a network to remote end-users, a network control program was developed and executes in the communications control unit. The network control program defines to the telecommunications access method the topology of the remote end-user network. The last and most important piece of SNA software to the end-user is the application subsystem. It executes on the main processor and uses the other two components of

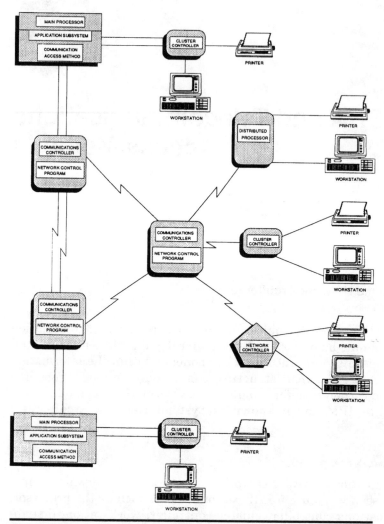

Figure 5.1   Network components.

SNA software to communicate to the end-user. The consistency of
data integrity between the end-user and the application subsystem
must be provided by an addressing scheme that addresses the individual end-user workstations and the corresponding requested information from the application subsystem.

5.1.2   Nodes

As we discussed, an SNA network consists of hardware and software
components. The hardware and its associated software components

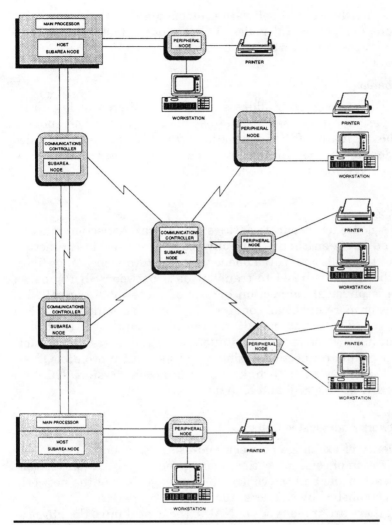

Figure 5.2 SNA nodes.

that implement the functions of SNA are called "nodes." In SNA there are three types of nodes: (1) host subarea nodes, (2) communications controller subarea nodes, and (3) peripheral nodes, as depicted in Figure 5.2.

Host subarea nodes control and manage the network. The hardware and software components that make up this node are the main processor and the telecommunications access method. The communications controller subarea nodes route and control the flow of data through a network. These nodes are associations of the communications controller and a network control program.

All other hardware and software components and their respective associations are peripheral nodes. This includes cluster controllers, distributed processors, workstations, and printers.

5.1.3 Subareas

A subarea is a designated address given to a subarea node and its attached peripheral nodes. This address informs SNA which subarea node is the originator of the information being transmitted and is also used to denote the destination subarea node of the transmission. Figure 5.3 diagrams subarea nodes in SNA.

5.1.4 Links

Since SNA nodes are separate entities, a physical connection must be established between them. This physical connection is the actual medium of transmission. A telephone wire, microwave beam, or fiber-optic medium can be used to transmit data between SNA subarea nodes. This physical connection by any of these media is called a "link." One or more links can connect adjacent subarea nodes, as seen in Figure 5.4. The media constitutes the link connection. Each link connection has two or more link stations. The link stations transmit the data over link connections using data link control protocols. SNA supports the following data link control protocols: System/370 data channel, SDLC, BSC, S/S, and X.25 interface.

5.1.5 Network Addressable Units

The efficiency of exchange between end-users is dependent upon the synchronization of communication, management of the resources that make up each node, and the control and management of the network. This is accomplished by Network Addressable Units (NAUs).

In SNA there are three types of NAUs: the logical unit, the physical unit, and the system services control point. The first allows the end-user access to the network. This NAU is a Logical Unit (LU). The LU manages the exchange of data between end-users. It is the port to the network. In other words, there is not a one-to-one connection between end-users and LUs.

For end-users to communicate with each other, a mutually agreed upon relationship must be established. This relationship is called a "session." When a session connects two LUs, it is called an "LU-LU" session. If multiple, concurrent sessions occur between the same two LUs, they are called "parallel LU-LU" sessions. LU-LU sessions can exist only between LUs of the same type. SNA defines nine types of LUs, as can be seen in Appendix C.

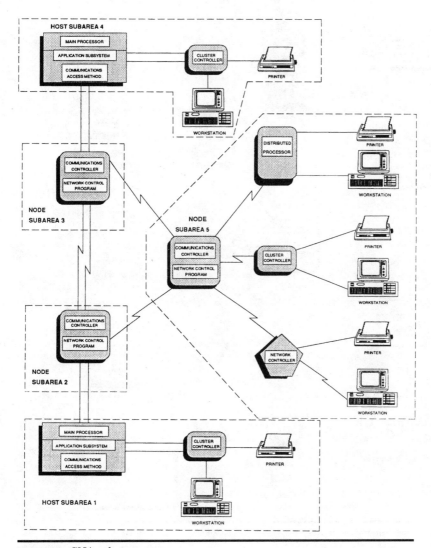

Figure 5.3 SNA subareas.

The LUs are managed by a Physical Unit (PU). The PU represents a processor, communications controller, or cluster controller and presents the LUs to the links that connect its node to adjacent nodes. As described in Appendix C, there are four types of physical units, one representing each type of SNA node. PUs are implemented by a combination of hardware and software components that describe to the PU its own node.

The third NAU, System Services Control Point (SSCP), lies within the telecommunications access method. The SSCP is found only in

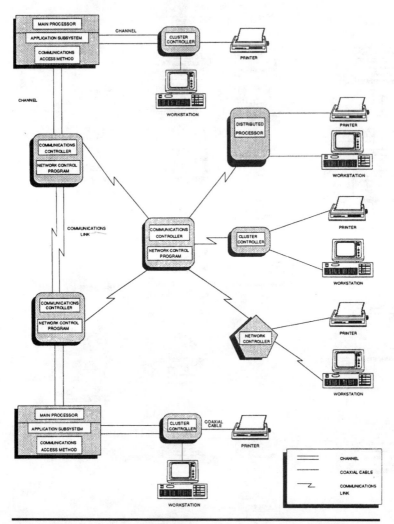

Figure 5.4 Transmission facilities.

host subarea nodes. Each SSCP in a network activates, controls, and deactivates network resources that have been defined to that SSCP as being in its domain. When there is one SSCP in a network, it is called a "single-domain" network. If there is more than one SSCP in a network, it is called a "multiple-domain" network.

5.1.6 Network addressing

This leads us to the actual assignment of addresses for network resources. Each NAU is assigned a unique address within a network based on the subarea to which the peripheral node is assigned and an element address for each NAU of that subarea node. The unique sub-area address is assigned to each host subarea node and communications controller subarea node during the telecommunications access method definition process. This subarea number is assigned to the subarea address field of the network address. All network resources in that subarea and any peripheral node attached to that subarea along with the link and link stations that connect adjacent subareas, will be assigned this unique subarea number.

Now that each node has a subarea address, each NAU within that node must have a unique address. This is the element address of the network address. The element address is assigned by the SNA access methods and Network Control Program during the resource definition process, dynamic reconfigurations, switched SDLC link resource activation, and the initializing of parallel LU-LU sessions. These element addresses are assigned in sequence according to the order in which the resource definition statements are read by the SNA access method or the Network Control Program.

Now that we have seen how the network address is derived, let's take a look at the three formats for the network address. In Figure 5.5, we see a 16-bit network address format and two extended formats, 23-bit and 31-bit network address formats. The 16-bit address format allows from 1 to 255 subareas. However, the length of the subarea address field is variable according to the maximum number of subareas that are defined to the SNA access method. For example, if we defined the maximum subarea address as 63 in our network, the subarea address field will use 6 of the available 16 bits for the network address. This limits the size of the element address field to 10 bits. The highest element address that can be assigned to any NAU in any subarea node for this example is therefore 1023. Likewise, if the maximum subarea address is defined as 31, 5 bits are used for the subarea address field, leaving 11 bits for the element address field. This translates into 2047 elements or NAUs available for any one subarea. This address split has to remain constant throughout the network, and many large shops quickly exhaust their element address limits. Because of this limitation of network resources on large networks, IBM announced Extended Network Addressing (ENA).

At first, ENA added 7 bits to the network address format and removed the network address split. This allowed for 255 subareas, which is the same as in the 16-bit format, but it created a constant element address field size of 15 bits. This provided each subarea node with a maximum of 32,768 addressable NAUs, thus allowing for virtually unlimited growth for large networks.

In September 1988, IBM announced network addressing enhancements that break all bounds by adding 8 more bits to the subarea

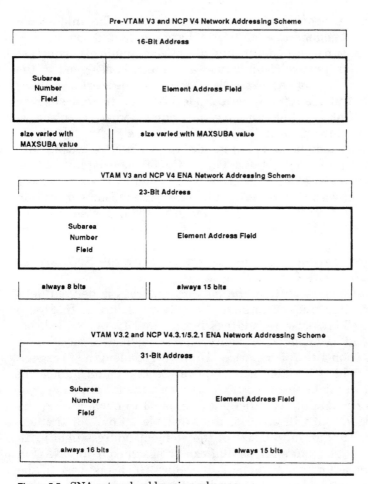

Pre-VTAM V3 and NCP V4 Network Addressing Scheme

Figure 5.5 SNA network addressing schemes.

addressing field of the network address. Now with VTAM V3.2, NCP V4.3.1, and NCP V5.2.1, we can have 65,535 subareas with 32,768 elements in each subarea.

Addressing within peripheral nodes is also necessary. Local addresses are assigned to each NAU in each peripheral node. These local addresses are unique only within that peripheral node. These addresses are translated into their corresponding network addresses by the boundary function of the host and communications controller subarea nodes.

During the access method definition process, the NAUs can be assigned network names. These network names must be unique within the network and are used by the network operator, workstation operator, and application programs. A directory is built during the definition process, and the SSCP uses this directory to translate the network name to the network address.

5.1.7 SNA routes

In the previous section we discussed how each NAU in a network is assigned a unique network address which consists of a subarea number and an element number. This addressing scheme has fostered the need for a technique to route information through the network. This routing technique is designed to maximize the amount of data being transmitted, data security, and route availability as well as minimize transmission time, transmission errors, traffic congestion, and cost. Although these are the objectives, not all can be equally satisfied. If one is favored, another will lose out. The choice of which objective takes priority depends upon what type of session is more important to the corporation's service goals.

As we discussed in Section 5.1.4, links connect adjacent subarea nodes. In SNA, if two or more links connect adjacent subarea nodes, they are called "parallel" links. Each link or group of links between two adjacent subarea nodes is called a "Transmission Group" (TG). A transmission group appears to the path control component of SNA as a single link. In order for a TG to be addressed, a transmission group number is assigned to the TG. A TG with parallel links has more availability than a TG with one link assigned. Usually, similar transmission characteristics govern which links are mapped to a TG. An example would be links that have the same transmission speed.

For SNA to follow a specific route, a path for that route must be defined. This path consists of the nodes, links, path control, and data link control components that are used for the transmission of data between two NAUs. A path is defined to SNA by an Explicit Route (ER) and a Virtual Route (VR) along with the TG.

An Explicit Route is the physical connection between two subareas. Each Explicit Route is mapped to a transmission group. However, you can define more than one ER to the same TG. This will increase the probability that a path will be available for transmission between two nodes. ERs are bidirectional. They have a forward and reverse path. The ER Number (ERN) assigned to the forward path can be different from the ERN assigned to the reverse path. The only requirement is that the ER pair must traverse the same set of subarea nodes and transmission groups.

The Virtual Route of a path is the logical connection between two subarea nodes. Each VR is mapped to an ER. One or more Virtual Routes can be assigned to an Explicit Route, again providing greater availability for transmission of data through the network. Each VR takes on the characteristics of the ER it is mapped to. The primary usage of VRs in SNA is the assignment of transmission priority to a session over an ER. The correlation of VR with a transmission priority is accomplished through the Class Of Service (COS) table. During an LU-LU session initiation, a COS is requested. It is this COS that determines the transmission priority of the session. More discussion about the definition and usage of the COS table will follow in Chapter 13.

One more important point of SNA routing should be mentioned. SNA distributes the responsibility of defining paths between subarea nodes to all the subarea nodes. Each subarea node needs only the information to pass data to the next adjacent subarea node, even if that adjacent subarea node is not the intended destination of the information. For example, Figure 5.6 shows data that needs to be routed from Node A to Node D. The path definitions in Node A define a route to Node D, but there are no physical links that connect Node A to Node D. However, Node A transmits the data to Node D via an Intermediate Networking Node (INN), which in turn transmits the data over TG 5 ER 2. Node D then sends the data to the destined logical unit.

5.1.8 SNA layers

Each of the seven layers of SNA is well defined to perform a specific SNA function in the architecture. The layers have been designed to be self-contained, allowing for autonomy between the layers. Although each layer is autonomous from the others, each layer performs services for the next higher layer, requests services from the next lower layer, and allows peer-to-peer communication between equivalent layers.

As seen by the end-user (Figure 5.7), the first layer involved with communications is the Transaction Services Layer. This layer allows the end-user access to the network by providing application services and requests services from the Presentation Services Layer. The Transaction Services Layer also has the function of providing services to control the network's operation through three components (Figure 5.8).

The Configuration Services component controls resources associated with the physical configuration during the communication access method's System Services Control Point (SSCP) to a Physical Unit (PU) session. This is known as a SSCP-PU session. These services are used for activation and deactivation of communication links, loading same-domain software, and assigning network addresses if dynamic reconfiguration is being used.

The establishment of an LU-to-LU session is a function of the Session Services component. It is during SSCP-SSCP and SSCP-LU

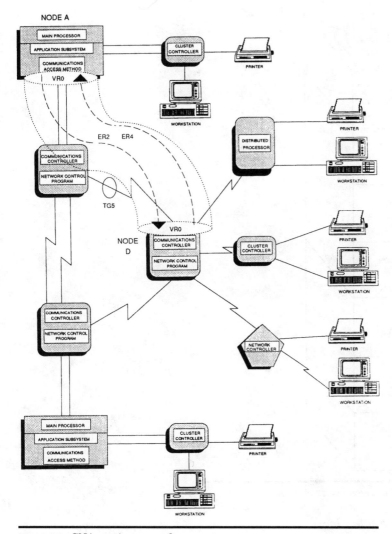

Figure 5.6 SNA routing example.

sessions that these services are invoked. Session Services perform translation of network names to network addresses. User passwords and user access authority are verified here, as is the selection of session parameters.

The third component that makes up the Transaction Services Layer operational control of the network is the Management Services component. These services perform the monitoring, testing, tracing, and recording of statistics for network resources for SSCP-PU and SSCP-LU sessions. SSCP, PU, and LU will be discussed in further detail in the following section.

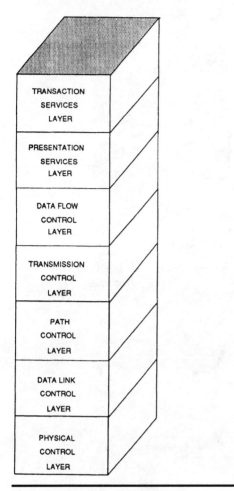

TRANSACTION
SERVICES
LAYER

PRESENTATION
SERVICES
LAYER

DATA FLOW
CONTROL
LAYER

TRANSMISSION
CONTROL
LAYER

PATH
CONTROL
LAYER

DATA LINK
CONTROL
LAYER

PHYSICAL
CONTROL
LAYER

Figure 5.7   The seven layers of SNA.

The Presentation Services Layer provides services for transaction programs by defining and maintaining the protocol of communications. Presentation Services also controls the conversational level of communications between transaction programs by providing the function for loading and invoking other transaction programs, enforcement of correct verb parameter usage and sequencing restrictions, and the processing of transaction program verbs.

LU-LU session flow is controlled by the next SNA layer, the Data Flow Control Layer. Communication of data between two LUs must be transmitted in an orderly and cohesive fashion. The services provided by this layer do just that. The units of data being transmitted are given sequence numbers, and an end-user request is correlated

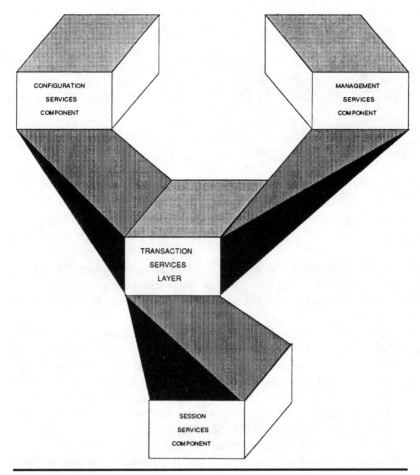

Figure 5.8 The three service components of the Transaction Services Layer.

with the transactions response. Related request units are grouped into chains, and related chains are grouped into brackets of data. This layer enforces the protocol and coordinates which session partner can send and which session partner can receive at any given moment.

The Transmission Control Layer is responsible for the synchronization and pacing of session-level data traffic. It does this by checking the session sequence numbers that were assigned to the request units by the Data Flow Control Layer. Another function is the enciphering and deciphering of end-user data.

The routing of the data units through a network is performed by the Path Control Layer. All types of sessions use this layer. Its main function is to route the unit of data to the desired destination in the

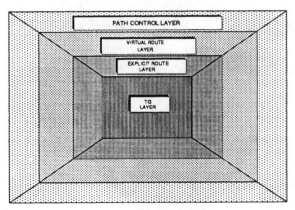

Figure 5.9 The sublayers of the Path Control Layer.

network, be it a terminal or an application program executing on a mainframe computer. The function is performed in subarea nodes. The Path Control Layer consists of three sublayers, as seen in Figure 5.9. The inner layer, the Transmission Group Layer, provides the transmission group connections between subarea nodes. The middle layer, Explicit Route Control, determines which of the transmission group connections information is to be passed between two end sub-area nodes of a path. Finally, the third sublayer is the Virtual Route Control sublayer. It is considered the outer sublayer of the three sublayers. Virtual Route Control provides the explicit route(s) that can be used to route information during a session.

The sixth layer of control in SNA is the Data Link Control Layer. This layer performs scheduling and error recovery for the transfer of data between link stations of two nodes. The Data Link Control Layer supports the link-level flow of data for both SDLC and System/370 data channel protocols.

The last layer in SNA provides the description of the physical interface for any transmission medium. It defines the electrical and transmission signal characteristics necessary to establish, maintain, and terminate all physical connections of a link. This layer is called the Physical Control Layer.

The seven layers of SNA are grouped into two distinctive functional groups, as seen in Figure 5.10. The first four layers are grouped into the Network Addressable Unit and Boundary Function classification. If you review the previous discussion, you will see that this classification is accurate. Remember, it is the first four layers that enable the end-users to send and receive data through the network by providing the definition of the logical unit functions in each node. The last three layers of SNA are grouped as the Path Control Network Functions.

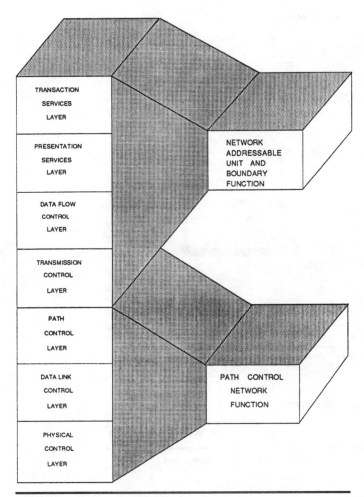

Figure 5.10   The two functional groups of SNA.

As discussed previously, it is in these three layers that the data is actually routed and transmitted between network addressable units.

5.1.9 SNA sessions

In SNA there are four session types that occur in the network: (1) SSCP-SSCP, (2) SSCP-PU, (3) SSCP-LU, and (4) LU-LU. Each session is used for a specific purpose in SNA. Figure 5.11 illustrates the four SNA sessions.

The SSCP-SSCP session is used by SSCP to communicate with another SSCP in another VTAM. This session is primarily used to set up cross-domain LU-LU sessions in a multidomain network and cross-network LU-LU sessions in an SNI network.

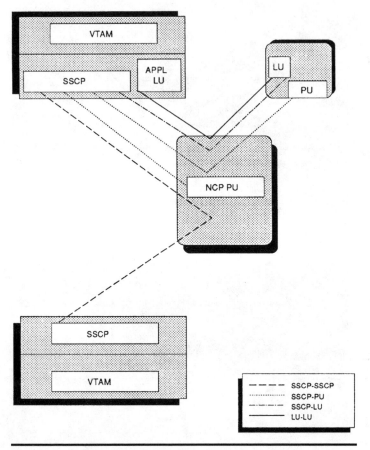

Figure 5.11 Diagram depicting the four SNA session types.

The SSCP-PU session determines the boundaries of the SSCP's domain. The PU is seen as being owned or controlled by the SSCP it is in session with. The session occurs when the SSCP receives the PU's acknowledgement of the SSCP's activation request. This session is used primarily for activating the LUs associated with the PU and for request and response network management data that flows on the SSCP-PU session.

The SSCP-LU session occurs when the LU acknowledges the SSCP's activation request. It is this session that allows an end-user to request a session with an application. The request for the application flows on the SSCP-LU session. The SSCP, in turn, passes the request to the application over the application's SSCP-LU session. Once the two LUs have agreed on session protocols, the LU-LU session is established. It is this session with which most end-users are familiar.

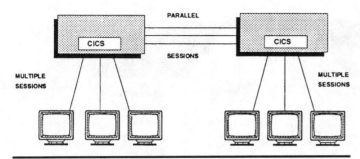

Figure 5.12   Example of multiple and parallel LU-LU sessions.

LU-LU sessions come in two types. Applications such as CICS and IMS can have multiple and parallel LU-LU sessions. As seen in Figure 5.12, multiple sessions can occur concurrently between CICS and several end-users. However, from the end-user's point of view, he or she is the only session partner. The distinction here is that each of the end-users is in session with the same network addressable unit. That is, CICS uses the same network address for all multiple sessions. Parallel sessions, on the other hand, can also occur concurrently between LUs. The distinction here is that the sessions are occurring between the same two LUs, but each session partner of each session is assigned a unique network address.

5.1.10   Open Systems Interconnection and SNA

So far we have discussed IBM's architecture for data communications, SNA. But other companies and organizations are in the process of developing, or have developed, alternative data communications architectures. One group, the International Standards Organization (ISO), has developed a standard for exchanging information between different architectures, Open Systems Interconnection (OSI).

Unlike SNA, OSI is designed to exchange information between autonomous systems by standardizing the communication protocols between the architectures. SNA, however, is designed solely to standardize communications protocols between IBM products. Although no direct correlation of the layers in the two architectures can be drawn, the functions provided by each layer are quite similar.

If we look at Figure 5.13, we can see the similarities of the two distinct architectures. The table shows that the SNA Path Control Layer functions are found in the OSI Network and Transport Layers. Likewise, the OSI Physical Layer and the SNA Physical Control Layer both describe to the architecture the physical characteristics of the transmission medium.

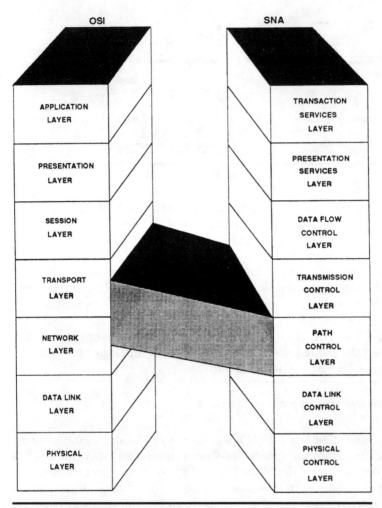

Figure 5.13 The OSI and SNA architectures.

It is important to remember that although there are many similarities between the two architectures, the purpose of each is distinct. SNA was designed specifically for IBM networks using IBM products. OSI was designed to provide communications between unlike network architectures. IBM has acknowledged the ISO OSI standard and is pursuing APPC/LU6.2 as the OSI standard for OSI's peer-to-peer communications.

5.1.11   SNA/LEN and APPC/LU6.2

The rapid expansion of distributed office processors dedicated to specific applications has brought a dramatic change in SNA. The distributed processors provide local access to the majority of the applications end-user community; however, there may still be end-users attached to the host processor that need periodic access to the data located at the distributed processor. The distributed processor and host processor are frequently hardware and software incompatible, and a means for sharing applications and data between the two end-user communities is necessary. For this reason, a new concept of "any-to-any" has come forth in SNA. This concept is implemented using Low-Entry Networking (LEN) and Advanced-Program-to-Program Communications using LU6.2 (APPC/LU6.2).

LEN implements what is known as SNA Node Type 2.1, from here on also referred to as PU T2.1. This special PU supports the unique capabilities of APPC/LU6.2. Each LU6.2 allows for peer-to-peer communications with another LU6.2, allowing for multiple and parallel sessions between LU6.2s. On each of these sessions a conversation can occur between application programs that use the LU6.2 session (Figure 5.14). LEN allows these peer-to-peer sessions to occur through an existing SNA network, provided that VTAM V3.2 and NCP V4.3 and/or V5.2 are implemented in the SNA network. LEN with APPC/LU6.2 allows any resource of any network to communicate with any resource of any other network.

The PU T2.1 node does not necessarily participate in an SSCP-PU session. This is because it supports two new types of LUs. Dependent LUs (DLU) require the assistance of an SSCP to establish sessions. The DLU uses the SSCP-LU session to request the SSCP services to establish an LU-LU session. These are historically the most common LU Types 0, 1, 2, and 3. In this case, the PU T2.1 supports an SSCP-PU session on

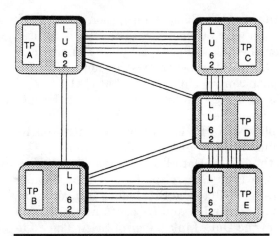

Figure 5.14 The multiple and parallel session capabilities of APPC/LU6.2. Note that VTAM is not involved in the session configuration.

behalf of the DLUs. However, the second type of LU, Independent LUs (ILU) can request and establish an LU-LU session without the services of an SSCP. LU Type 6.2 is an ILU. The PU T2.1 that is supporting ILUs therefore does not participate in an SSCP-PU session. See Chapter 10 for more information on Low-Entry Networking (LEN).

5.2   SNA/SDLC Format Summary

From the previous discussions, we now know how SNA addresses and routes data to different types of processors, communications controllers, cluster controllers, workstations, and printers. In this section we will see how this information is relayed to the nodes and NAUs of the network by using SNA frames. These SNA frames are created using the SDLC protocols. There are three types of frames. The information frame passes SNA commands and responses, end-user data requests or responses, and acknowledges information frames (I-frames). The supervisory frame acknowledges I-frames and reflects the status of the NAU as being Ready-to-Receive (RR) or Receive-Not-Ready (RNR). The final frame is the unnumbered frame, which passes SDLC commands and is used in data link management.

5.2.1   Message unit formats

In SNA, each message unit along the route is given a specific format. This format is directly related to the location of the message unit on the SNA route. If we look at Figure 5.15, there are three types of message unit formats. Each one correlates to a specific layer within the SNA architecture.

Network Addressable Units use the Basic Information Unit (BIU) format to exchange information with other NAUs. The BIU is created as a function of the LU when the LU prefixes a Request Header (RH) to the Request Unit (RU). Only NAUs use request headers. The LU then passes the BIU to the path control element for that NAU.

The path control elements for NAUs prefix a Transmission Header (TH) to the BIU. This transmission header denotes a Path Information Unit (PIU). As its name implies, the PIU is used by the Path Control Layer to route the message unit to the appropriate destination.

Path Control then passes the PIU to Data Link Control for preparation to transmit the PIU. Data Link Control prefixes a Link Header (LH) and appends a Link Trailer (LT) to the PIU before transmission. As is suggested from their names, the LH and LT contain information that concerns the functions of the Data Link Layer of SNA. The addition of LH and LT to the PIU makes up a SDLC frame, also known as a Basic Link Unit (BLU).

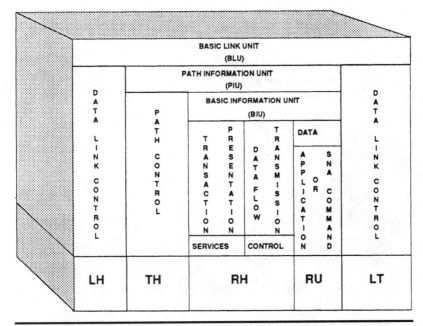

Figure 5.15   Message units and their relationship to the SNA layers.

5.2.2 Link header

The format of the link header can be seen in Figure 5.16. The link header contains three fields. The first field, the flag, is always a hexadecimal value of 7E. This notifies the data link control element that this is the start of a new SDLC frame.

The second field is the SDLC station address. This address can be a specific station address or a group address. It can also reflect a broadcast address. This broadcast address is always a hexadecimal value of FF and is mostly used by link-level protocol for NAU notifications. One more address may be found in this field, a "no stations" address, which is reserved and is a hexadecimal 00 value.

The third field of the link header is the control field. This field describes to the data link control elements the frame format being used. The frame format may be an unnumbered, supervisory, or information frame. The unnumbered frame dictates the SDLC link-level command being issued. The supervisory frame denotes whether the station is in the RR or the RNR state or if the previous frame transmitted was rejected. The information frame basically keeps track of the sequence frames sent and frames received. This is used for synchronization of BLUs.

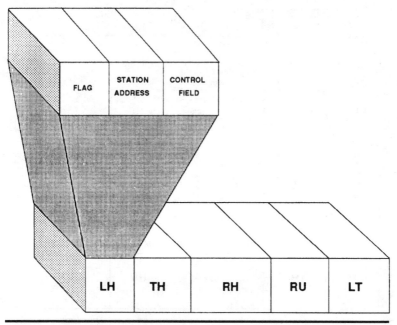

Figure 5.16   Format of the SDLC link header.

5.2.3   Transmission header

Because SNA supports different types of nodes in a network, the transmission header has five formats. These formats are identified by the first byte in the transmission header. It is the Format Identification Field, or FID. The FID type depends on the type of node involved with the transmission. If we look at Figure 5.17, we can see the relationship of FID type to the corresponding node type. We can also see that the length of the TH varies with the FID type. The evolution of SNA networks and the dominance of node Types 2 and 4 being used as the backbones of these networks has promoted the use of FID Types 2 and 4 in the majority of transmission formats.

It is important to remember that the purpose of the TH is path control (Figure 5.18). In the TH we can find the origin address field, the destination address field, the sequence number of this PIU in relation to other PIUs for this request or response, the length of the RU, and whether it is the only, first, middle, or last in the transmission. For FID Type 4, explicit and virtual route information is found along with the originating subarea and element addresses and the destination subarea and element addressees. The FID Type 4 is also used to indicate if the PIU is from a non-SNA device (FID Type 0) by setting a bit in the transmission header. The diagram in Figure 5.19 depicts the flow of FID types through a network.

NODE TYPE	FID TYPE	TH LENGTH	USAGE
Non-SNA	0	10 bytes	Non-SNA traffic between adjacent subarea nodes not supporting explicit and virtual route protocols
PU5 PU4	1	10 bytes	SNA traffic between adjacent subareas not supporting explicit and virtual route protocols
PU5 PU4 PU2 PU2.1	2	6 bytes	SNA traffic between a subarea node and an adjacent peripheral node
PU4 PU1	3	2 bytes	SNA traffic between an NCP subarea node and a peripheral node type 1
PU5 PU4	4	26 bytes	SNA traffic between adjacent subarea nodes supporting explicit and virtual route protocols
PU5 PU4	F	26 bytes	Specific SNA commands between adjacent subarea nodes supporting explicit and virtual route protocols

Figure 5.17   Relationship of SNA PU node types to SDLC FID types.

5.2.4   Request/Response Header

Following the transmission header is a 3-byte header that describes the actual information being transmitted (Figure 5.20). This header is called the Request/Response Header (RH). The basis of the RH is to describe the data that follows. The header type is indicated by bit 0 of the header. If bit 0 is a 0, the header is a Request Header; if it is a 1, it is a Response Header.

The Request Header provides information on the format of the data and the protocol that governs the session. It describes to the PU if a definite response is indicated for this transmission, if brackets are being used, where in the chain of data this BIU is positioned, and if a pacing request is indicated for this transmission.

The Response Header provides the appropriate information to the requester in regard to the requested protocol. The Response Header is chiefly responsible for returning a positive or negative response to the requester. If a negative response is transmitted, a sense data indica-

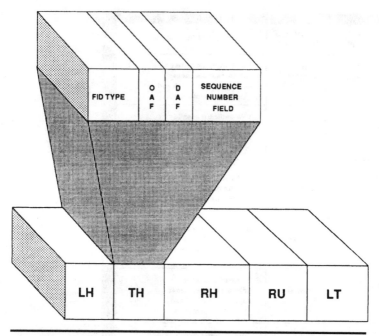

Figure 5.18   The common fields of transmission header FID types.

tor bit is set to 1, which indicates to the requester that sense data follows that explains the negative response to the request.

5.2.5 Request/Response Units

Request Units follow Request Headers (Figure 5.21). These units are variable in length and may contain end-user data or an SNA command. The data RUs contain the information that is to be exchanged between end-users. The command RUs control the operation of the network by issuing appropriate SNA commands. Although the Request Unit is variable and in theory is infinite, the restriction of a 2-byte field in the TH, the Data Count Field (DCF), allows for a maximum RU size of 64 kbytes. However, most logical units support only an RU size of 256 bytes. This is because the data link buffer in most PUs is only 256 bytes. The number 256 has been chosen as the prime size for a link buffer because of tests that were performed on telephone lines. These tests showed that 256 bytes can be transmitted with the lowest percentage of transmission error. The Response Unit in the case of a positive response to a data Request Unit is null; there is no response. However, in the case of an SNA command response, the response unit is generally 1 to 3 bytes in length. Negative Response Units are usually 4 to 7 bytes long. The first 4 bytes indicate sense code data that describes the reason for the negative response. This could be a protocol

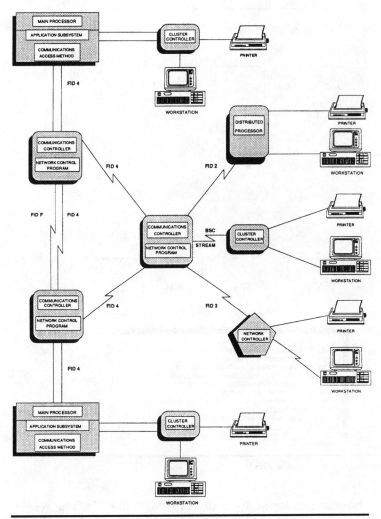

Figure 5.19   FID flows through the SNA network.

violation, transmission error, or a possible path outage. Three more bytes may be sent to identify the rejected request.

5.2.6   Link Trailer

The final field in the SDLC frame format is the Link Trailer (LT). The link trailer has two fields: the Frame Check Sequence (FCS) field and the link trailer flag as seen in Figure 5.22.

The FCS field is used to check the received frame for transmission errors. The transmitting link station executes an algorithm based on Cycle Redundancy Checking (CRC). The data for the computation is

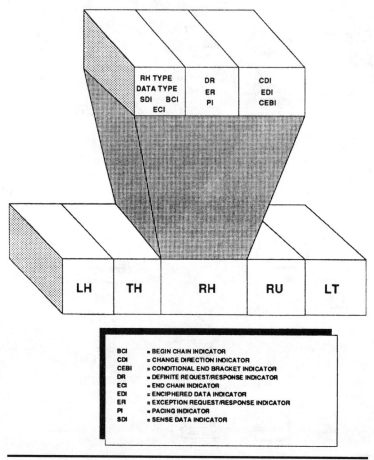

Figure 5.20   Request and response header fields.

inclusive of the link header address field through the RU. The receiving link station performs the similar computation and checks its results against the FCS field. If the results are not acceptable, a transmission error sense code is sent to the originating link station stating that retransmission is necessary.

The link trailer flag indicates to the receiving data link control element that the frame has ended and a new frame should be expected. The flag is also defined as a hexadecimal 7E.

5.3   Basic Telecommunications Access Methods

One of the first telecommunications access methods is Basic Telecommunications Access Method (BTAM). This provides the user

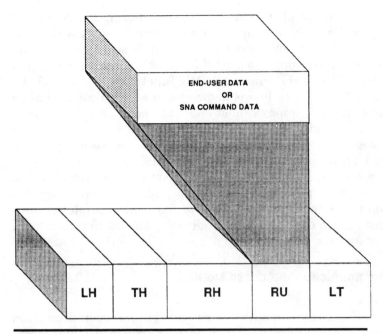

Figure 5.21 Request and response unit.

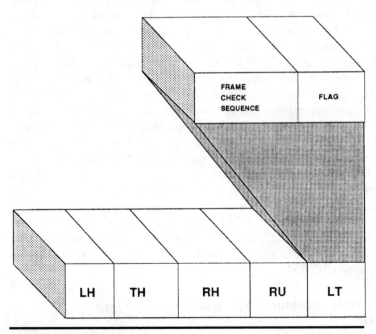

Figure 5.22 Link trailer format.

with support for input operations from the end-user. BTAM can be regarded as the basis for subsequent access methods. It provides routines to manage buffers and to detect and handle some transmission errors, switched line support, leased line support, and control of network resources. However, since BTAM's function is to control the lines to the end-user, each line is dedicated to an application; unlike on workstations, applications cannot share the same line. The user is responsible for providing code to BTAM to perform queuing and routing of messages, any additional error analysis and recovery above that which BTAM supplies, and system security and integrity. The biggest drawback to BTAM is that the user must supply the control program. This includes keeping lines active, adding or removing line control characters, assembling blocks of data into completed messages, translating messages to terminal code, and editing messages. This is extensive and complex coding.

5.4 Telecommunications Access Method (TCAM)

TCAM is a queued access method. It has a high-level control program called Message Control Program (MCP). By using supplied TCAM macro instructions, the user can direct the MCP to queue messages received from or sent to remote terminals. TCAM can transfer messages from one remote terminal to another and between terminals and application programs. TCAM supports SDLC, BSC, and Asynch terminals. The advantage TCAM has over BTAM and VTAM is its message control and queuing abilities. These two functions greatly increase the user's security and data integrity by preventing loss of data. Most brokerage houses, banks, and other institutions whose loss of data can cause a loss of millions of dollars use TCAM's message control and queuing to the fullest extent. With TCAM came full control of network resources. In conjunction with TCAM is the IBM Network Control Program, which extends the access method's ability to control and maintain telecommunications lines. TCAM operates under an MVS or OS/VS1 operating system environment. It supports both SNA and non-SNA devices concurrently.

The most recent version of TCAM, V3, has been stripped of access method responsibilities but has kept the message control and queuing capabilities intact. This version executes on a host processor as an application under VTAM.

5.5 Virtual Telecommunications Access Method (VTAM)

VTAM is an access method that controls communications between logical units. It directly controls the transmission of data to and from channel-attached devices and uses the IBM Network Control Program (NCP) to forward and receive data from remotely attached devices over telecommunication links. Unlike TCAM, VTAM is not a queuing access method. Its main function is to route data to its desired destination. The VTAM user is concerned only with logical units. Non-SNA devices must be handled by an emulation program or a protocol converter of some type, be it hardware or software. The only exception to this is that VTAM does support 3270 BSC devices. An emulation program or protocol converter can be included in the NCP as a partitioned emulation program (PEP). We may also use IBM's Network Terminal Option (NTO) or Non-SNA Interconnect (NSI) in conjunction with the NCP. VTAM operates in MVS. OS/VS1, VM, and VSE operating system environments.

5.6 VTAM Extended (VTAME)

VTAM Extended operates only under a VSE operating system environment. It provides all the functions of VTAM plus support for non-SNA devices, other than BSC 3270 terminals, attached by a loop adapter. VTAME also provides intermediate-node routing functions that versions of VTAM prior to VTAM version 2 did not provide.

5.7 Summary

In this chapter we learned about various network components like nodes, subareas, links, and NAUs. Considerations for routing data such as Explicit Route, Virtual Route, and Transmission Group were discussed. Functions of various SNA layers and their relationship to the OSI model were considered. SNA sessions consist of four different types: SSCP-SSCP, SSCP-PU, SSCP-LU, and LU-LU. APPC/LU6.2 forms the basis of distributed processing in an SNA environment. LEN provides for peer-to-peer sessions without the participation of SSCP. Finally we learned about various headers such as RH, TH, LH, and LT and different information unit formats such as BIU, PIU, and BLU. Now that we feel comfortable with SNA components, let's move on to an extension to SNA. That extension is known as Advanced Peer-to-Peer Networking.

Advanced Peer-to-Peer
Networking

Early computing systems were dominated by batch-oriented processing. This means that several programs were grouped together to be processed by the mainframe computer. These programs were designed for noninteractive execution; hence, they could be grouped or "batched" for processing. Following batch processing was interactive processing. Information systems were developed that allowed almost real-time access to data. The LANs of the 1980s brought this data closer to the end-user and hence called for a new networking architecture, an architecture that provides peer-to-peer communications rather than a hierarchical architecture.

6.1 Why Peer-to-Peer Networks?

The early communications networks made access to corporate data available on an almost timely basis, meaning that the new data was not presented until the batch processing was completed. In fact, many of the early networks were used to support remote batch processing. Remote batch processing allowed remote corporate sites to enter data and then transmit the entered data to the mainframe in a file for further batch processing. This transmission of data on the early networks was slow and unreliable.

The cost of maintaining the early networks was high, and they were therefore within the reach of only large corporations, the government, and educational institutions with grant money from the government. The deployment, operation, and management of these early networks was nonstandardized. These computing networks required a hierarchical architecture so that data processing professionals could manage them with greater control. IBM's Systems Network Architecture (SNA) was created to answer this need.

SNA established the first real successful commercial network architecture and has dominated corporate networks for approximately 10 years. Although SNA was announced in 1974, it did not really become widely used until the early 1980s. By this time SNA had evolved to support multisystem networking and SNA Network Interconnection (SNI). Both these functions allowed networked terminals to access multiple applications residing on any VTAM host anywhere in the SNA network. The growth of SNA networks accelerated with the advent of larger mainframe computers, the addition of front-end processors running Network Control Program (NCP), and the higher-speed communications facilities provided by the common carriers. The networks quickly evolved from simple to highly complex, requiring extensive definitions of each network resource attached.

SNA's hierarchy established four types of SNA nodes to provide the hierarchical or "master-slave" architecture necessary at the time to manage these extensive networks. VTAM sits at the top of the hierarchy. VTAM controls and manages all the resources in the network. It is called a Physical Unit (PU) Type 5. NCP is a slave to VTAM and controls the peripheral devices attached to its front-end processor; it is known as a PU Type 4. Cluster controllers attached to VTAM's mainframe and the NCP's front-end processor are known as PU Type 2.0 nodes. The terminals and printers attached to the cluster controller are defined as Logical Units (LU) and are categorized into several different types, most notably LU Type 2 for 3270 terminal displays, LU Types 1 and 3 for printers, and LU Type 6 for program-to-program communication. In each instance the LU could not begin a session with another LU unless VTAM facilitated the session establishment. PU Type 2.0 nodes are limited to a single link and act as secondary link stations. The LUs attached to a PU Type 2.0 are limited to a secondary logical unit (SLU) with a single-session capability to the VTAM application which acted as the primary logical unit (PLU). This hierarchy provides manageability and a sound networking structure but allows the LUs to participate in LU-LU sessions only with VTAM applications and not with other LUs attached to PU Type 2.0 nodes. This hierarchy also provides detailed knowledge of the network configuration and the painstaking definition of each individual resource in the network.

6.2 Peer-to-Peer Networking Concepts

In the early 1980s, a new age of information processing was taking hold. The impact of microprocessors and digital communications created a new type of networking paradigm. That paradigm was peer networking. The processing power available to the end-user at the desktop now matched that of many mainframes. These powerful workstations could run applications that were once run on the mainframe. Local area networks provided the media and architecture these workstations needed to communicate with each other for sharing data, programs, and services. A hierarchical network architecture had become outdated. Any computing system, small or large, now needed to communicate with any other computing system on a peer level without the assistance of a "master."

The low cost of workstations and the ease of installing them on a LAN increased the need for peer networks. Therefore, many departments went out and built their own networks without regard to compatibility across the corporation. Eventually, the individual departments found that they needed their computing systems to communicate. The result of this growth and need for interdepartmental communication fostered the following peer networking requirements:

- A peer network must be easy to use, change, manage, and grow.

- A peer network must have decentralized network control with the ability for centralized network management.

- A peer network must support any type of networking topology.

- A peer network must have flexible support for physical-level attachments.

- A peer network must be able to internetwork and interoperate with subarea SNA.

- A peer network must be simplistic in nature and low in cost.

- A peer network must provide for continuous operation.

Providing solutions to these requirements has taken some time. Any peer network must specify services to meet these needs. Configuration services will specify a means for simple physical attachment to the peer network. A directory services function will be needed to locate any resource in the peer network dynamically. The provision for the most efficient route and the ability to provide an alternative route between peer network resources can be provided by a topology and routing service. The need for reliable transport through the use of flow control, the prevention of deadlock, and the ability to handle variable network packet sizes is handled by a data transport service. Control and management of a peer network is provided by a management service.

6.3 APPN Architecture

IBM's Advanced Peer-to-Peer Network (APPN) was developed to meet the needs of peer-to-peer networks as described above. APPN is an extension to the PU Type 2.1 node architecture. It has enhanced the

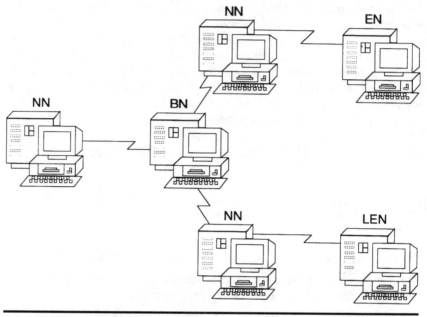

Figure 6.1   The few types of APPN nodes.

PU T2.1 node architecture to support directory functions and dynamic update of network topology, routes, and directories. The APPN architecture defines four types of nodes as shown in Figure 6.1.

The previously defined Low-Entry Networking (LEN) T2.1 node is the first APPN node to be supported. Recall that the LEN T2.1 node does not support a control point function and hence does not have within it the APPN enhancements but is defined as a supported node in APPN. The APPN End Node (EN) is a T2.1-based node that supports many of the APPN T2.1 node service enhancements. As its name implies, it must be the end of a route. The APPN Network Node (NN) T2.1 resource provides all the functions of an end node, but it also assists ENs and LENs to locate APPN resources throughout the network. The Border Node (BN) is really a function of a NN. The BN serves a purpose much like that of SNI for SNA networks. In APPN, a BN acts as a NN to its native network and as an EN to the nonnative network. In this way, unique APPN networks can be joined, allowing for topology isolation while integrating the directory services of the two autonomous networks.

All the APPN nodes have a function called a Control Point (CP). Figure 6.2 diagrams the use of the control point. The CP is a program that provides the different services for the node. The CP for a NN is much more enhanced than that for a LEN. LENs do not establish CP-CP sessions. NNs can establish CP-CP sessions with direct NNs or

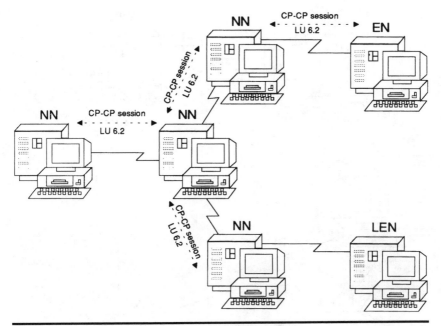

Figure 6.2 APPN CP-CP sessions between NNs and ENs.

with direct attached ENs. The CP-CP communication is established using LU6.2 sessions. A CP is then by definition an LU6.2 application. As such, the CP must have a fully qualified network name. A network node may also take on additional services for its ENs and is then called a Network Node Server (NNS). Each NNS and the ENs it has established a CP-CP session which defines the domain of the NNS. Finally, the physical links attaching APPN nodes are known as Transmission Groups, just as in SNA. However, at present only a single link TG is allowed between APPN nodes, although there can be several TGs off of a node. For instance, in Figure 6.3, NN1 has multiple TGs, one to each adjacent NN, and NN2 has parallel TGs to NN3. Each TG is uniquely identified by a TG number and the name of the partner CP. This makes the TG unique between the adjacent partners and allows for the reuse of TG numbers between adjacent nodes because of the unique fully qualified CP name.

6.3.1 LEN Node

The LEN CP provides the five major services of APPN, with limitations on directory services and topology and routing services as shown in Figure 6.4. LEN is actually the base T2.1 node architecture without the APPN extensions. LEN nodes can attach to an APPN network through an adjacently attached NN. The LEN provides no network

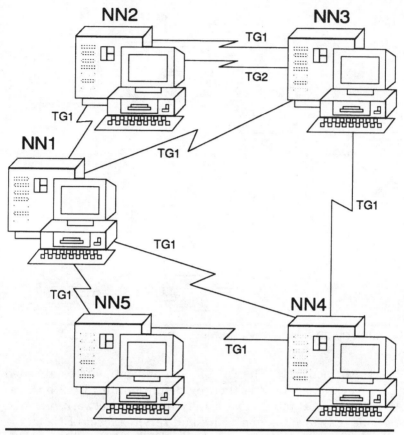

Figure 6.3 Example TG configuration illustrating single-link TGs and parallel TGs between APPN nodes.

services to other nodes, since it does not support a CP-CP session. The directory services on a LEN node are limited to local definitions. These definitions include both local and remote resources that need to communicate. A LEN node will provide session services for its own LUs only and does not provide TG link topology other than its own links. LENs do support multiple TGs but not parallel TGs. IBM product offerings that support the LEN node are listed in Figure 6.5.

6.3.2 APPN End Node (EN)

The end node in APPN implements PU T2.1 architecture with limited support of directory and topology and routing services. The end node in APPN is where LUs are located. ENs participate in an APPN network through an adjacent NNS. An EN can also be directly attached to another EN for LU-LU session support, as shown in Figure 6.6. The EN determines the whereabouts of partner LUs for its own

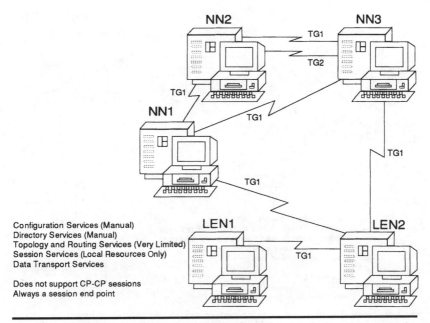

Figure 6.4   LEN node services under APPN.

Products Supporting Low Entry Network (LEN) Node	
AS/400	VTAM V3R2+
RS/6000	NCP V5R2+
OS/2 Extended Services	Network Services/2
OS/2 Extended Edition	Network Services/DOS
Transaction Processing Facility (TPF)	DOS APPN/PC
System/36	System/38
System/88	RT/PC
Series/1	DOS APPC/PC

Figure 6.5   IBM product offerings supporting LEN node functionality.

resources either by local definitions or by sending a LOCATE request to its NNS. Local definitions are required for ENs directly attached to another EN. ENs support CP-CP sessions only with the adjacent NNS. CP-CP sessions with the NNS occur in single-pair LU6.2 sessions, one for sending and one for receiving. When an EN starts its links, it will establish the CP-CP session with its NNS, exchange link information for topology services, and register its LUs with the NNS. More than one NN can be attached to an EN, but only one can be the NNS at a time. The CP of an EN is responsible for session services on its LUs only. Figure 6.7 contains a table listing IBM products that support APPN EN.

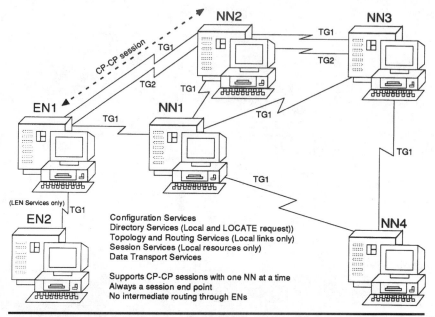

Figure 6.6 APPN EN services under APPN.

Products Supporting End Node (EN)	
AS/400	VTAM V4R1
RS/6000	Network Services/2
OS/2 Extended Services	Network Services/DOS
OS/2 Extended Edition	DPPX/370

Figure 6.7 IBM product offerings supporting EN functionality.

6.3.3 APPN Network Node (NN)

Although APPN is being represented as a peer networking architecture, it does have its roots in SNA, and hence it has the flavor of a hierarchical structure. The APPN NN implements all of the APPN services. The NN defines a domain just as VTAM defines a domain. A NN domain is characterized by its own resources, ENs that participate in a CP-CP session with the NN, and these EN resources. LEN nodes and their LUs, adjacently attached to the NN, are not part of the NN domain because a LEN does not participate in a CP-CP session; however, a NN will perform directory services for the LEN node. Communication between the NN and the EN is accomplished through the pair of CP-CP sessions. A NN domain is diagramed in Figure 6.8.

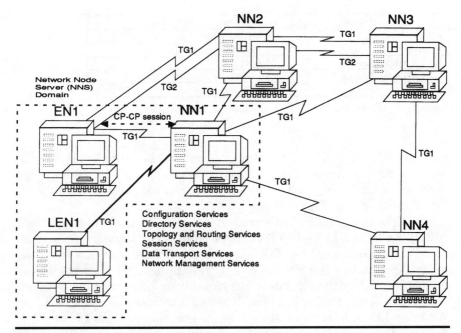

Figure 6.8 Diagram depicting a Network Node domain.

A network node acts as the entry point to the network for APPN nodes attached to it. The APPN NN provides

- LU-LU session services for its own LU
- Session routing functions
- Network directory searches and route selection
- Management services as a focal point or relay point for network problem management

The NNS maintains directory entries for resources located within itself and manually defined for those that are found in attached LEN nodes. This directory will also contain entries from ENs that have registered their resources automatically with the NNS. Figure 6.9 contains a table listing the IBM products that support APPN NN.

6.3.4 Link and transmission groups

A link is defined as the physical transmission medium and the two end points. The end points are known as link stations. As shown in Figure 6.10a, the link can be point-to-point or over a shared transmission medium like token-ring. Token-ring is really a multipoint connection. However, APPN does not currently support multipoint con-

Products Supporting Network Node (NN)	
AS/400	VTAM V4R1
RS/6000	Network Services/2
OS/2 Extended Services	6611
OS/2 Extended Edition	DPPX/370
	3174

Figure 6.9 IBM product offerings supporting NN functionality.

nections. The APPN architecture overcomes this with the implementation of Shared Access Transport Facility (SATF). SATF presents the nodes on a token-ring to APPN as if they were all point-to-point links. This can be seen in Figure 6.10b.

The ENs in a SATF configuration define a Virtual Routing Node (VRN). As shown in Figure 6.11, EN1 defines two network nodes and two TG connections. One TG attaches the EN to the VRN, and the other TG attaches the EN to NN1. Likewise for EN2. The route vectors generated by the ENs describe two routes for connection between them, one route through a real NN and one route through a VRN. When an LU on EN1 wants to establish a session with an LU on EN2, NN1 route selection services will indicate to EN1 that the VRN route should be used, since the resources are on the same shared medium. In this case the NN does not have to be involved with the transport of data between the two EN LUs. This type of configuration is also called a connection network.

6.3.5 Comparing APPN to SNA layers

APPN encompasses the SNA Path Control and Transmission Control functions. The Physical and Data Link Control Layers of SNA are fully supported under APPN. APPN will support any physical connection currently provided today. Session services in APPN are provided by LU6.2, which takes on the SNA Data Flow Control and Presentation Services Layers. Resources are located through the directory services functions of the control point. The control point in APPN resides at SNA Layer 7, Transaction Services. This comparison can be seen in Figure 6.12.

6.4 APPN Services and Functions

APPN is a sophisticated architecture that provides dynamic facilities. Coordination of definitions between resources is significantly reduced thanks to this sophistication. Location of resources, network change adaptation capabilities, learning network topology, route optimization, class of service, and transmission priorities are all brought about through the four services defined for the APPN architecture. These

(a) Link Configurations

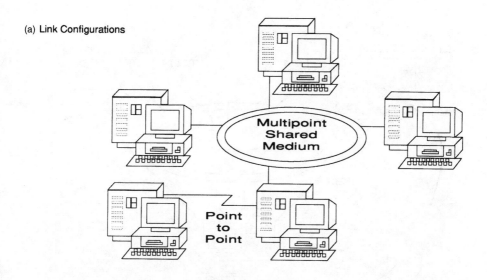

(b) Shared Access Transport Facility (SATF)

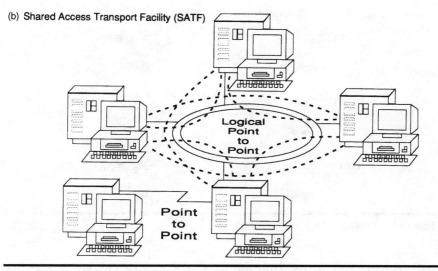

Figure 6.10 (a) Sample configuration of point-to-point and shared media multipoint link configurations. (b) Logical point-to-point configuration over shared media.

dynamics are functions of the configuration services, topology and route selection services, directory services, and session services.

6.4.1 Configuration services

Configuration services is a component of the control point in an APPN node. An APPN node is responsible for its own definition. Each

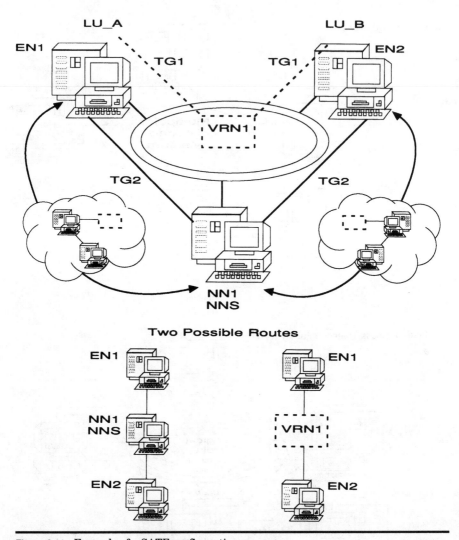

Figure 6.11   Example of a SATF configuration.

APPN node has a function called the Network Operator Facility (NOF). The NOF initializes configuration services. Through the NOF, an operator can define, start, stop, and query configuration services components. It is through the use of the NOF that node definitions and characteristics are specified. These definitions are

- Data link control
- Ports
- Links and link stations

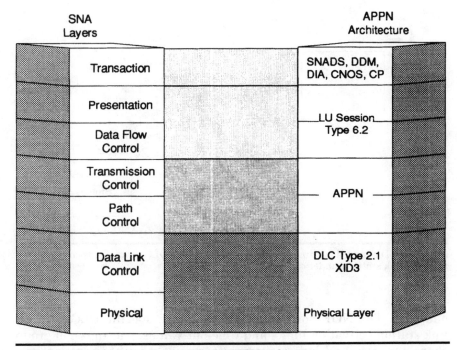

Figure 6.12   SNA layers as compared to the APPN architecture.

■ Attached connection network (virtual routing node)

Data link control.   Just as in SNA, the DLC layer ensures reliable delivery of information between a pair of nodes through the use of a protocol that provides for frame sequencing, acknowledgment, error recovery, and the establishment and synchronization of the information between the node partners over a common communication medium. The configuration services CP creates a DLC manager and a DLC element for each type of DLC in use. Each manager and element may support one or more ports. Figure 6.13 depicts DLC protocols and the relationship between DLC, ports, and link stations. The DLC manager is responsible for the activation and deactivation of DLC elements and links, acts as coordinator between the DLC element and the CP, and notifies the CP when a station or port is operative or inoperative. The DLC element is responsible for exchanging data with adjacent DLC elements and performs the physical transfer of data to the communications medium.

Ports.   A port is associated with a DLC process and acts as the physical connection to the link hardware. Ports are synonymous with

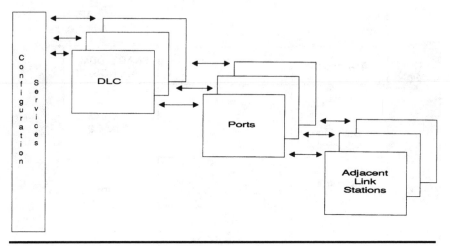

Figure 6.13 Diagram depicting the relationship between DLC protocols, ports and adjacent link stations under configuration services.

adapters. The NOF ports are defined with a specific DLC process, link station activation limits and time-out values, transmission group characteristics and buffer sizes, information about connection networks, and virtual routing nodes defined to this port.

Links and link stations. A connection between two nodes is made up of a local link station and an adjacent link station. Each link station is assigned a name. The role of a link station may be predefined or negotiated at link activation time. In a predefinition scenario, each definition on the adjacent nodes of the link must agree. If both nodes have their link station defined as primary or secondary, then the link activation will fail. To avoid this, the link station role can be defined as negotiable. If a link station is defined as negotiable and its adjacent link station is defined as primary, then the negotiable link station will take on the role of a secondary link station. The opposite holds true if the adjacent link station is defined as secondary. In the case where both link stations are defined as negotiable, the roles are determined at link activation time based on fields in the exchange identification passed between the two link stations. The exchange identification format 3 (XID3) is used between two adjacent APPN nodes over the link stations. The following information is passed between the adjacent nodes to resolve configuration definitions:

- Adjacent link station name.
- Control point capabilities.

Network node provides services over this link.

Network node does not provide services over this link.

End node supports CP-CP sessions over this link.

End node does not support CP-CP sessions over this link.

End node supports and requests CP-CP sessions over this link.

- CP name.
- Link characteristics.
- TG number.
- A subarea PU name, if applicable.
- The Product Set ID (PSID).
- The capabilities of the node.
  Parallel TG support.

  The type of DLC support.

  The BTU size.

Connection Network (CN) and VRN.   Through NOF, configuration services can define a VRN and a CN. A CN is composed of a SATF and the VRN representing the SATF. The CN must be supplied with a fully qualified network name. This is the CP name associated with the VRN. Hence, if several APPN nodes want to communicate over a token-ring network, they must all have definitions for the same CN name. Since a VRN is not a real node, it cannot support CP-CP sessions. Recall that a CP-CP session is required for session establishment and routing of data between adjacent nodes. Any-to-any connectivity is provided using a VRN by each end node that has at least one CP-CP session with the same NNS, or a CP-CP session between two NNSs that have CP-CP sessions with different end nodes (Figure 6.14). Without the use of a CN and VRN, each individual APPN node would need a definition to all of the other ENs on a token-ring. CN and VRN reduce the definition requirements and the amount of Topology Data Units (TDUs) flowing over the SATF and reduce the overhead on NNSs, since the ENs communicate directly with one another by using the VRN.

6.4.2   Topology and Route Selection (TRS) services

In each APPN node resides a topology and Class-Of-Service (COS) database. TRS is responsible for the creation and maintenance of a COS database and a network topology database. In LENs and ENs

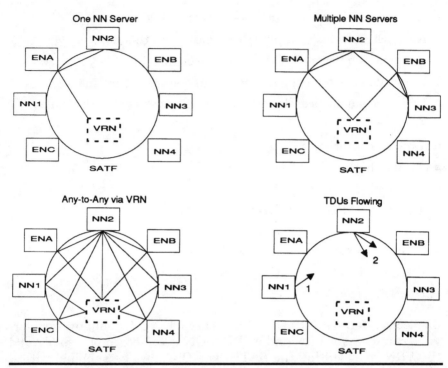

Figure 6.14   Use of the VRN and TDU flows.

this database is reflective only of its local resources. The APPN NN is the only APPN node that can contain a full network topology database. These two databases provide the necessary information for selecting end-to-end routes between nodes. There are three components that make up TRS; these are topology database manager, Class-Of-Service Manager (COSM), and Router Selection Services (RSS).

Figure 6.15 illustrates the relationships between the three functions of TRS and the relationships between the services. Through NOF, pertinent parameters are passed to the TRS at initialization. These parameters are

- Type of node (NN, EN, LEN)
- CP name of this node
- Network ID of the node
- Support for Class-Of-Service/Transmission Priority Field (COS/TPF)

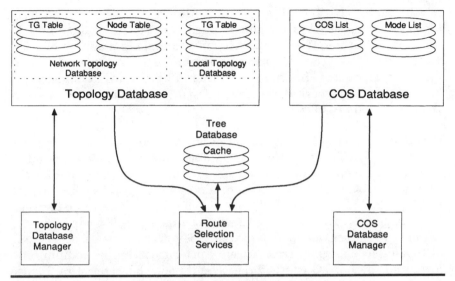

Figure 6.15 The three components of TRS and their relationships.

- COS database file name
- Topology database file name

The main function of TRS is to provide the optimal route between two APPN nodes over the network. RSS needs information from several databases. The topology database is actually broken down into two types of databases and two types of information.

Topology database. APPN networks are defined by TGs. TGs that connect NNs are called intermediate routing TGs, and TGs that connect ENs to NNs, VRNs, or other ENs are called endpoint TGs. The information about the TGs is kept in two databases managed by the TDM. These are the network topology database and the local topology database. The network topology database resides only on NNs and contains information about intermediate routing TGs. This database does not contain TG information about LENs, ENs, or the TGs attached to them. It includes only information about the TGs that interconnect the NN to other NNs or to VRNs. The network topology database is replicated on all NNs and is kept across device IPLs. The local topology database resides on both NNs and ENs and maintains information about end-point TGs. The local topology database, however, is not kept across device IPLs.

Within the network topology database is a node table. The node table specifies the characteristics of a node and associated definitions for the node. The TG table is found in both the network and local topology databases. As the name implies, this table contains information specific to the TGs. Each entry in the table is made up of a TG vector and a TG record. Two entries exist for each TG, since the TG is direction dependent. Information gathered by a NN that changes the topology of the network is transmitted to other NNs through Topology Database Updates (TDUs).

COS database. The COS database provides predefined definitions for prioritizing traffic over TGs. The database itself is made up of information from configuration services definitions for the TGs and the COS definitions. This information, along with data provided by the topology database, allows RSS to decide on the optimal route between two end points. The database itself contains a list of mode names, a list of COS names, and a weight index. The COS name associates one of three user-assigned transmission priorities—high, medium, and low—and rows of TG characteristics and node characteristics. Figure 6.16 lists the TG characteristics and the node characteristics used in the COS database. The mode name table contains a list of mode names. The mode name identifies session characteristics and points to a COS name to use for LU session requests. This is quite similar to the mode table found in VTAM. The COSM uses the mode name pro-

Mode Name	Corresponding COS Name
Default	#CONNECT
#BATCH	#BATCH
#INTER	#INTER
#BATCHSC	#BATCHSC
#INTERSEC	#INTERSEC
CPSVCMG	CPSVCMG
SNASVCMG	SNASVCMG

TG Characteristics	Node Characteristics
Cost per Byte	Route Additon Resistance
Cost per Connect Time	Congestion
Effective Capacity	Weight Field
Propagation	
Security Level	
User Defined-1	
User Defined-2	
User Defined-3	
Weight Field	

Figure 6.16 TG and node characteristic parameters used in the COS database.

vided by the LU to select a COS name for the COS database. The weight index structure of the COS database allows for a mechanism for computing the TG weight only once and storing it for future reference rather than computing the weight each time.

RSS. Optimal route selection is performed by the RSS of TRS. Using the network topology and COS databases, RSS calculates the optimal route for each LU-LU session supported by the NNS. The current route selected by RSS is a static route. This means that if any piece of the physical connection that makes up the selected route fails, the LU-LU session will fail, and it will then have to be reestablished, at which time RSS will determine a new optimal route for the session. Dynamic routing—that is, the ability to retain LU-LU sessions during a network resource outage—is made available in APPN High-Performance Routing (HPR).

Routes in APPN are made up of an ordered sequence of nodes and TGs found between two end nodes. RSS processes the required route characteristics on a session-by-session basis. These requirements are defined by the mode name supplied by the LU issuing the session request. Recall that the mode name will point to a COS name. RSS will then calculate all possible routes between the two end points based on the COS and TG requirements. This information is pulled from the network topology database and the local topology database for each of the end points. The optimal route selection begins with the exclusion of all nonacceptable routes at the time of the session establishment request. RSS can exclude potential routes by examining the current TG characteristics and network topology for congestion on intermediate nodes, depletion of resources on the intermediate nodes or the end point itself, a quiesced node along the route, an infinite weight assigned to the route by TRS, and user-definable values for adding route resistance. To speed up the calculations, RSS can cache a routing tree. The routing tree is computed from the original NNS and a given COS. Each NN calculates and maintains its own routing trees. The tree represents the optimal route between the original NNS and any other NN in the network. These routes are unidirectional and include the network nodes and the TGs that interconnect them. Each COS used needs it own routing tree. The tree is cached by RSS so that it need be calculated only once unless the status of a network resource changes. If multiple routes have the same weight, one is randomly chosen and an aging algorithm is used to cause the routing tree to be deleted and then recalculated. Once the optimum routes have been determined, RSS will select the route with the least weight.

6.4.3   Directory services

Logical units initiate a session request to their CPs. The CP must determine where and on which node in the network the LU resides. After this determination, a route can be selected by TRS. The directory services component of APPN performs the resource locate function.

Recall that the network topology database is replicated across all NNs in the network. Directory databases, however, are distributed. Each node in APPN maintains a local directory database. Local resources are entered into the local directory through configuration services. These resources include the control point LU and other user application LUs. The directory may also contain file names as well.

The LEN node does not support CP-CP sessions and therefore will not participate in a distributed directory search. All LUs remote to a LEN node that LEN LUs need to establish sessions with must be predefined on the LEN local directory database. This can be seen in Figure 6.17, where LEN_B has predefined LU_A and LU_1 as being reachable at NN_1 using TG2. The reverse holds true for the APPN node adjacent to the LEN node. The APPN NN node has an advantage over the LEN node because it can use a wild-card character to define all the resources on the adjacent node.

APPN ENs can either have their resources manually entered on the NNS like the LEN or use a registration process with the NNS. In

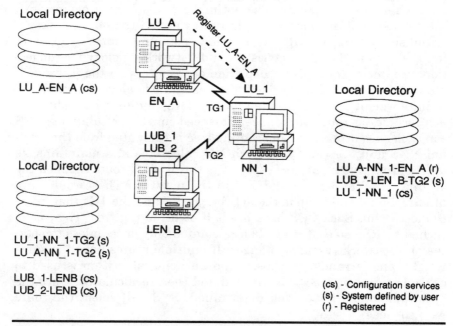

Local Directory

LU_A-EN_A (cs)

Register LU_A-EN_A

LU_A

EN_A

TG1

LU_1

Local Directory

LUB_1
LUB_2

TG2

NN_1

Local Directory

LEN_B

LU_A-NN_1-EN_A (r)
LUB_*-LEN_B-TG2 (s)
LU_1-NN_1 (cs)

LU_1-NN_1-TG2 (s)
LU_A-NN_1-TG2 (s)

LUB_1-LENB (cs)
LUB_2-LENB (cs)

(cs) - Configuration services
(s) - System defined by user
(r) - Registered

Figure 6.17 Registration and static definitions of LU resources residing on EN and LEN node.

Figure 6.17, EN_A registers its resource LU_A with NN_1 after the CP-CP session is established. EN_A can perform a registration process only if it is an authorized EN. An authorized EN can also allow the NNS to query it for LU resources that have not been registered with the NNS.

When a request for a Destination LU (DLU) is sent to the CP, the node will first look at its own local directory database. If the resource is found, the session establishment can begin. For LEN nodes, if the DLU is not found in the local directory database, then the session is not established. However, if the DLU is properly defined but is not local to the LEN, then the LEN can issue a BIND command to the owner of the resource or to a NN. An APPN EN can request directory assistance from its NNS by issuing a LOCATE request and forwarding it on the CP-CP session between the EN and the NNS.

Directory services uses three types of network searches to discover the location of a LU resource. These are one-hop search, broadcast search, and directed search. Figure 6.18 diagrams the three types of searches.

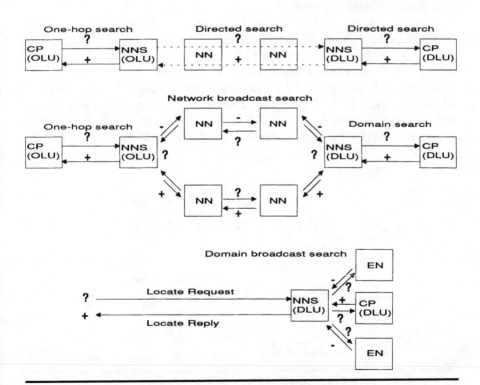

Figure 6.18 Illustration of the three types of APPN LU search processes.

One-hop search. This type of search is found only between an EN and its NNS. The EN issues a Locate search request to the NNS. The response is based on the then-current network topology database on the NNS for its domain. On behalf of the requesting EN, however, the NNS may issue a directed or broadcast search to locate the desired DLU.

Broadcast search. This type of search is used by an NNS when the DLU is not found on its local directory database. The broadcast is a Locate(broadcast) request that flows to all NNs and ENs adjacent to the NNS. Locate(broadcast) requests are sent to adjacent ENs only if the EN had indicated that it would entertain Locate(broadcast) requests from the NNS during the CP-CP session establishment. This type of broadcast search is also called a "domain search." A Locate(broadcast) request propagated throughout the network to other NNs is called a "network search." An NN attached to the EN that contains the resource sends the Locate(Find) request to verify that the DLU still exists on the EN. The EN then returns a Locate(Found) response to the originating EN through the same path used to discover the resource. At this point, TRS can begin determination of the optimum route.

Directed search. Instead of the Locate request being transmitted to all NNs in the network, a predefined path is used for known resources. This reduces the amount of broadcasts needed to locate a resource in the network. For example, in Figure 6.19, the path to LU_C in NN_1 is still in the RSS cache. LU_A on EN_A requests a second session with LU_C on EN_C. NN_1 sends a Locate(directed) request over the path that resides in the RSS cache. The NNS for

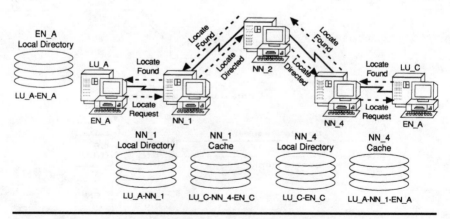

Figure 6.19 Example of a directed search over an APPN network.

EN_C verifies that LU_C still exists on EN_C, which then returns a Locate(found) response to EN_A. Note that this involved only two Locate requests through the network, rather than a broadcast which could potentially flood a network.

Directory services also supports a Central Directory Server (CDS). NNs can forward their own directories to a centralized server. This central directory server becomes the repository for the location of all the resources in the network. VTAM is currently the only CDS provided with APPN.

6.4.4   Session services

At initialization the NOF provides the necessary information to session services for the establishment of sessions between EN resources and CP-CP sessions between ENs and NNs. Session services is responsible for generating unique session identifiers and for the activation and deactivation of CP-CP sessions, and it assists in LU-LU session initiation. Session services involves directory services to locate the DLU partner, invokes topology and routing services to determine the optimum route between end points, indicates to management services the status of CP-CP sessions, and requests configuration services to activate TGs.

Each session initiated by session services is assigned a unique network Fully Qualified Procedure Correlation Identifier (FQPCID). This identifier is used to correlate requests and replies between the APPN nodes, recovery procedures, problem determination, accounting, auditing, and performance monitoring. The originating node specifies the FQPCID. The FQPCID is composed of the CP name of the originating APPN node and a 4-byte sequence number.

Sessions between LUs that are not found on adjacent nodes are facilitated by the intermediate routing functions of APPN NNs. Once the NNs have determined the optimum route, the originating LU can issue a BIND request to the DLU to establish a connection. This BIND is similar to the SNA BIND but has been extended to support APPN and LU6.2. The BIND now contains such information as a Route Selection Control Vector (RSCV), the FQPCID, the COS/TPF value, and a BIU indicator to determine whether the RUs sent on the session can be segmented.

Each session traversing a TG is assigned a unique identifier by the address space manager function of session services. The identifier is called a Local Form Session Identifier (LFSID). The LFSID is used in an extended ID Type 2 transmission header. This extended ID Type 2 header is used for communications between APPN nodes. The LFSID value changes after each intermediate NN found in the selected path. Figure 6.20 illustrates the format of the LFSID in the extended ID

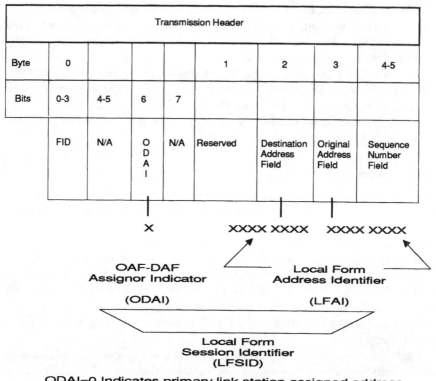

Figure 6.20 Illustrated format of the LFSID field.

Type 2 TH. In SNA Type 2.0 nodes, the address field is 8 bits in length. This address field is used to identify the PU of the sending and receiving LUs. Because of this 8-bit limit, the maximum number of sessions allowed on a PU is 255. In a Type 2.1 node, the origin and destination fields are combined to create a Local Form Address Identifier (LFAI) providing for up to 65,024 (64K) unique sessions that can travel over the TG. Bit 6 of the TH is also used to identify the session and is called the Origin Address Field (OAF)–Destination Address Field (DAF) Assignor Indicator (ODAI). This bit partitions the address field. Since each address space manager on each endpoint APPN node can define 65,024 addresses, the total number is the aggregate of 130,048 addresses that can be used to identify unique sessions over a TG.

Intermediate NNs found on the route between the two end points create Session Connector Blocks (SCBs). The SCB is used to map the incoming LFSID to an outgoing LFSID. The BIND used in session establishment is forwarded to the next adjacent node determined from the route found in the RSCV. The SCB also stores pacing values, transmission priorities, and whether the RU can be segmented.

6.5   VTAM/NCP Support for APPN

VTAM and NCP are the cornerstone of SNA. As such, these two communications applications must support both the hierarchical structure of SNA and the horizontal structure of APPN. VTAM/NCP began its support for APPN as a LEN node. This support allowed VTAM LU6.2 applications executing on the mainframe under CICS or APPC/MVS to establish sessions with LU6.2 resources on "right-sized" platforms like OS/2 servers or UNIX-based workstations. LEN nodes do not participate in CP-CP sessions with other APPN nodes, VTAM/NCP has limited APPN functionality and responsibilities. Since the introduction of LEN support, VTAM/NCP have been enhanced to support pure APPN nodes along with the ability to provide a migration path to a pure APPN network from a subarea-based SNA network.

6.5.1   LEN support on VTAM/NCP

LEN support was made available by VTAM V3R2 and NCP V4R3 on the IBM 3725 Communication Controller and by NCP V5R2 on the IBM 3745 Communication Controller. Figure 6.21 diagrams a LEN configuration. APPN networks made up of AS/400s and OS/2 servers were able to connect to the SNA subarea network for LU6.2 sessions. VTAM and NCP combined to create a LEN node called a Composite Network Node (CNN). Since there is no CP-CP session used to transfer information, there is no distributed directory service to exchange resource and routing data. Resources on VTAM had to be manually defined on APPN nodes, and APPN resources had to be manually defined on VTAM. The LU6.2 sessions established between APPN networks attached to an SNA subarea network used dynamic APPN routing within their networks and static SNA subarea routing within the SNA network. Although LEN support enhanced connectivity for VTAM/NCP, it did not provide the dynamic benefits of APPN. LEN support required the coordination of LU and routing definitions.

6.5.2   VTAM/NCP support for APPN

APPN NN and EN support are provided by VTAM V4R1 and NCP V6R2. NCP V6R2 executes only on an IBM 3745 Communication Controller. This is the first base implementation of APPN on VTAM. VTAM V4R1 supports traditional SNA subarea functionality with

Composite Network Node

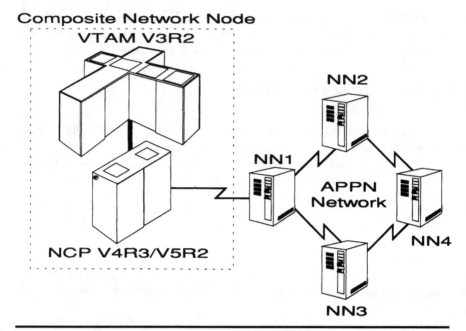

Figure 6.21 LEN configuration support with VTAM and NCP.

additional support for APPN NN, EN, and CDS. Figure 6.22 illustrates VTAM V4R1 and NCP V6R2 support for APPN.

The dual role supported by VTAM and NCP requires extensions to the APPN architecture to allow for migration from a dominated subarea architecture to a peer architecture. These extensions allow VTAM to isolate subarea specifics from APPN nodes but provide the ability to allow both APPN and subarea LU6.2 sessions to flow through and/or into APPN and subarea networks.

Now that VTAM V4R1 can be a NN or an EN, it can participate in CP-CP sessions. CP-CP sessions from VTAM to other APPN nodes can be established either through the boundary function of NCP or a connection network. These CP-CP sessions integrate the APPN and subarea directories that enhance resource search functions, allowing APPN network resources to gain connectivity into and through the subarea network. These sessions make VTAM V4R1 a good choice to act as a CDS. VTAM V4R1 must be defined as a NN to act as the CDS. The use of a CDS greatly reduces the number of APPN broadcast searches.

VTAM V4R1 also enhances connectivity options for APPN-to-subarea connection. VTAM now supports parallel and multitail TG connections to APPN networks. Parallel TG support allows two APPN nodes to be connected through multiple physical paths. Multitail sup-

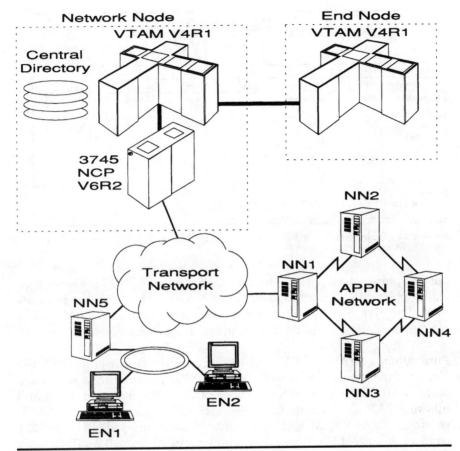

Figure 6.22 VTAM V4R1 and NCP V6R2 APPN support.

port allows independent LUs to establish sessions to or through a subarea network using multiple physical connections. Multitail support was previously available on VTAM V3R4.

VTAM's support for SSCP takeover has been extended to APPN nodes attached through an NCP boundary function. This takeover is nondisruptive to both the SNA and APPN sessions flowing through the NCP. VTAM issues a new nonactivation XID exchange that provides the new CP name of the recovering VTAM.

6.5.3 VTAM node types to support APPN

VTAM's ability to support both subarea and APPN network functions is provided by defining VTAM as one of five different types of nodes. Various combinations of the NODETYPE and HOSTSA VTAM start

Node Type Functional Summary						
Node	Node Type	HOSTSA	CP-CP Sessions	SSCP-SSCP Sessions	NCP Ownership	Inter-Change Function
Subarea	Not coded	Coded	No	Yes	Yes	No
Interchange	NN	Coded	Yes[1]	Yes	Yes	Yes
Migration Data Host	EN	Coded	Yes[1]	Yes	No[2]	No
Pure Network Node	NN	Not coded	Yes[1]	No	No[2]	No
Pure End Node	EN	Not coded	Yes[1]	No	No[2]	No
Note: 1. Node level default for CP-CP start option is YES. 2. Activation of an NCP is not allowed.						

Figure 6.23 Table identifying VTAM functionality with APPN node type and functional SNA and APPN support.

options parameters determine the functionality of VTAM. Figure 6.23 contains a table listing the various combinations of these two parameters. The five different functions are pure subarea VTAM, interchange node, migration data host, pure EN, and pure NN.

Pure subarea VTAM. VTAM will take on only traditional SNA PU Type 5 functionality when the NODETYPE parameter is not coded and the HOSTSA parameter is coded. This is the default setting and allows VTAM to participate in SSCP-SSCP sessions but not CP-CP sessions. Since VTAM acts as a subarea node only, all PU Type 2.1 connections are LEN-type CNN connections. Figure 6.24 illustrates the subarea-only functionality with LEN.

Interchange node. An InterChange Node (ICN) is defined in VTAM when the NODETYPE parameter specifies the value NN and the HOSTSA parameter is coded with a valid SNA subarea number. An ICN is a single APPN VTAM node with one or more NCPs owned by the ICN VTAM and running NCP V6R2. Figure 6.25 illustrates an ICN subarea domain configuration. ICN nodes provide FID 2 and FID 4 support for both SSCP-SSCP and CP-CP sessions. The ICN configuration can effectively replace a Communication Management Configuration (CMC) VTAM by providing ownership of the dependent LUs.

An ICN facilitates the routing of sessions from APPN nodes into and through the subarea network using SNA subarea routing. As shown in Figure 6.25, routing from the APPN NNs to the ICN subarea domain is performed by APPN Type 2.1 connectivity, and within the subarea domain SNA ERs and VRs are in use.

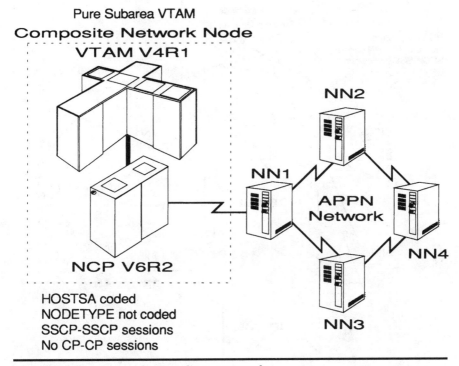

Figure 6.24 SNA subarea functionality support only.

A main function of an ICN is to transform APPN Locate requests to SNA subarea Direct Search List (DSRLST) or Cross-Domain Initiate (CDINIT) requests and SNA subarea DSRLST and CDINIT requests to APPN Locate requests.

An NCP can belong to several ICNs because of the ability of multiple VTAMs to activate the NCP and their ability to own link stations attached to the NCP's communication controller.

ICNs can be interconnected by using an APPN TG connection between the NCPs of the ICN or by using a subarea routing connection between the ICN NCPs. The subarea connection can be just an SNA INN link, or it can be an SNA/SNI link. Figure 6.26 diagrams these two possible ICN configurations.

Migration Data Host (MDH). In current CMC configurations, VTAMs that do not own NCPs but do support activation of the channel and link station to the NCP's communication controller are known as data hosts. The MDH node function of APPN was created to support a data host's migration to APPN. An MDH function is defined by supplying an SNA subarea number, and the NODETYPE parameter indicates that the VTAM will act as an EN. As shown in Figure 6.27, the MDH

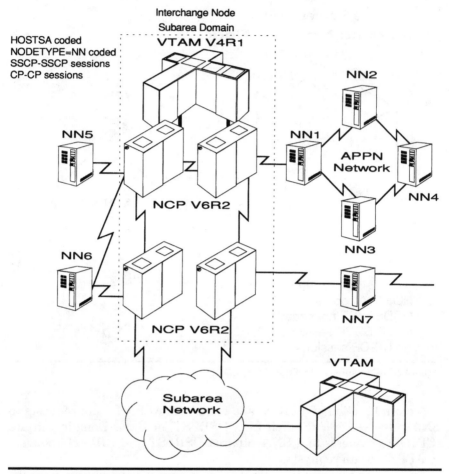

Figure 6.25   SNA subarea and APPN functionality support under ICN.

host can connect to an ICN VTAM through a channel connection, or alternatively through a channel connection to an NCP that is participating as a component of the ICN configuration. Since the MDH is an APPN EN, it does not provide for APPN protocol to SNA subarea protocol transformations. This must be done at the ICN on behalf of the MDH. An MDH can provide automatic registration of its resources when the TG between it and the adjacent ICN becomes active and the CP-CP session is established. The APPN connection between the ICN and MDH can be either over a channel-connected communication controller executing NCP V6R2 or through a token-ring using an IBM 3172 Interconnect Controller. MDH does support both CP-CP sessions and SSCP-SSCP sessions; however, these sessions cannot traverse the same physical link, and hence two links must be connected.

(a) Interchange Nodes with Full APPN Connectivity

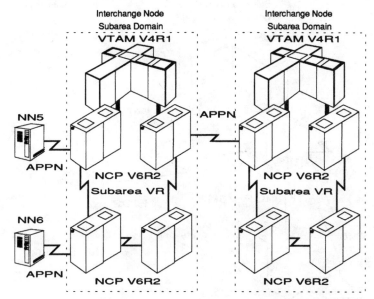

(b) Interchange Nodes without Full APPN Connectivity

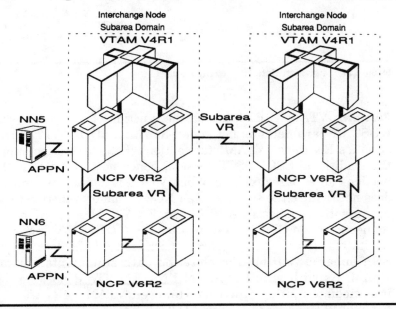

Figure 6.26 Complex ICN configurations.

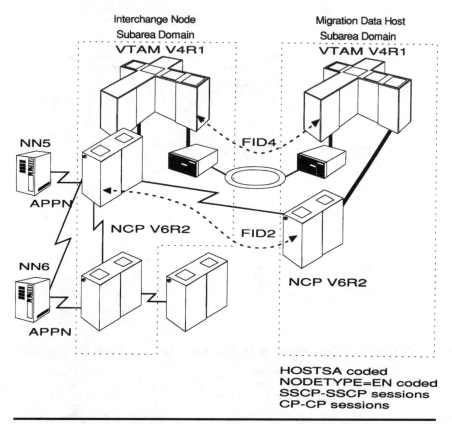

Figure 6.27 Migration Data Host configuration.

Pure EN. VTAM V4R1 takes on the function of a pure EN when the HOSTSA parameter is not coded and the NODETYPE parameter specifies EN in the VTAM start options list. There is no support for the traditional SSCP-SSCP session, hence no support of FID 4 connections for subarea routing. When VTAM is defined as a pure EN, it does not require the services of an NCP boundary function to communicate with an adjacent APPN node. EN VTAM supports only FID 2 connections and CP-CP sessions. Like the MDH, EN VTAM can register its resources with an NNS or CDS. EN VTAM does not support SNA subarea routing; therefore, no path statements are coded in VTAM. As with all other APPN ENs, EN VTAM does not support intermediate routing functions. However, VTAM V4R1 will continue to support dependent LUs like LU Type 2 that will be serviced with a future release of VTAM. LU Type 2 sessions will be transported over LU6.2 sessions through the APPN network to the dependent LU's partner. The dependent LUs can be owned by VTAM on its own node

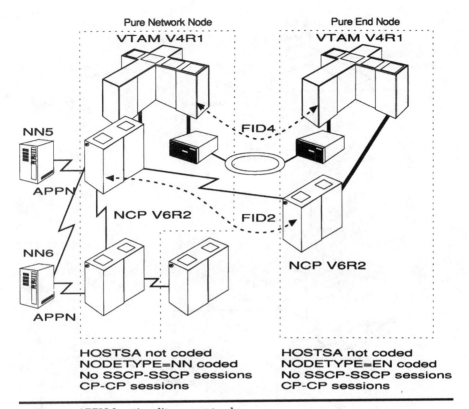

Figure 6.28   APPN functionality support only.

or on nodes adjacent to the VTAM boundary function. Figure 6.28 depicts both a pure EN and a pure NN for VTAM V4R1.

Pure NN. VTAM V4R1 can be configured as an APPN network node by defining the NODETYPE parameter with a value of NN and not coding the HOSTSA parameter. Again no subarea number is defined, so VTAM does not participate in SSCP-SSCP sessions and maintains a CP-CP session with adjacent network nodes and optionally with adjacent end nodes. NN VTAM can be configured as a CDS for all the APPN nodes to utilize. Like the EN VTAM, a NN VTAM can own dependent LUs on itself or on nodes adjacent to the VTAM or NCP boundary function.

6.5.4   Central Directory Server (CDS)

One of the benefits of APPN is the reduced definition requirements for network resources on the various nodes in an APPN network. VTAM V4R1 can act as a NN and therefore a CDS. Using VTAM as the CDS greatly reduces network search broadcasts, thereby improving network

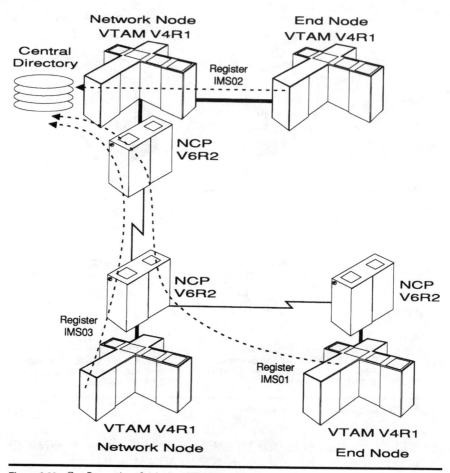

Figure 6.29 Configuration depicting VTAM as the CDS.

performance. Network performance is improved because the NNs on the APPN network will issue a directed search to the CDS on VTAM rather than a network broadcast. Figure 6.29 shows a CDS configuration. Multiple CDSs may be defined in an APPN. A multiple CDS configuration can be used for backup as well as for distribution of requests.

6.5.5 Connection network support

VTAM V4R1 does not define connection networks. But the TRS function in VTAM V4R1 does support connection networks and virtual routing nodes. Figure 6.30 diagrams a connection network configuration for VTAM V4R1. In the figure, both EN_A and EN_B are capable of defining connection networks and as such define the VRN connection as well

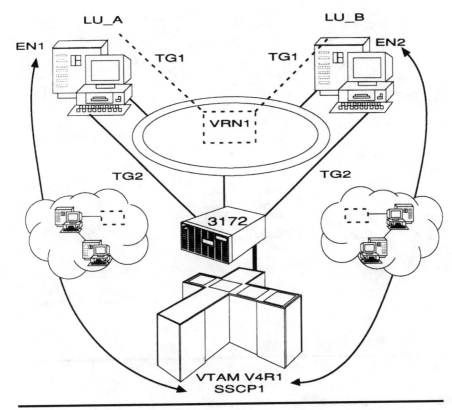

Figure 6.30   A connection network configuration.

as the TG connection to the VTAM by specifying the MAC address and
SAP address of the IBM 3172 and that VTAM_A is the preferred NNS.
VTAM defines the EN dynamically once the line and port definitions of
the physical connection to the ring through the IBM 3172 have been
entered on VTAM. CP-CP sessions between the two end nodes and
VTAM_A are established, and the TG tail vectors sent by the end nodes
will identify to VTAM that the direct route is through the VRN. The two
end nodes could be other VTAMs and provide for LU-LU sessions
between them only if the route used went through the NN VTAM or if a
direct connection were predefined in the two partners.

6.6   SNA-to-APPN Migration Considerations

There are several ways in which to approach a methodology of
migrating from an SNA subarea-based networking scheme to APPN.
Figure 6.31 will be used to flow through the five possible stages of
migrating such a network to APPN.

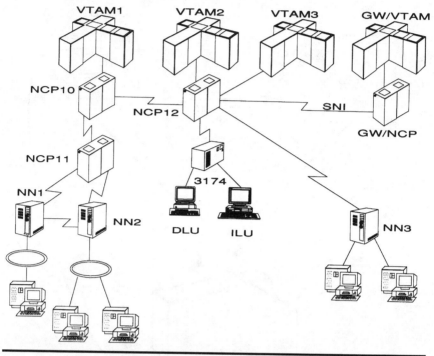

Figure 6.31 Traditional SNA subarea network.

6.6.1 Current SNA subarea/APPN network

In Figure 6.32 we have four distinct networks. Two are SNA networks, and two are APPN networks. The SNA networks are connected by an SNI link, and the two APPN networks are connected through the SNA subarea network NET_A. Note that the VTAM and NCP versions executing in the SNA subarea networks currently support LEN nodes. The two APPN networks appear to NET_A as two LEN nodes.

6.6.2 Create an APPN interchange node

Prior to VTAM V4R1, a LEN consisted of VTAM and NCP together. In such a configuration they are called a Composite Network Node (CNN). For APPN capabilities, the VTAM V4R1 and NCP V6R2 CNN is an APPN interchange node. Figure 6.33 shows the second phase in SNA-to-APPN migration. In this phase, VTAM_1 and NCP_11 are defined as part of the ICN and become a NN in appearance to NN_1 and NN_2. Note that NCP_10 is not part of the ICN configuration. NCP_10 still supports the SNA subarea flows, and thus path definition statements must still exist in VTAM_1 and NCP_11 as well as NCP_10 to support the SNA subarea flows. NCP_11, however, uses

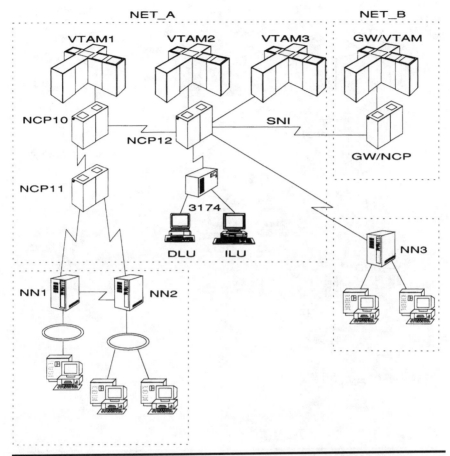

Figure 6.32 Illustration depicting the two SNA subarea networks and the two APPN networks.

FID 2 flows to NN_1 and NN_2. At this stage all traditional VTAM definitions are still required; however, VTAM_1 will participate in dynamic routing and topology updates with NN_1 and NN_2.

6.6.3 Migrate second VTAM domain to ICN

In the third phase of the migration to APPN VTAM_2, NCP_12 and the link-attached IBM 3174 with APPN support, along with NN_3, become a new APPN network. This can be seen in Figure 6.34. Note that the 3174 has both a DLU and an ILU attached to it. The ILU is found on a workstation that appears to the 3174 NN as an EN. The 3174 therefore must support both SNA Type 2.0 for the DLU sessions and Node Type 2.1 for the EN sessions. Again we see here that traditional SNA subarea flows are found between NCP_10 and NCP_12, NCP_12 and VTAM_3, and NCP_12 and NET_B through the SNI

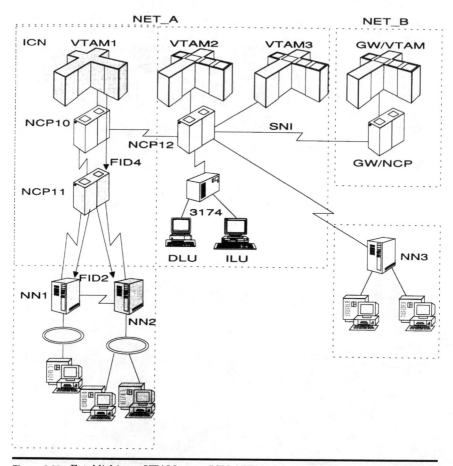

Figure 6.33 Establishing a VTAM as an ICN APPN component.

link; and APPN network flows are found between NCP_12 and NN_3. LU6.2 sessions flowing from APPN_B to APPN_A must have their FID 2 format converted to FID 4 format and vice versa as the data flows between NCP_10 and NCP_12. It may be good practice to keep this type of connection until the data flowing between the NCPs amounts to more APPN traffic than traditional SNA traffic.

6.6.4 Convert data host to pure EN

In Figure 6.35, VTAM_3 is defined as a pure APPN EN. As such, VTAM_3 uses only FID 2 formats and APPN protocols. This phase will now let us eliminate the traditional subarea definition that was found in VTAM_3. SNA definitions found in NCP_12, VTAM_2, NCP_11, and VTAM_1 can now eliminate the SNA subarea defini- tions that pertained to VTAM_3. VTAM_2 can now automatically reg-

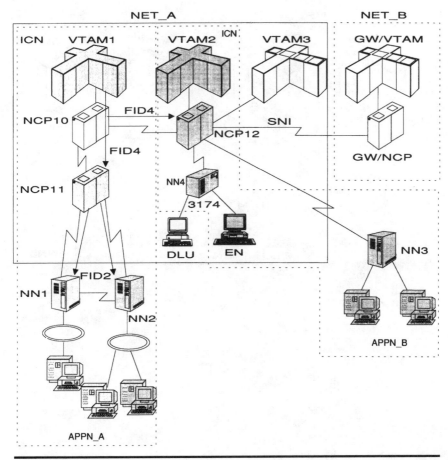

Figure 6.34 Selection of a second VTAM to participate as an ICN on the migration path to APPN.

ister its resources with NN VTAM_2. Suppose that much of the data from VTAM_3 was destined for NET_B. In such a scenario, it may be wiser to convert VTAM_3 to a migration data host instead of a pure EN. Even though the HOSTSA parameter is not defined in VTAM_3, a subarea number of 1 is assumed. This is to support subarea sessions between LU Type 2 resources in VTAM_3's domain and applications owned by a VTAM in another domain. However, only FID 2 format is routed over the APPN connection to the destination VTAM.

6.6.5   Single APPN network

At some point in time, NCP_10 was converted to an NCP level that supports APPN CNN. The link between NCP_10 and NCP_12 can now be defined as an APPN TG connection, creating one large APPN network configuration. This final configuration can be seen in Figure

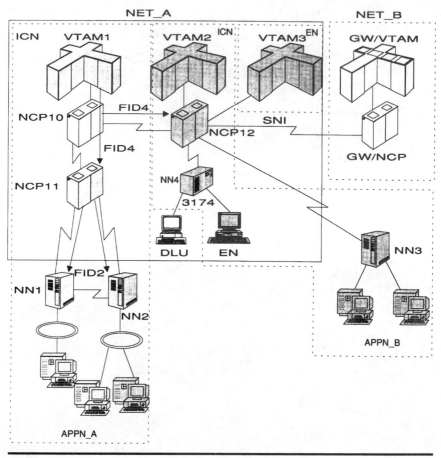

Figure 6.35 Convert any SNA Data Host to a pure APPN EN.

6.36. SNA subarea definitions in VTAM_1, NCP_10, and NCP_11 for connectivity to VTAM_2 and NCP_12 can be eliminated. The reverse is true for VTAM_2 and NCP_12. However, VTAM_1, NCP_10, and NCP_11 still need a path definition statement for connection with each other, as does VTAM_2 for the connection to NCP_12. In addition, VTAM_2 and NCP_12 must retain their SNA subarea definitions for connectivity to NET_B through SNI. All previous LU-LU sessions that were available prior to the migration to APPN are still fully supported and have not been affected.

6.7 Summary

In this chapter we discovered how the impact of powerful microprocessors and the wide acceptance of local area networking led to the

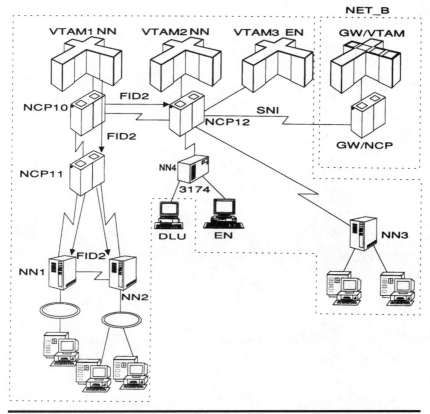

Figure 6.36 A complete APPN network with support for traditional SNA/SNI communication.

need for peer-to-peer networking architectures. IBM's APPN is its answer to this peer-to-peer networking architecture requirement. APPN is an enhancement of Low-Entry Networking, which was first introduced on IBM System/36 devices. APPN is based on the LEN Type 2.1 node, with extensions to create a network node and an end node. Each node has a control point that provides configuration, topology and routing, directory, session, and management services. We also learned how VTAM and NCP have incorporated APPN node functionality and extended it to provide for SNA subarea connectivity and APPN network connectivity concurrently. This ability to provide concurrent support for both networking schemes allows for a migration methodology that enables your corporation to run its business while moving the network from a hierarchical one to a peer network structure.

Communications Protocols

Protocols define a set of rules which two or more participating entities must follow for proper exchange of information among them. In a computing environment, such protocols become more important because of the complexities of modern-day networks. The protocols must ensure that the information is accurate and exchanged in an orderly fashion. In this chapter we will first discuss the link-level protocols, such as SDLC and BSC. Thereafter, we will talk about the session-level protocols, such as half duplex and full duplex.

7.1   Link-Level Protocols

In wide area networks, most of the communications takes place over long distances. The physical media to carry information from one point to the other are provided by common carriers such as AT&T, MCI, Sprint, and local telephone companies. Such physical media could also be located in the same building or campus and may be owned by the enterprise itself. One of the inherent shortcomings of long-distance data communications is the erroneous information which could be transmitted to the receiving node. Such errors could be caused by communications equipment, electromagnetic interference, weakening of signal, and many other reasons. The common carriers cannot totally eliminate such communications errors. It is the responsibility of the Data Terminal Equipment (DTE) to ensure that the transmitted and received data are the same. DTEs can be the communications controllers and cluster controllers.

The protocols which ensure *accurate and orderly* exchange of information between two or more DTEs are the link-level protocols. Such protocols logically connect the Data Link Control Layers (DLC) of SNA. The DLC Layer is one level above the Physical Control Layer. One such protocol that was developed in 1965 was Binary

Synchronous Communications, or BSC for short. Although still in use, it is being rapidly replaced by a more advanced protocol, Synchronous Data Link Control (SDLC). SDLC was introduced in 1973 and is the recommended protocol for SNA networks.

7.1.1 DTEs and link-level protocols

When the DTEs are two communications controllers, the preferred link-level protocol is SDLC. Since such devices exchange relatively large amounts of data, they need to exchange information simultaneously. Later in this chapter, we will learn that such simultaneous exchange is called "full duplex," which is supported only by SDLC.

If one of the DTEs is a communications controller and the other one is a cluster controller, the protocol can be BSC or SDLC. The most widely used remote cluster controller is the IBM 3274-41C. Any cluster controller model having a suffix of C is a remote one. A remote cluster controller can be BSC or SDLC depending upon the microcode disk which is used to configure it. Examples of other remote cluster controllers are 3274-51C (8 ports) and 3274-61C (16 ports).

Remember that link-level protocols ensure end-to-end data integrity between DTEs only. Although there are also DCEs (modems and DSU/CSUs) involved in the middle, they do not participate in such protocols. However, SDLC and BSC protocols do require synchronous modems. Asynchronous modems *cannot* be used for BSC or SDLC protocols and devices.

7.2 Synchronous Data Link Control (SDLC)

SDLC is the preferred link-level protocol for SNA networks. Introduced in 1973, it is very widely used in the IBM networking environment. It is equivalent to a similar protocol developed by the International Standards Organization (ISO) and adopted by many non-IBM vendors. ISO's link-level protocol is called "High-level Data Link Control" (HDLC).

SDLC protocol ensures that the DTEs exchange information accurately and in an orderly fashion. There are two entities involved in data transmission—the sender and the receiver. Communications controllers and cluster controllers can be both senders and receivers in a data exchange.

7.2.1 Primary and secondary stations

In a point-to-point configuration, there is one primary and one secondary station. In a point-to-multipoint configuration, there is still one primary but many secondary stations (Figure 7.1).

In other words, there is one and only one primary station on an SDLC line. The primary station is aware of the transmission status of

a) Point-to-Point Configuration

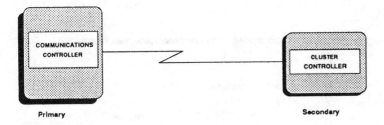

b) Point-to-Multipoint Configuration

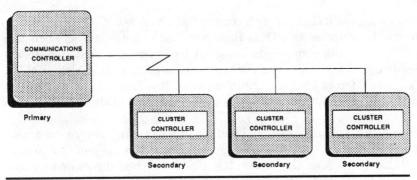

Figure 7.1   Primary and secondary stations in an SDLC environment.

different secondary stations all the time. A primary station can send data to a secondary station at any time, but the reverse is not true. A primary station has to *invite* the secondary station to send data.

In a point-to-multipoint environment, a primary station usually operates in full-duplex mode, while the secondary stations operate in half-duplex mode.

7.2.2   SDLC frames

DTE breaks down the transmission data into a number of small pieces of information. Each chunk of information is enclosed within transmission control fields before it is sent over the communications lines. Such a transmission unit is called an "SDLC frame." Figure 7.2 gives various fields constituting an SDLC frame. Let's look at the contents and functions of each field in the frame.

Flag field.   There are two flags in an SDLC frame—at the beginning and at the end. Each flag is 8 bits (1 byte) in size and always contains

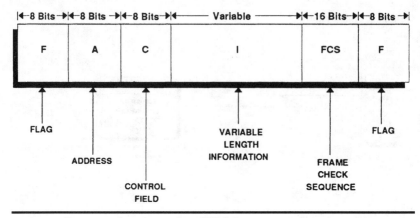

Figure 7.2   Format and contents of an SDLC frame.

a binary value of 01111110 or hexadecimal 7E. A set of six 1s following a binary 0 indicates an SDLC flag. Any data contained between the flag fields may not contain six consecutive 1s because that implies an end-of-frame status. Since in reality an information field may contain any combination of bits, a technique called "bit stuffing" is used to get around this problem. Bit stuffing is discussed later in this section.

Address field.   *This field always identifies the secondary station on a line.* When the primary station is inviting the secondary station (polling) to send data, it identifies the station being polled. When the secondary station is sending data to the primary station, it identifies the sender's own address. In a point-to-point configuration, this field may contain only one address because there is only one secondary station. In point-to-multipoint, it can identify up to 254 secondary station addresses. Although an 8-bit address can identify 256 different addresses, the remaining two bit configurations are used for different purposes as follows:

- All binary 0s are used for testing only.
- All binary 1s are used to send a broadcast message from the primary station to all secondary stations.

It is also worthwhile mentioning that the address hexadecimal FD is used by IBM modems. Therefore, if multiple cluster controllers are multidropped on a line, the logical limitation is 253 such devices.

Control field.   A control field is an 8-bit field which can have three different formats. Figure 7.3 gives the contents for each format. Notice that the leading bit of the transmission has been shown in the extreme right position. The logic for determining what format the control field belongs to is as follows:

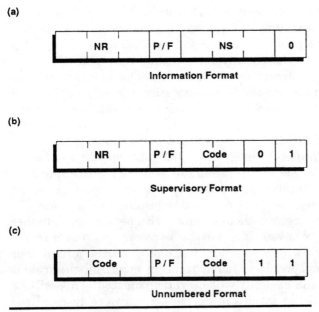

(a)

| NR | P / F | NS | 0 |

Information Format

(b)

| NR | P / F | Code | 0 | 1 |

Supervisory Format

(c)

| Code | P / F | Code | 1 | 1 |

Unnumbered Format

Figure 7.3 Format and contents of the control field in an SDLC frame.

- If the leading bit is 0, it is an information format.
- If the leading bit is 1, it is a supervisory or unnumbered format. If the second bit is 0, it is a supervisory format; otherwise it is an unnumbered format.

The significance of each format is explained as follows:

- The *information format* shown in Figure 7.3a, indicates that the frame contains application-specific data. The source of such data could be VTAM applications such as CICS/VS, IMS/DC, TSO, VM/CMS, etc.
- The *supervisory format,* shown in Figure 7.3b, contains a 2-bit code field which can yield four different combinations. A value of 00 indicates that the station is ready to receive data (RR). A value of 01 means it is not ready to receive data (RNR). Finally, 10 indicates a reject (REJ) and requests retransmission of the frame.
- The *unnumbered format,* shown in Figure 7.3c, contains a 5-bit code field and can have up to 32 different values. So far, only 15 different commands have been devised and used. Readers are advised to refer to the appropriate IBM literature to get details about such commands.

Now let's talk about NR and NS. Simply put, NS indicates the number of frames sent, and NR represents the number of correct frames received. The primary and secondary stations exchange NR and NS values to know how many frames they have sent the other and what is the last correct frame received. The primary station maintains the NS value for each secondary station and keeps track of NR information coming from every secondary station based on the address field.

The 3-bit NR and NS fields can contain a maximum value of 7. Therefore, in SDLC protocol the sending node (station) can send up to seven frames (NS = 7) before expecting a response for the first one. In contrast, the BSC protocols would not allow transmission of a second frame unless a response has been received regarding the first one.

There is another thing you should know. NR specifies the number of error-free frames received. *If a particular frame has had a transmission error, every frame following that one is retransmitted.* For example, if the second frame is found to be in error, all the frames from the second to the seventh will be retransmitted. This will be done even though all the frames after the second were transmitted without errors.

Because of the advent of high-speed communications links such as DDS lines and T1 lines, it was found that the maximum limitation of seven transmissions was inadequate. The satellite-based network further aggravated the problem with long transmission delays. With the release of NCP Version 3, you have a choice of keeping the limit at 7 or increasing it to 127. It is controlled by the MAXOUT parameter (MAXOUT = 127) during NCP generation. When the network is handling the capability of transmitting up to 127 unconfirmed frames, NS and NR values are contained in 7 bits each. Thus, the control field under those circumstances is a 16-bit field. This feature also permits the retransmission of a selected erroneous frame rather than having to retransmit all the frames following the frame in error.

Finally, let's talk about the P/F bit in the control field. When the frame is being transmitted from the primary to the secondary station, this bit means a poll. It indicates that the secondary station can send more data. When the frame is being transmitted from the secondary to the primary, it means the final frame. In other words, the secondary station indicates that this is the final frame and it has nothing more to send.

Information field. The information field is a variable-length field and contains application-specific information. In most cases, this is the real data you intend to transmit. It is always a multiple of 8 bits. It is an optional field, since control fields containing unnumbered commands do not transmit application data in the frames.

Frame Check Sequence (FCS). The FCS is a 16-bit field and is used to verify whether the data is error-free or not. It contains the *Cyclic Redundancy Check (CRC)* value. The sending station computes the FCS and appends it as a 2-byte field after the information field. The receiving station uses the same formulas to calculate the CRC constant and compares it with the value it received. If they are identical, the data is correct. The fields used to calculate the CRC constant are address, control, and information.

7.2.3   Bit stuffing

Bit stuffing is also referred to as "zero insertion." Since the presence of six 1s indicates a flag field, the presence of this bit combination can give a false indicator of the end of frame. To overcome this problem, bit stuffing is used to ensure that the address, control, information, and FCS fields do not contain six contiguous 1s anywhere. Figure 7.4 shows how this is done. Whenever the sending station detects five contiguous 1s in an SDLC frame, it inserts a 0 after it. This way, you will not see six contiguous 1s except in the flag fields. The receiving station takes out the single zero after it detects five 1s. Although bit stuffing increases transmission volume, it ensures data transparency for any bit combination. In general, bit stuffing adds about 5 percent overhead to the transmitted data.

7.2.4   Other characteristics

SDLC is a full-duplex protocol. The primary and the secondary stations can send data to each other at the same time. Each station can

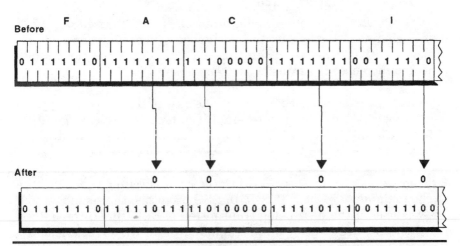

Figure 7.4   An example of bit stuffing in an SDLC frame.

send up to 7 (or 127) frames before waiting for an acknowledgment. Since SDLC is a bit-oriented protocol, it is insensitive to code, which may be ASCII or EBCDIC. The acknowledgment for the data is usually sent with the data itself, while in other protocols, such as BSC, it is a separate transmission. This feature of SDLC reduces unnecessary transmissions between the stations. In general, SDLC is a better protocol, with more efficient transmissions and lesser amounts of overhead.

7.3   Binary Synchronous Communications (BSC)

BSC protocol, announced in the mid-sixties, was rapidly replaced by the new and powerful SDLC protocol. We will give minimal coverage to this protocol because of its antiquity. The following are some of the highlights of its characteristics.

Batch orientation.   When this protocol was designed, there were very few on-line applications. Most of the transmissions were batch-oriented and involved the use of remote job entry stations. The terminals which used BSC for batch transmissions were IBM 2770, IBM 2780, and more recently the IBM 3780. Although these batch terminals are still in existence, the new RJE/RJP stations use SDLC protocols.

Propagation delays.   While SDLC protocol is full duplex, BSC supports only half-duplex transmissions. The sending station must get an acknowledgment for the data it sent before sending another batch. Therefore, BSC is not suitable for those environments in which the physical transmission media can cause long propagation delays. An example of such media is satellite transmission.

Although BSC is a half-duplex protocol, it can be used over full-duplex channels. In the case of full duplex, synchronization is maintained in either direction at all times. This reduces the turnaround time and increases the transmission efficiency.

Point-to-point versus multipoint.   As in the case of SDLC, BSC supports point-to-point as well as point-to-multipoint transmissions. Point-to-point transmissions can occur over leased as well as dial-up (switched) lines.

Error checking.   In the case of SDLC, the error checking is done by the Cyclic Redundancy Check (CRC). In BSC protocols, the same methodology is used if the data being transmitted is EBCDIC. If the data is ASCII, there are two techniques applied to do the error checking. Vertical Redundancy Check (VRC) is applied to check each ASCII character, while Longitudinal Redundancy Check (LRC) is used to check the entire block.

Byte-oriented protocol. While SDLC is a bit-level protocol, BSC deals at the byte level. In SDLC, all the control information, flags, and data fields are dealt with at the bit level. The bit stuffing takes into account only consecutive 1s, irrespective of the fact that they may span 2 bytes. In contrast, BSC deals with the transmission at the byte level. The control characters are all interpreted at the byte field only. Individual bits within the characters do not hold any special significance.

7.3.1 SDLC versus BSC

Figure 7.5 outlines the comparison between SDLC and BSC protocols. Most of the characteristics have already been discussed in the previous sections. In SDLC protocols, an invitation from the primary station (polling) to the secondary station to send data is included in the control field of the frame. In BSC, it is a separate sequence with its separate transmission. Similarly, acknowledgment for the data is included in the SDLC frame itself and is represented by NR. Again, in BSC it is a separate sequence. In other words, SDLC protocol pro-

CHARACTERISTIC	SDLC	BSC
Duplex	Full Duplex	Half Duplex
Polling	In All Frames	Separate Sequence
Maximum Message Count	128	2
Protocol	Bit Level	Byte Level
Addressing	In All Frames	Separate Sequence
Good for Satellite Transmission	Yes	No
Textual Transparency	Yes	Yes (through a special character)
Acknowledgment	May Be Included with Data	Separate Sequence
Code Sensitivity (ASCII/EBCDIC)	No	Yes

Figure 7.5 Comparison between SDLC and BSC link-level protocols.

vides for performing multiple functions in the same transmission. In BSC, each transmission usually performs only one function.

The BSC protocol is ill-suited for physical transmission media having substantial propagation delays. Thus BSC is not recommended for satellite-transmission-based networks.

SDLC protocol guarantees textual transparency and deals with the occurrence of end-of-frame flags by using bit stuffing techniques. In BSC, you have an *option* to use the transparency feature with the use of a Data Link Escape (DLE) character before the occurrence of an End of Text (ETX) character within the text. In EBCDIC transmissions, ETX is a hexadecimal 03 while the DLE is a hexadecimal 10.

Since SDLC is a bit-level protocol, it is basically code-insensitive. It does not care whether the information being transmitted is ASCII or EBCDIC. Since BSC is a byte-level protocol, it is code-sensitive. There are different sets of control characters for ASCII and EBCDIC transmissions.

SDLC protocols are the same irrespective of the type of device used. The only difference in an SDLC environment is whether the maximum message count value before acknowledgment is 127 or 7. In the first case there will be 2 control bytes, while in the latter case there will be 1. In contrast, there are many types of BSC protocols. Two of the most popular BSC protocols are 2780 BSC and 3270 BSC. The first is oriented toward batch transmissions, while the latter is designed for interactive communications.

7.4   Asynchronous Communications

Asynchronous communications was developed in the 1950s and is a primitive protocol compared to SDLC and BSC. It is still a very widely used mode of communications for the PC environment. Its widespread use can be attributed to the low-cost asynchronous modems that support this kind of communications. On the other hand, SDLC and BSC protocols require synchronous modems, which are more expensive.

Another name for this protocol is start-stop communications. According to this protocol, only one character is transmitted at a time. In SDLC and BSC communications, a block of data is transmitted at a time. Even the error detection is done at the character level in asynchronous communications.

7.4.1   Data stream

Figure 7.6 shows the bit structure of a single character transmission in asynchronous communications. The first bit is called the "start bit."

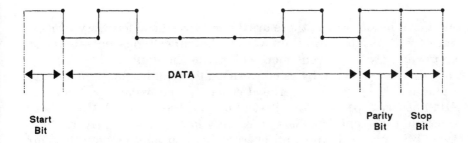

- Data Bits Transmitted: 7
- Total Bits Transmitted: 10
- Parity Can Be Odd or Even

Figure 7.6   Asynchronous communications.

The next 7 bits include the real data that is to be transmitted. Since each ASCII character set consists of 7 bits, each transmission can send one ASCII character. The eighth bit is used for error checking and is generally referred to as the parity bit. Parity can be odd or even. If the parity is even, all the binary 1s in the data and the parity bit must be an even number. If the parity is odd, the number must be odd. Since the communications software at either end can choose even or odd parity, you must ensure that both ends use the same parity. The last bit in the transmission is a stop bit.

7.4.2 Error checking

The odd or even parity checking methodology employed by asynchronous communications does not ensure complete data integrity. If 1 of the 7 data bits in an ASCII character is altered, the parity bit can detect the error. If 2 bits are altered during transmission, it is not possible to detect the error. Two such scenarios are given as follows:

- If a binary 1 is altered to a 0 and a binary 0 is changed to a 1, parity will remain the same.
- If two binary 1s are both changed to 0s or two binary 0s are both changed to 1s, parity will still remain the same.

In either example, erroneous data will be transmitted, but the error detection procedures will not be able to detect it. For these reasons, asynchronous communications is not recommended for applications where data integrity is a must.

7.4.3 Performance

As shown in Figure 7.6, there are 3 bits of overhead for every 7 bits of data transmission. In other words, the superfluous data bits add approximately 43 percent more volume to the data. In SDLC protocols, there are about 6 bytes of overhead for each frame. If an average frame is 200 bytes, the overhead comes to approximately 3 percent. Add another 5 percent for bit stuffing and the total SDLC overhead comes to 8 percent. This is significantly lower than 43 percent, however. BSC protocols have an overhead which may be slightly more than that of SDLC.

7.4.4 Eight-bit ASCII code

Seven-bit ASCII code allows for a maximum character set of 128. Some vendors use an extended ASCII character set in which 8 bits are used to represent each character. In such cases, it is not possible to use the eighth bit for error checking. Parity checking is turned off under these circumstances.

Asynchronous communications is generally used by devices which have no buffers. It is also used for reasons of economy because asynchronous modems are inexpensive. However, it is pertinent to add that some communications protocols use asynchronous modems but use better error checking techniques to ensure high data integrity. Some of these protocols are Xmodem, Kermit, and Crosstalk.

7.5 Token-Ring Protocol

In the winter of 1980, the Institute of Electrical and Electronics Engineers (IEEE) Computer Society established the data link and physical standards for local area networks. Figure 7.7 shows the IEEE 802 standard that implements local area networking in the two lower layers of SNA and OSI. The Data Link Layer for the IEEE 802 LAN standard is subdivided into two sublayers. These are the IEEE 802.5 standard for token-ring Medium Access Control (MAC) and the IEEE 802.2 standard for Logical Link Control (LLC).

7.5.1 Medium Access Control (MAC) sublayer

Communications between stations on a token-ring requires an addressing mechanism that will guarantee that each station address on the token-ring is unique. This is needed to ensure receipt and delivery of information to and from the source and destination stations on a token-ring. The MAC sublayer provides this addressing mechanism to control the transmission of data so that only one sta-

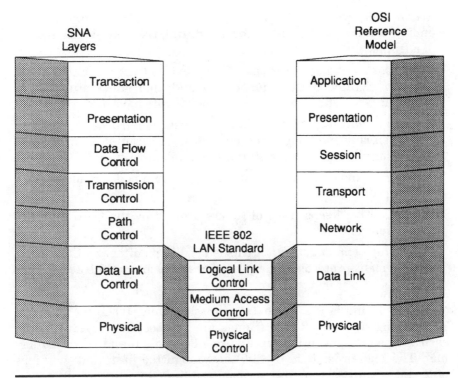

Figure 7.7 Comparison of IEEE 802 LAN standard with SNA and OSI architectures.

tion is transmitting at any given time. The MAC determines whether a station on the token-ring is in transmit or receive state and controls the routing of data over the LAN. The main functions of the MAC are:

Addressing.   The MAC address is the physical address of the station's device adapter on a LAN. Recognition of the station's address is found in the physical header of a MAC frame. Each station on a token-ring must be able to recognize its own MAC address and an all-stations (broadcast) address or a null address for frames which are not to be received by the station. The MAC address identifies the physical destination and source of any frame transmitted over the token-ring.

Frame copying.   After MAC has recognized its own address, meaning that the frame received has this device's MAC address as the destination address, MAC uses this function to copy the frame from the token-ring into the device adapter buffers.

Frame recognition.   This function determines the type of frame received and the frame's format—for instance, a system or user frame.

Frame delimiter.   The MAC must determine the beginning and ending of a frame. This is performed during transmission or receipt of a frame.

Frame status and verification.   This provides the checking and verification of frame check sequence bits and the frame status field in each frame to determine if transmission errors have occurred.

Priority management.   This function ensures fairness of access to the token-ring medium based on priority by issuing priority-level tokens for all participating stations on a token-ring.

Routing.   This determines which function in a node should process the frame.

Timing.   This keeps track of timers utilized by the MAC management protocols.

Token (LAN) management.   This is used to monitor the LAN using management protocols to handle error conditions at the access control level.

A MAC frame is the Basic Transmission Unit (BTU) for the IBM Token-Ring network. These frames are composed of several fields, each 1 byte or more in length. Figure 7.8 details the MAC frame format. The high-order byte (byte 0) is transmitted first, and the high-order bit within each byte is transmitted first. In other words, transmission is from left to right. The frame is composed of two sections with an optional section for information. The first section, the physical header, contains the Starting Delimiter (SD), the Access Control (AC) field, the Frame Control (FC) field, the destination and source address fields, and an optional Routing Information (RI) field. The second section of the MAC frame, the physical trailer, contains the Frame Check Sequence (FCS), the Ending Delimiter (ED), and the Frame Status Field (FS).

The SD is a single-byte field of the MAC frame. Token-ring frames and the tokens themselves are valid only when the right combination of bits that defines the byte as the starting delimiter of a frame is present.

The AC field is a 1-byte field to denote the access required by the frame on the LAN. The access defines the priority as well as the type of BTU (i.e., frame or token).

The next single-byte field is known as the FC field. The FC field identifies the type of frame, including specific MAC and information frame functions. There are currently two types of frames that can be identified with the FC field.

The 6-byte Destination Address (DA) field allows for a 48-bit representation of the MAC address for the receiving station. This address can be an individual or a group address. The individual address can

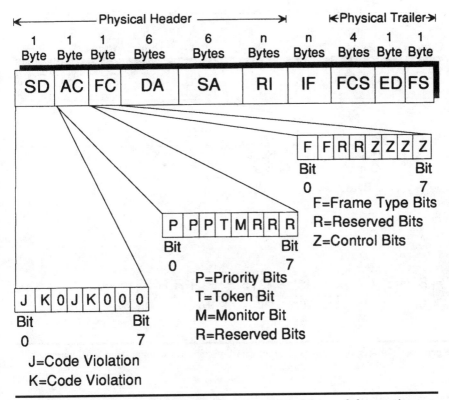

Figure 7.8 The token-ring MAC frame format and breakdown of the starting delimeter, Access Control and Frame Control fields.

be assigned by a LAN administrator or by the adapter manufacturer. A value of hexadecimal FFFFFF means that this frame is an all-stations broadcast and that every station on the network will make a copy of the frame and interpret its information.

Like the destination address, the Source Address (SA) is 6 bytes in length (48 bits). The first byte of the SA field indicates whether the routing information field is present and whether the address is provided by a LAN administrator or by the manufacturer.

The RI field is an optional field that is used only when the frame is going to leave the originating token-ring or source ring. There are two fields that make up the RI field. These are the route control field and the route designator field. Each is 2 bytes in length. The route control field is followed by up to eight route designator fields. Routing performed in this fashion is also called "source routing." The format of the RI field is detailed in Figure 7.9.

When a station contacts another station on the token-ring, it sends out a broadcast or test command. Token-ring uses Source-Route

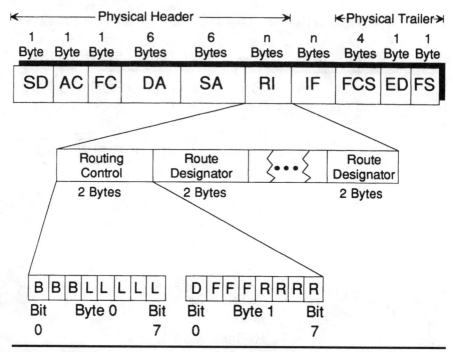

Figure 7.9   The format of the routing information field in the token-ring MAC frame.

Bridging (SRB) to determine the physical path through the token-ring network between the station partners.

Since the total number of route designators in the route information field is eight, a frame cannot traverse more than seven LAN segments before reaching the destination address.

The route designator field is made up of 2 bytes. The 2 bytes indicate the ring number and the bridge number. Each ring is assigned a networkwide unique number. Any bridge attached to the same ring can have the same ring number. Bridges attached to different rings have different ring numbers. Each bridge is assigned a number. The bridge number can be the same for bridges attached to the same ring. However, bridges attached to the same two rings must have unique bridge numbers. Bridging in this manner is called "parallel bridging." It is the ring number that guarantees a unique route designator. When a bridge receives a frame from the ring, it interprets the route designator field. The bridge compares the route designator field values with its attached ring number and its own bridge number. The last route designator field in the routing information field will not contain a bridge number. This is because the end of a route is the target ring number.

The information field follows the route designator field. The information field for a MAC frame will contain MAC control information. If this frame had been denoted as an LLC frame, then the information field would contain end-user data. The LLC information is also called a Logical link control Protocol Data Unit (LPDU).

The FCS field is 4 bytes in length. Its value is created using an algorithm that covers the FC field, the destination and source addresses, the optional RI field, the information field, and the FCS itself. The algorithm results in a 4-byte cyclic redundancy check value placed in the FCS field by the originating station. The frame type determines the position of the FCS field and therefore the protection guaranteed by cyclic redundancy checking.

The ED will indicate code violations, errors, or that this frame is one of many associated frames being transmitted.

The FS field indicates whether the frame has been copied or the DA has been matched or an error occurred.

7.5.2 Logical Link Control (LLC) sublayer

The LLC sublayer is IEEE 802.2. The LLC has two accepted types of operational procedures and a proposed third:

- Type 1—connectionless
- Type 2—connection-oriented
- Type 3—acknowledged connectionless

Together these three types of operations provide link-level services for applications.

Connectionless mode of operation indicates that a logical data link connection is not established between the LAN stations before transmitting information frames. The LLC does not guarantee delivery of the information unit. Using this mode of operation, there is no flow control, correlation between frames, error recovery, or acknowledgment of receipt of the information frame. These services must be provided by upper-layer network services.

Connection-oriented operation requires that a logical data link be established prior to transmitting an information unit. This operation creates an LLC Type 2 control block. The control block in association with delivery and error recovery services constitutes a link station. LLC Type 2 service provides sequence numbering of information frames at the Data Link Layer, error detection, and basic recovery and flow control, including acknowledgment. The LLC Type 2 acknowledgment allows for up to 127 frames to be sent before an acknowledgment is expected from the receiving station. This is also known as modulus 128.

Acknowledged conectionless is the proposed method of operation. This mode of operation does not require a connection before transmitting information units. However, it does expect link-level acknowledgments from the destination station. This type of operation is particularly useful in LANs that may have high bursts of traffic, such as those used with file servers and backbone connectivity.

There are two classes of LLC operations defined in the IEEE 802.2 standard:

2. Class I—connectionless only

2. Class II—connectionless and connection-oriented

All stations support conectionless operations. Only class II stations support Type 1, Type 2, and Type 3 modes of operation.

The LLC sublayer has three main functions:

1. A specification for interfacing with the network layer above

2. Logical link control procedures

3. A specification for interfacing with the MAC sublayer below

The interface specification to the network layer defines the calls for unacknowledged connectionless service. This type of service allows stations to exchange information units without establishing a connection or acknowledgments. This is also known as a datagram. The LLC sublayer also provides a connected service as an option to the network layer.

Service Access Points (SAP) provide the interface between the application and the logical link control. Each SAP is uniquely architected for an application existing on a specific device type. A main function of the SAP is to allow multiple applications executing on a device to access the token-ring network through a single connection or adapter. SAPs support connectionless and connection-oriented transmission.

The logical link control frame, shown in Figure 7.10, is made up of four fields. These fields together are called the Logical link control Protocol Data Unit (LPDU).

The first field of the LPDU is the Destination Service Access Point (DSAP) address. This field is 1 byte in length and identifies the value of the SAP to which this LPDU is destined. The next field in the LPDU is the Source Service Access Point (SSAP) address field. This field identifies the SAP address of the originating SAP. This field is also 1 byte in length and is used in much the same manner as the DSAP. The third field of the LPDU is the control field. This field has three formats: the "information format" (I-format), the "supervisory format" (S-format), and the "unnumbered format" (U-format). The I-format contains application-level data. The S-format LPDUs are used

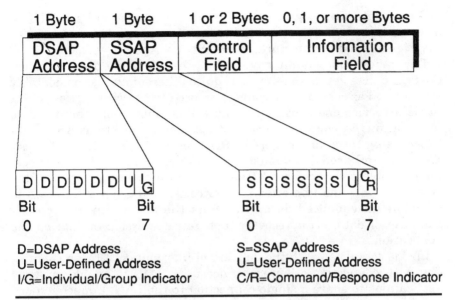

Figure 7.10   The format of the Logical Link Control frame and breakdown of the Destination and Source Service Access Point fields.

to acknowledge I-format LPDUs, request retransmissions, and temporarily suspend transmission of I-format LPDUs. The S-format LPDUs do not have an information field, since they are really denoting the status of the link station. The S-format of the LPDU control field is 2 bytes in length. The final LPDU control field format is the U-format. This format is used to send additional control functions and data transfer functions. The control field is 1 byte in length.

Two important control commands used in token-ring are the Exchange Identification (XID) and the test command. The XID command is used to carry identification and characteristics of the sending link station, causing the remote link station to respond with an XID response LPDU. This command was included in the IBM Token-Ring network implementation of the IEEE 802.2 standards to support SNA. The remote link station's response to this XID command must always be an IEEE 802.2 XID response containing the above information.

The test command is used by the link station to perform a basic test of link station–to–link station connectivity. The test command can contain an optional information field of test characters that are returned by the remote link station to determine connection solidity in a test response. This command was also included by IBM in its implementation of the IEEE 802.2 standards to support SNA.

The final field of the LPDU is the information field. This field, found in the I-format, will contain, if present, higher-layer protocols and user data. For example, an SNA Path Information Unit (PIU) is inserted into the information field when token-ring is being used as the transmission architecture over an SNA WAN.

The prior discussion covered connection-oriented service. Connectionless service, however, is quite different and is considerably streamlined as compared to connection-oriented service. Since connectionless service does not require data link connection, there are no link stations in connectionless service. Communications are established using three unnumbered LPDU formats. These are XID, test, and Unnumbered Information (UI) commands and responses. Connectionless service uses XID and test in the same manner as was described for connection-oriented service. The UI command is used to transmit unsequenced data. This puts the responsibility for resequencing of data, error recovery, and retransmission of data on the application.

LLC is concerned with the delivery of information. LLC in conjunction with connection-oriented service uses SAPs to enable multiple applications to share a single connection to the LAN. Connection-oriented service utilizes link stations to manage the logical connections, providing an extensive error recovery mechanism for maintaining data integrity. Connectionless service, on the other hand, provides no error recovery or guarantee of successful data transport.

7.6 Ethernet Protocol

The Ethernet protocol was developed by Xerox Corporation, and the final frame format was codeveloped by Digital Equipment Corporation (DEC), Intel Corporation, and Xerox. The popularity of Ethernet quickly grew, and the IEEE architected an Ethernet standard for all vendors. This standard is known as IEEE 802.3. IEEE 802.3 and Ethernet frames are quite similar, and the majority of vendors utilize the 802.3 standard. Unlike token-ring, Ethernet LANs use a contention-based bus architecture. This can also be considered a probabilistic architecture. In a bus architecture environment, it is possible for two or more stations to simultaneously send data onto the LAN. This causes collisions on the LAN, hence a standard protocol to sense for these types of errors was developed. The standard is called Carrier Sense Multiple Access/Carrier Detect (CSMA/CD). Stations on an Ethernet LAN see all transmissions as a result of the bus architecture. This is the multiple-access function of CSMA/CD. Each station analyzes the destination field for the destination address of the frame. If the address matches the station's address, the frame is copied. If the destination address does not match, the station discards the frame. In token-ring, the frame is sent back out onto the LAN. In an Ethernet environment, the frame is "thrown away." The time of the transmission of an Ethernet frame onto the LAN is determined by a station "listening" to the wire. If the wire is busy, then the station waits to send its data. If the wire is "silent," the station transmits its data. This is the carrier sense function of CSMA/CD. If two or more stations have a frame to send and each senses that the wire is silent, each will send a frame out onto the wire. This will result in collisions. Collision detection causes a station to abort transmission and sends a jamming signal to the other stations on the LAN. This jamming signal is a random pattern, usually 63 bytes in length. Stations on the Ethernet LAN will wait a random time after receiving a jamming signal before retrying transmission. This random wait time helps to avoid collisions. The number of collisions on an Ethernet LAN is a factor of the number of stations attached to the LAN.

7.6.1 Ethernet frame format

The Ethernet frame format depicted in Figure 7.11 shows the simplicity of Ethernet. The first field is called the "preamble." The preamble is 8 bytes in length and is a pattern of alternating bits. The preamble provides synchronization for Ethernet. The next field is a 6-byte field that contains the physical address of the Ethernet adapter of the station. This field is the destination address field. It contains the target address of the partner station. The following 6-byte field is the source address. This is the physical address of the originating station's Ethernet adapter. The type field is a 2-byte field that indicates the upper-layer protocol found in the data field. The more commonly found types are x'805D' for IBM SNA over Ethernet, x'8138' for Novell IPX, and x'0800' for TCP/IP. The data field in an Ethernet frame is variable in length and can range from 46 to 1500 bytes. The data found in this field is provided by the higher-layer protocols. The last field is the Frame Check Sequence (FCS) field. This is a 32-bit field that is used for Cyclic Redundancy Checking (CRC) using the destination and source address fields and the type and data fields.

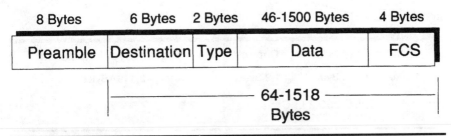

Figure 7.11 The format of the Ethernet frame.

Frames in Ethernet can range from 64 to 1518 bytes in length excluding the preamble. Frames outside of this range transmitted onto the LAN using Ethernet protocols are considered invalid. Frames shorter than 64 bytes are called "runts." Frames longer than 1518 bytes are referred to as "giants."

7.6.2   IEEE 802.3/Ethernet

Figure 7.12 illustrates the IEEE 802.3 frame format for Ethernet. Note that this is a MAC frame format which will allow IEEE 802.2 LLC data units to be encapsulated. Here again the Ethernet frame begins with a preamble, but in this format it is only 7 bytes in length. The preamble is followed by a Start Frame Delimiter (SFD), a single-byte field. These two fields together create the identical pattern required for the preamble field of the Ethernet frame discussed above. The SFD is followed by the destination address field. Again, the address found here is the address of the target station on the Ethernet LAN. However, instead of the address field's being 6 bytes in length, it can be 2 or 6 bytes in length; 6 bytes is more common. In the IEEE 802.3 specification, the address field will indicate whether the address is an individual station address or a group address and whether the address is universally assigned or locally administered. The source address follows the destination address, and it too can be either 2 or 6 bytes in length. The requirement is that it must match the length used in the destination address field. Following the source address field is the length field. Note that in the IEEE 802.3 specification, the length field replaces the Ethernet frame type field. The length field provides the length of the data unit field in bytes. The

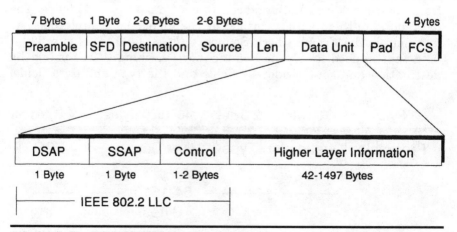

Figure 7.12   The IEEE 802.3 Ethernet frame format.

data unit field is an IEEE 802.2 LLC protocol data unit. IEEE 802.3 requires that the data unit be a minimum of 46 bytes. If the data unit field is smaller than 46 bytes, the pad field is used to make up the difference. The data unit field and the pad field combined cannot exceed 1500 bytes in length. Again the last field is the FCS field. This 32-bit CRC value is based on the contents of the two address fields and the length, data, and pad fields.

7.7 Transmission Control Protocol/Internet Protocol (TCP/IP)

The call for a simplified networking architecture that provided basic services was developed by the Advanced Research Projects Agency (ARPA) of the U.S. Department of Defense in the late 1960s. The goal of this network was to investigate any-to-any connectivity for computer network technologies by creating an internetwork between universities and research organizations. The resulting network, called ARPANET, was the first packet switching technology for data communications. ARPANET provides peer-to-peer communications between unlike computing devices. In its first incarnations, ARPANET protocols were slow, and networks were unreliable.

TCP/IP has its roots in a design specification proposed in 1974. This new protocol specification provided a robust, versatile protocol that is not dependent on the Physical and Data Link Control Layers of such networking architectures as SNA and OSI. The TCP/IP architecture has accounted for this protocol's longevity and durability in data communications. Since its protocol design is independent of hardware, operating systems, and telecommunications facilities, TCP/IP has seen commercial success in recent years since the advent of right-sizing MIS organizations and their applications. Therefore, it is appropriate to address the TCP/IP architecture and its role in information networks.

7.7.1 TCP/IP architecture

The TCP/IP architecture is a layered architecture that provides a rational, simple, and easy-to-modify structure; this has added to TCP/IP's success. The architecture is often referred to as a five-layer architecture, but it actually defines only the three higher layers.

As seen in Figure 7.13, the two lower layers, Physical and Data Link Control, can be defined by different networking architectures, such as token-ring, Ethernet, frame relay, ATM, SNA, and OSI. At these layers, the physical attributes of the communications network are defined and adhered to. Data units at this level are called

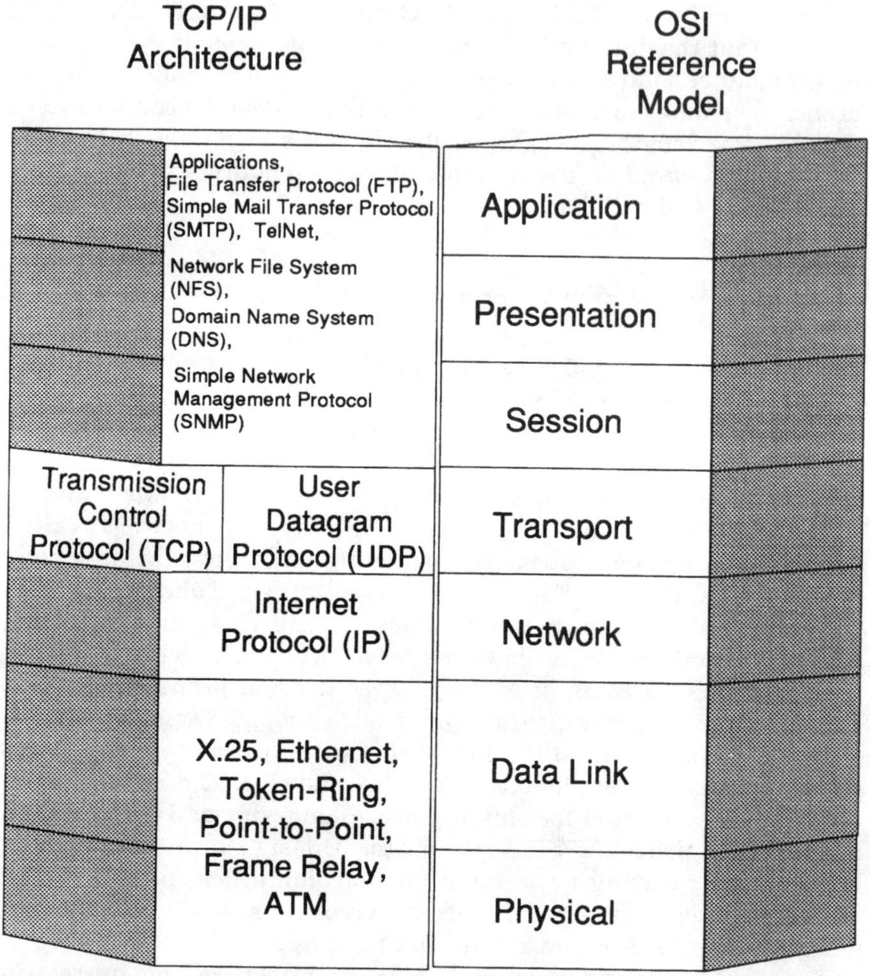

TCP/IP
Architecture

OSI
Reference
Model

Applications, File Transfer Protocol (FTP), Simple Mail Transfer Protocol (SMTP), TelNet,	Application	
Network File System (NFS), Domain Name System (DNS),	Presentation	
Simple Network Management Protocol (SNMP)	Session	
Transmission Control Protocol (TCP) / User Datagram Protocol (UDP)	Transport	
Internet Protocol (IP)	Network	
X.25, Ethernet, Token-Ring, Point-to-Point,	Data Link	
Frame Relay, ATM	Physical	

Figure 7.13 Comparing the TCP/IP architecture with the OSI Reference Model.

"frames." Moving up to the third layer, we find the Network Layer, where Internet Protocol (IP) resides.

IP's single main function is to route data between hosts on the network. The data is routed in datagrams. IP uses connectionless network service for transmitting the datagrams. Since it uses connectionless service, IP does not guarantee reliable delivery of the data or the sequencing of the datagram. Hence, the upper layers (e.g., TCP) must provide this function. What IP does provide is the ability to route traffic independent of available physical routes regardless of the application-to-application requirements. IP allows datagrams to be rerouted through the network if the current physical route being used becomes inoperative. This is a valuable feature built into IP that SNA does not possess.

The Transport Layer, layer 4 of the OSI architecture, is where TCP resides. TCP connection-oriented services, as such, provide reliable delivery of data between the host computers and requires that a "connection" between the applications exist prior to sending data. TCP guarantees that the data is error-free and in sequence. Data units at this layer are called "segments." TCP passes its segments to IP, which then routes the datagram through the network. IP will receive incoming datagrams and pass the segment to TCP. TCP will analyze the TCP header to determine the application recipient and pass the data to the application in sequenced order. Standard applications provided with TCP/IP that use TCP are File Transfer Protocol (FTP), Simple Mail Transfer Protocol (SMTP), and Telnet terminal services.

At the Transport Layer, another protocol is found in the TCP/IP architecture that provides connectionless-oriented services. This protocol is called User Datagram Protocol (UDP). UDP is used by various applications to send messages to other applications where the integrity of the data is not as important. UDP passes user datagrams to IP for routing. UDP is used by the Simple Network Management Protocol (SNMP), Domain Name System (DNS), Network File System (NFS), and other applications. These user datagrams are usually small single messages, and hence it was thought that to resend such a message would not cause a network impact. This is in contrast to SNA, where every frame, no matter how small or large, is provided with guaranteed delivery.

7.7.2 IP addressing

Internet Protocol has an addressing scheme that is based on the Network Layer. It uses a logical address rather than the address found in the Data Link Control Layer, which is a physical interface address. Every host in an IP network is assigned an IP address. A TCP/IP host can be a mainframe computer, a UNIX workstation, or an OS/2- or DOS-based personal computer. The address is 32 bits (4 bytes or 4 octets) in length. This allows for a total of 2^{32} addresses. That's over 4 billion hosts on an IP network! The addresses are used in binary form but represented in a decimal form. This representation is called "dot notation." Since 8 bits make up an octet or byte and there are 32 bits to an IP address, there are 4 octets or bytes that must be represented. Hence in dot notation four values represent the IP address of an IP host. For example, the address

```
130.11.31.132
```

is the decimal representation of the binary value

```
10000010 00001011 00011111 10000100
```

Obviously the decimal representation is easier to comprehend, and this is the reason why the dotted decimal notation is used to define the IP address. If all the binary values are 1s, then the IP address is 255.255.255.255 and is considered a broadcast address which every IP host will receive.

The IP address format is composed of a network address and a local or host address. The network address identifies the network to which this host is attached. The local address is the unique local address of the host within the network.

The IP address formats use the first 4 bits of the first octet to further subdivide the IP addressing scheme into classes. As shown in Figure 7.14, there are five IP address classes. The classes are denoted by the starting bits of the address. For example, if the first bit of the address has a 0 value, then it is a Class A address, with address numbers starting from 0 and continuing through to 127. A Class A networking scheme is used for large IP networks; it provides over 2^{24} or 16,777,216 addresses. Class B, with an address space of 2^{16} or 65,536 addresses, is used for medium-sized networks, and Class C, using only 2^8 or 256 addresses, is used for small networks. In TCP/IP a network consists of all the addresses within a given network. For instance, a Class A address of 56 is a different network from a Class A address of 15.

Class A, B, and C addresses are used by the user community to build networks. Class D addresses are used by IP for a mechanism called "multicasting." Multicasting allows for the distribution of a single message to selective hosts on a network. The Class E addresses are used for experimental purposes and are therefore reserved.

The IP address schemes Class A, B, and C can be further subdivided by partitioning the IP address to include a subnet address. Subnetting provides a sliding mechanism to extend the network address into the host address field. For instance, in Figure 7.15 we have extended the network address by creating a subnet. In the figure, a Class A network of 34 (which is denoted as 34.0.0.0) has been subnetted by applying a "mask" to the bits in octets 2 and 3. This mask is denoted in decimal form as 255.255.255.0. The value of the mask sets the bits used in the subnet and network addresses to 1 and the host address bits to 0. So, in our example, the network is 34, the subnet is 128.80, and the host field is the last octet. This type of subnetting will provide 254 host addresses for the network 34.128.80.0. It provides only 254 because the 0 host address is reserved to represent the network in the host and the value of 255 is reserved for

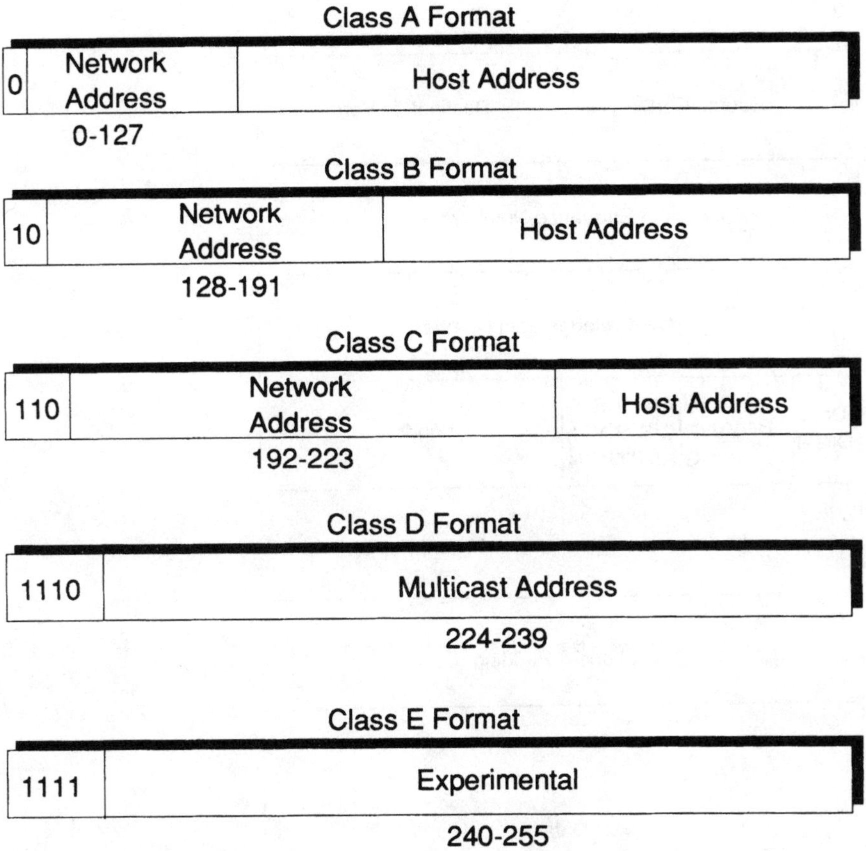

Class A Format

0	Network Address	Host Address

0-127

Class B Format

10	Network Address	Host Address

128-191

Class C Format

110	Network Address	Host Address

192-223

Class D Format

1110	Multicast Address

224-239

Class E Format

1111	Experimental

240-255

Figure 7.14 Use of the first 4 bits to divide the IP addressing scheme into five IP address classes.

Class A network address:	34.0.0.0	00100010.00000000.00000000.00000000	34.0.0.0
16-bit subnet mask	255.255.255.0	11111111.11111111.11111111.00000000	0.128.8.0
Class A subnet address:	34.128.8.0	00100010.10000000.00001000.00000000	34.128.8.0

Figure 7.15 Example illustrating the use of IP addresses and subnets.

broadcasting. By applying a mask of 255.255.252.0, the number of hosts can be increased to over 4000 for the subnet.

7.7.3 TCP segment

Transmission Control Protocol formats are called segments. Each TCP segment is made up of a TCP header and a data field. Figure 7.16 illustrates a TCP segment. The segment begins with the source and destination port identifiers.

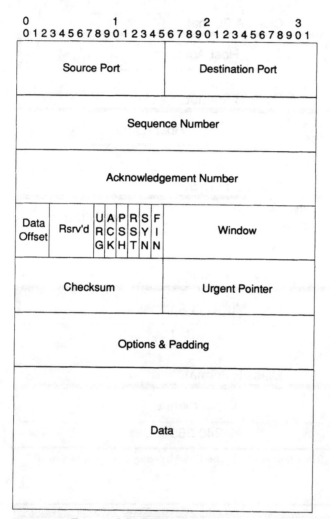

Figure 7.16   Format of the TCP segment.

Ports in TCP can be viewed as identifiers for applications. Each TCP application is assigned or selects a port number. In TCP/IP the standard or well-known applications like File Transfer Program (FTP) are given standard port numbers that no other application can use. For example, FTP is always port number 21. Each port number field in the TCP header is 2 octets in length. As can be seen from Figure 7.16, the source and destination port values take up the first 4 octets of the TCP header. Following the port number identifiers is the sequence number field. Just as in SNA, the sequence number identifies the position of this TCP segment in relation to the other outgoing segments found in the data stream. The sequence number field is 4 octets in length. The next field is the Acknowledgment (ACK) field. The ACK identifier indicates the sequence number of the expected next incoming segment. If the ACK number field is entered, then the ACK indicator field is set to a 1. The sequence and ACK number fields are equivalent to the NS and NR fields in SNA/SDLC.

The data offset field contains the length of the TCP header. This field is 4 bits in length and measures the TCP header in 32-bit increments. The header must end on a 32-bit boundary, and hence the options and padding field found at the end of the header is used to pad the header out to the 32-bit boundary.

The Urgent field (URG) is set to a 1 when the segment contains urgent data for the TCP application. The TCP applications locate the urgent data by the value placed in the urgent pointer field. This value points to the last octet of the urgent data. Any data found after this location is ignored. A terminal break or interrupt is often sent as urgent data.

The window field is the number of bytes, beginning with the acknowledgment number field, that can be received by this host. This field is continually updated as the host determines the number of bytes that it can receive into its buffers.

The PUSH flag, which follows the ACK flag, when set to a 1 indicates to the receiving TCP host that once TCP receives the segment, it is to deliver the data to the application identified by the destination port number field in a timely manner.

Connections are aborted when the REST flag is set to a 1. It is also sometimes used in response to a segment error.

Connections are closed by setting the FIN bit to a 1 in disconnect messages.

The value in the checksum field in the TCP header is determined by computing the length of the TCP header and the length of the data. Incoming segments are validated by using this computation and then verifying the results against the value found in the checksum field. If the values do not match, then the segment is discarded. Note that the checksum field itself is also included in the calculation, since it resides in the header. Also realize that this checksum value is at the Transport Layer. IP also performs a checksum at the Network Layer, and the Data Link Layer also performs a checksum as well.

7.7.4 IP datagram

The IP uses connectionless service to send data between TCP/IP hosts. Hence it does not guarantee delivery or the sequence of the

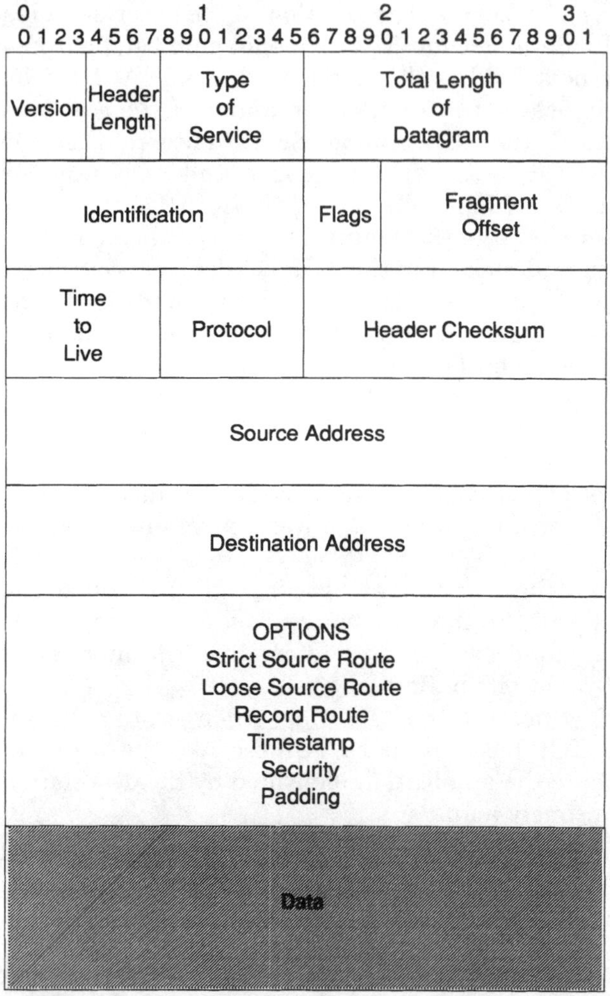

```
0                   1                   2                   3
0 1 2 3 4 5 6 7 8 9 0 1 2 3 4 5 6 7 8 9 0 1 2 3 4 5 6 7 8 9 0 1
```

Version	Header Length	Type of Service	Total Length of Datagram	
Identification			Flags	Fragment Offset
Time to Live		Protocol	Header Checksum	
Source Address				
Destination Address				
OPTIONS Strict Source Route Loose Source Route Record Route Timestamp Security Padding				
Data				

Figure 7.17   Format of the IP header.

datagrams. IP was designed in this way to provide internetworking for any device. As you can see from Figure 7.17, the IP header is quite large. In fact, the header can reach as much as 60 octets in length.

The first octet in the header contains two fields. The version field identifies the IP version being used for this datagram. This value is always 4. Datagrams received with a previous version number are discarded by the receiving device. The header length field of the first octet contains the length of the IP header in 32-byte boundaries. The padding field is used here just as in TCP to ensure this requirement.

If no options are included in the datagram, then this header length field will always indicate a length of 20 octets or 5 words.[1]

The type of service field is used to give precedence and priority to different types of IP traffic flowing through the network. For the most part this field is ignored by most implementors of TCP/IP and is a serious drawback to the current IP implementation. However, some vendors, like the router provider Cisco Systems, are implementing this type of service in their routing software in a manner analogous to providing a class of service to the various protocols and applications.

The length of the datagram is calculated by adding together the length of the IP header and the data fields. This 2-octet field architecturally provides for a datagram to be up to 65,535 bytes in length. However, networking restraints and host computer buffer constraints doe not make this a reality.

Though IP does not support sequencing, it does support fragmentation. Fragmentation occurs when the data field makes the overall length of the datagram larger than any node can accept on the way to the destination host. The Identification field is used to assist in determining which datagram fragments belong together for reassembly at the receiving host. The Flags field indicates whether a frame can be fragmented and whether it is one of many fragments or the last fragment. The Fragment Offset field indicates the number of 8-byte units that this datagram's data is from the start of the first datagram. This 8-byte unit is called a "fragment block."

The Time-To-Live (TTL) field is a measurement in seconds. This field is decremented by 1 by each router along the path to a destination station. The value for the field is set by the sending host and can be set as high as 255. If the TTL field value reaches zero before the datagram reaches the destination host, the datagram is discarded. In reality it is not a timer at all but a hop count mechanism. Once the hop count is exceeded, the datagram is removed from the network by the router that decrements the TTL field to zero.

The Protocol field indicates which upper-layer protocol is to receive the data field of the datagram. For instance, a value of 6 indicates that TCP is to receive the data.

The Header Checksum field of the IP datagram is a value based on the 16-bit one's complement of the one's-complement sum of all 16-bit words in the header. The checksum value itself is set to all zeros prior to this calculation. This checksum calculation is performed by the sending host and every router traversed in the network because of the TTL decrementing and fragmentation and the addition of option fields.

---

[1]A word is 4 octets in length. An IP header has a minimum length of 5 words and a maximum length of 15 words.

The Source and Destination address fields are the 32-bit addresses used for the IP address. Recall that these are not the hardware addresses but the Network Layer addresses of the devices. The hardware addresses, MAC addresses, are found in the Data Link Layer.

The Options field contains several different options that can be implemented by the sending host. For the most part these are self-explanatory and will not be discussed here.

The last field in the datagram is the data field. The data field of an IP datagram is the TCP segment or UDP segment.

This ends our discussion of TCP/IP. It may have seemed exhausting, but it was included to give the reader an understanding of the TCP/IP protocol so that the reader may make comparisons to SNA and APPN.

7.8 Full Duplex and Half Duplex

There are two ways in which data can be transmitted between nodes. In the full-duplex mode, two nodes can send and receive data at the same time. In half-duplex mode, only one node can transmit at one time. Only after one node is done with transmission is the second node allowed to initiate a transmission back to it.

7.8.1 Transmission versus application data flow

The terms "full duplex" and "half duplex" are generally misunderstood. When these terms are used by communications engineers, they are usually associated with the *transmission* media. Physical transmission between two devices can be half or full duplex. Such transmission is associated with the data communications equipment. When they are used by communications systems programmers, they refer to the data flow between applications. *One has nothing to do with the other.* While a physical transmission could be taking place in full-duplex mode, the applications could be communicating in half-duplex mode.

The term "duplex" for transmission flow is link-based and refers to the Data Link Control Layer of SNA. This layer is the second layer of SNA, as shown in Figure 5.7. It is controlled by the data link control protocols such as SDLC and BSC. The term "duplex" for flow control refers to the Data Flow Control Layer of SNA. This layer is the fifth layer of SNA, as shown in Figure 5.7.

7.8.2 Full Duplex

Figure 7.18a illustrates a full-duplex environment. It represents data flow control between Logical Units (LUs). An example of an LU in this case is an LU6.2, which could be a CICS running as a VTAM

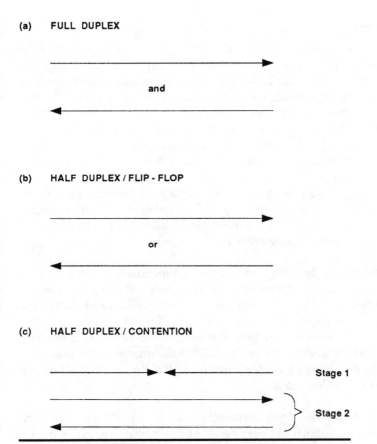

Figure 7.18 Flow control protocols. (a) Full duplex; (b) half duplex/flip-flop; (c) half duplex/contention.

application in a host. Such a CICS could be communicating with another CICS running in a different host. In this case, either CICS could send a message at any time. Therefore, both the participating LUs should be ready to receive and send data at any time. Usually, recovery in a full-duplex data flow control environment is more complex. This is because various LUs (CICS in our example) should be able to correlate the requests they send and the responses they get from multiple sources on an ongoing basis.

Data flow control between Network Control Programs (NCPs) running in different communications controllers is also full duplex. However, the transmission between the NCPs controlled by data link control could be half duplex. Notice the difference. In the first case it is the logical transmission of data between two entities, and in the latter case it is the physical transmission of data between the same two entities.

7.8.3 Half Duplex

Figure 7.18b and 7.18c illustrates the concept of half duplex. Each of the LUs in a session takes turns being the sender and later on being the receiver of data. But an LU cannot be both at the same time. This protocol is relatively less complex and ensures easy recovery.

Data flow control between an intelligent LU (e.g., CICS) and a terminal LU (e.g., 3278) is half duplex. In this case it is an LU6.2 and LU2 communicating with each other. Also, the link-level communications between NCP and a single 3274 cluster controller is also half duplex. This is irrespective of whether you define it as full or half duplex. The only time you can take advantage of full-duplex communications in an NCP-to-cluster-controller setup is when these are multiple 3274s on a line. As we already learned, such a setup is called a "point-to-multipoint" configuration.

Flip-flop half duplex. In a flip-flop mode of half duplex, the LUs take their turns in sequence. The sender of the message gives a change of direction indication to tell the receiver that it can send data now. When a session is established between two LUs, the BIND indicates who will initiate the first dialogue. For example, in the CICS environment, the INVITE parameter of the SEND command is equivalent to the change in direction indicator. Figure 7.18b illustrates half-duplex protocol in the flip-flop mode.

Contention half-duplex. In a contention mode, either LU can send data at any time. If both of them do so at the same time, this results in a contention. At the time of establishing the session, the session parameters indicate who will win in a contention situation. Figure 7.18c illustrates half-duplex protocol in the contention mode. In stage 1, the contention occurs. In stage 2, a smooth flow of data occurs depending upon who the first speaker will be as determined by session parameters. These session parameters are referred to as the bind image.

7.9 NRZI and NRZ

7.9.1 NRZI

Non-Return-to-Zero-Inverted (NRZI) is a technique used to maintain bit synchronization for modems. The maintenance of bit synchronization requires that the polarity of the modem signal be changed periodically. In the absence of polarity change, a long string of the transmission of the same signal can put the modems out of synchronization. Put it simply, NRZI works as follows: *Invert polarity when you see a zero.*

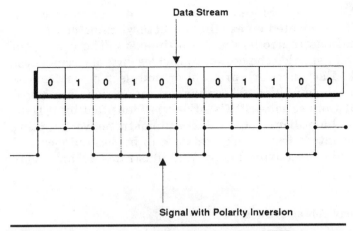

Figure 7.19   An example of polarity inversion using NRZI.

Figure 7.19 illustrates the use of NRZI. Notice that every time a zero is encountered, the polarity is changed. Different combinations of 1s and 0s work as follows:

- If a 1 is followed by a 1, keep the same polarity.
- If a 1 is followed by a 0, change polarity.
- If a 0 is followed by a 0, change polarity.
- If a 0 is followed by a 1, keep the same polarity.

Basically, NRZI is used for those modems that do not provide timing. In other words, if the DCE (modem) cannot provide its own timing, the DTE (communications controller/cluster controller) must provide it. An example of such a modem is the IBM 3872. For those modems which do provide timing signals, NRZI need not be used. However, you must ensure that if NRZI is used on a single DCE on a data link, it is used by all the DCEs on the same data link.

7.9.2   NRZ

Many people confuse NRZ with NRZI and think that it is a different kind of communications protocol. Remember that NRZ has nothing to do with data communications. NRZ is used for tapes.

7.10   Frame Relay

Frame relay is an interface definition with its origins in Integrated Synchronous Digital Network (ISDN). However, frame relay does not require an ISDN network. Since frame relay is a definition specification, it can be implemented on existing packet switching equipment like communications controllers and multiplexers. The concept of frame relay is very simple. The network provides virtual circuits over a physical link. Each network protocol or end-user device can be assigned a virtual connection. These virtual connections are called Private Virtual Connections (PVCs). Each is identified by a Data Link Connection Identifier (DLCI) value. Automatic network routing can be designed into a frame relay network to bypass outages and ensure end-to-end connectivity. Figure 7.20 illustrates a frame relay connection.

7.10.1 Frame relay characteristics

A PVC exists in a point-to-point connection across the network. Many PVCs share a single physical port. This saves the cost of the many physical connections that would be needed for point-to-point lines. Frame relay does not do any polling, thus saving bandwidth on the link. But every nonerror frame is relayed through the network. Frames found to be in error are discarded by frame relay, since error recovery is the responsibility of the link control protocol. Frame relay can use any link control protocol that defines error recovery mechanisms. This saves more overhead in bandwidth and processing. For the most part, frame relay vendors implement High-level Data Link

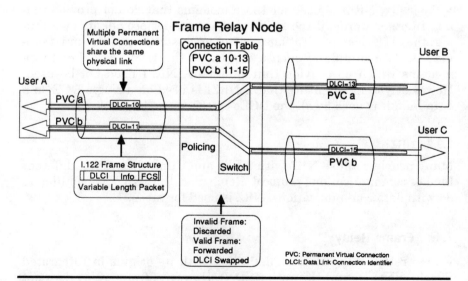

Figure 7.20 Illustration of a frame relay connection.

Control (HDLC). Although frame relay uses HDLC, it does not provide for class-of-service, as it is known in SNA. Each frame is equal in service requirements. Frame relay networks use a signaling mechanism to identify flow control congestion. This mechanism is called Forward Explicit Congestion Notification (FECN) and Backward Explicit Congestion Notification (BECN). Two bits in the frame relay frame, as shown in Figure 7.21, determine which type of congestion control is in use. The FECN bit is set to indicate that nodes along the path in the direction of the frame are congested. The BECN bit, when set, indicates that nodes behind the frame are congested. Congestion can be due to low-speed physical links between switches and/or to many different frame lengths. Frame relay allows for any length frame up to a length of 8 KB. Most frame relay vendors, however, support a maximum frame size of 2 KB. The characteristics of supporting variable-length frames may impose erratic response times on the network. To remove this potential problem, higher-speed links up to T1 may be required.

7.10.2 Frame relay device types

Frame relay nodes in the network act as switching devices and are referred to as Frame Relay Switching Equipment (FRSE). Devices that enter a frame relay network are called Frame Relay Terminal Equipment (FRTE). The PVC is defined between the FRTE. The function of the FRSE is to map the DLCI of the FRTE end points. This connection table consists of entries that define the node ID, the link ID, and the DLCI. Data delivered through the network with a given DLCI of a specific port is delivered to the appropriate virtual connection. The DLCI is swapped at the FRSE with the DLCI of the destination port. The IBM 3745 with NCP V6.2 can act as a FRSE.

The implementation of frame relay provides gains in network efficiency through trading network complexity for link capacity. The higher-speed lines are required to make efficient protocols stable.

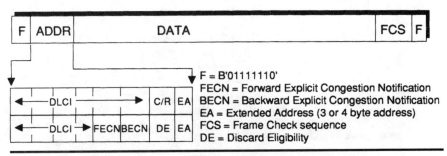

Figure 7.21 Break out of the bits in the address field for frame relay.

However, these higher-speed lines result in low delay through the network, thereby increasing productivity. The higher-speed lines in concert with the low overhead of frame relay can provide an improvement in throughput of 10:1 to 30:1 over previous network transmission standards on existing communications equipment.

7.11   Asynchronous Transfer Mode (ATM)

ATM is based on the concept of cell switching. Cells are set at a fixed length. Large blocks of user data are broken down to fit into the data area of the cell. The data area of a cell is called "payload." In ATM a cell is 53 bytes in length. The header takes up 5 bytes, and the payload uses 48 bytes (Figure 7.22). Even if the user data is only 1 byte in length, the cell will be 53 bytes long. Short fixed-length cells simplify the requirements on the switching hardware and therefore allow for higher speeds. Small cells result in shorter transit delay through a multinode network. The nodes on the network can handle cells more efficiently in their queues, reducing the variation in transit delays that results from variable-length frames. Finally, cell-based networks like ATM do not use link-level error recovery. Only the header field of the cell is checked for errors in ATM, and it is up to the higher-layer protocols to check for errors in the payload portion of the cell. All this adds up to the theoretical support of 2 gigabits per second (Gbps)! Cell switching is the most suitable for concurrently carrying voice, video, data, and image over the same physical link.

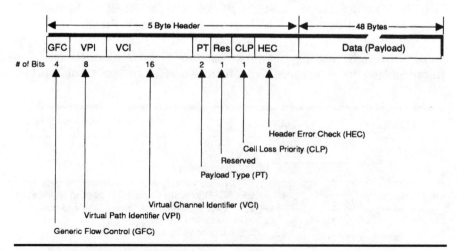

Figure 7.22   Format of the ATM cell.

7.11.1   ATM network architecture overview

An ATM network is composed of nodes, interfaces, links, and virtual resources, just like many of the other network schemes discussed. Figure 7.23 illustrates an ATM network. Network Nodes (NN) perform the switching functions and data transport within the network. Note that the ATM NNs are not the same as the APPN NNs. At the end points of an ATM network are the Customer Premise Nodes (CPN). This equipment is owned by the end-user of the ATM network. Within each node there is an interface. On NNs the interface is called the Network Node Interface (NNI). CPNs have a User Network Interface (UNI). The NNI provides network signaling procedures for operation and control of the network besides the data transfer functions found in the UNI. Physical links between nodes exist for the NNI and UNI to utilize. The physical links are divided into virtual paths, virtual connections, and virtual channels. A Virtual Path (VP) is a route through the network between two end points with the same quality requirements. The end-to-end connection over which data is transmitted is called a Virtual Connection (VC). VCs transmit data in only one direction; hence two VCs are needed for sending and receiving data. Each VP is subdivided into virtual channels. Virtual connections are provided by the interconnection virtual channels. Up to 256 VPs can be defined for each physical port and 4096 virtual channels for a single VP.

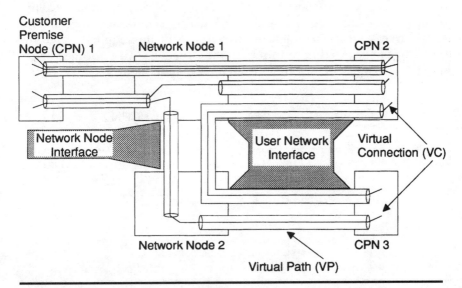

Figure 7.23 ATM network and connectivity illustration.

7.11.2 ATM layers overview

At the Physical Layer of ATM, the medium data rates and transmission technique are defined. The ATM specifications include T3, SONET, IEEE 802.6, Fiber Channel, FDDI, and most recently T1. The next layer in ATM, the ATM Adaptation Layer (AAL) (Figure 7.24), transforms the incoming signals into ATM cell format. This layer is subdivided into two levels. The first level breaks up the incoming data into manageable pieces. The segmentation and assembly level breaks up the user data into 48-byte fields for transmission and reassembly at the receiving end point. ATM has specified several AALs to support different traffic requirements. These include voice, video, image, and data for leased-line-, connection-, and connection-less-oriented sessions. Support for the various traffic requirements over ATM results in data rates of up to 2 Gbps.

The third layer to ATM is the ATM Layer. This layer defines the 53-byte cells that are transmitted through the network. A 5-byte header is added to the 48-byte information field received from the AAL to form the ATM cell.

7.12 Summary

In this chapter we discussed link-level protocols such as SDLC, BSC, LLC-2, and MAC. SDLC, SNA's architected protocol, has replaced BSC for the most part, and IBM's token-ring protocol using IEEE 802.2 and IEEE 802.5 is making large inroads in corporate networks as the data link transport for SNA. We also discussed Ethernet and its format as it relates to LLC-2. Ethernet is used largely by academic, research, and government organizations for its simplicity and low cost. It was reviewed here to provide the reader with an understanding of its format because of the increased functional features of the

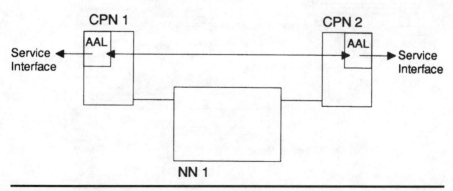

Figure 7.24 ATM adaptation layer illustration.

new IBM controller offerings that support Ethernet connectivity to the mainframe computer. We also learned about asynchronous communications, which is heavily used in the personal computer community. Full-duplex and half-duplex protocols were discussed for the Data Link and Data Flow Control Layers of SNA. It is emphasized that the Data Link Layer deals with the physical exchange of data and the Data Flow Layer is concerned with the logical exchange of data between two SNA logical units. The NRZI technique was also discussed; it provides timing signals in DTEs if they do not have this capability. Finally we introduced you to frame relay and ATM networks. Both of these networking schemes will eventually dominate wide area networking, with frame relay being used as the migration path to ATM.

Communications Software — VTAM and NCP

Designing an SNA-based communications network requires various considerations of hardware, protocols, and software. We discussed hardware in Chapters 2 and 3. We also learned about various communications protocols in Chapter 7. In this chapter we will talk about the software products which are SNA compliant and therefore help in the implementation of an SNA network.

8.1   SNA Software Products

SNA software products can be divided into two broad categories:

- Base software for SNA networks
- Application software for end-user applications

The *base software* is the required software which helps in defining and implementing an SNA network. Examples of IBM's base software products are ACF/VTAM and ACF/NCP; we will call them VTAM and NCP. VTAM runs in the host computer, and NCP runs in the communications controller. Examples of host processors that run VTAM are IBM 3084, 4381, 3090, and ES9000. The communications controllers that run NCP are IBM 3705, 3720, 3725, and 3745. The base software does not perform any application-specific function, but teleprocessing applications (e.g., CICS) need the services of such base software.

The *application software* for teleprocessing applications uses the services of base software to create an environment for end-user applications. Examples of such software are CICS/VS, IMS/DC,

TSO, and VM/CMS. They provide the framework for creating real end-user applications such as accounts payable, airline reservation systems, etc.

Since different terms are used by various communications professionals for the same product, let's summarize them to avoid confusion. VTAM, the foundation of SNA software, runs in the host processor as an operating system (e.g., MVS) application. CICS runs in the host processor as a VTAM application. VTAM provides the necessary communication services to CICS to create a networking environment. The end-user applications run within the CICS address space and are the real business applications. VTAM/NCP systems programmers generally use the term "application" to refer to a CICS region or any other teleprocessing software which uses the services of VTAM. CICS systems programmers use the same term to refer to business applications running within the CICS address space. So be aware that these terms may have different meanings depending upon who is using them. Even the IBM manuals use them inconsistently.

Before getting into more details about various communications software, we need to understand the concept of Logical Units (LUs) and Physical Units (PUs). In the SNA world, various functions and responsibilities are assigned to the logical and physical units rather than to hardware and software. It is the characteristics of those LUs and PUs which are implemented in the software and hardware. But first of all let's see what these LUs and PUs really are.

8.2   Physical and Logical Units

8.2.1   Physical unit

An SNA network is composed of various logical and physical units. A physical unit may be a physical device, but it does not have to be. As a matter of fact, a PU may be a combination of hardware, software, and microcode. A PU may use and handle many resources. In that respect it may be called a "resource manager." If we visualize a network as consisting of a collection of interconnected nodes, a PU is the operating system that runs in those nodes.

SNA defines various types of physical units. Figure 8.1 lists all the physical units that may make up SNA nodes. Notice that VTAM running on an IBM (or compatible) host is a Physical Unit (PU) Type 5 or PU Type 2.1. NCP running in a 36X5 or 3720 communications controller is a PU Type 4. An SNA cluster controller is a PU Type 2. Be aware that PU Type 3 is not defined by SNA, at least not yet.

SNA-, VTAM- and NCP-related literature makes extensive use of the acronyms PU and LU. Getting used to this terminology will make you feel more comfortable in the subsequent chapters.

SNA NODE TYPES				
PU 1	PU 2	PU 2.1	PU 4	PU 5
3271	3174	S/36	NCP	VTAM
6670	3274	AS/400	37X5	4300
3767	3276	PC	3720	308X
	PC	TPF		3090
	3770	6611		ES/9000
	AS/400	308X / VTAM V4		
		3090 / VTAM V4		
		ES/9000 / VTAMV4		

Figure 8.1   Physical units defined by SNA.

8.2.2 Logical unit

Logical units are the entities through which an end user can communicate with another end user. The end user does not have to be a human being. It can be a terminal, a piece of code, a printer, or another entity. As a matter of fact, the real purpose of an SNA network is to facilitate and manage interaction among various logical units. Physical units provide the means, but the logical units provide the end purpose of networking.

Figure 8.2 shows various LUs supported by SNA. Note that a logical unit is further split into two categories—Primary Logical Unit (PLU) and Secondary Logical Unit (SLU). When a session is established between two LUs, one of them is called a PLU and the other is called a SLU. It is possible for two LUs to have the same processing intelligence. For example, a CICS (LU6.2) can communicate with another CICS (LU6.2). However, although both CICSs have the same rank, one has to be designated a PLU and the other a SLU at session establishment. Whether an LU will be primary or secondary is determined during the BIND process when a session is established between two LUs. The PLU is usually responsible for recovery management in the case of a session failure. In Figure 8.2, the logical units 0 through 7, as shown in the box on the top of each chart, represent the SLUs on the right-hand side of that chart. Products listed on the left side of the chart represent the PLUs that can establish sessions with the respective SLUs.

Now let's talk about some of the more popular LUs. A 3278 is an LU2. A printer attached to a cluster controller (PU T.2) can be LU1 or LU3. A CICS looks like LU6.1 or LU6.2. Notice that LU5 has not been defined by SNA—again, at least not yet.

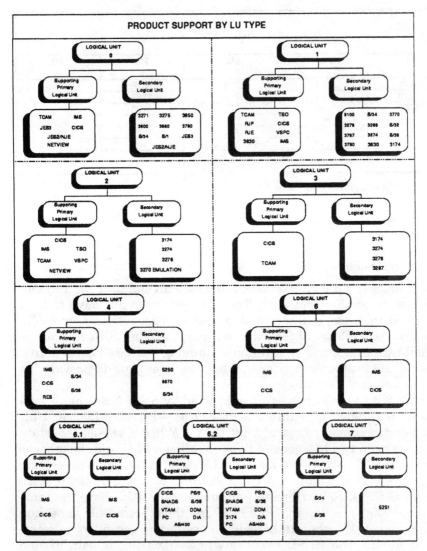

Figure 8.2 Logical units defined by SNA.

8.3 VTAM

Virtual Telecommunications Access Method (VTAM) is the strategic and most versatile communications access method of IBM. It runs in the host processor in a separate address space. It is supported to run under various IBM mainframe-based operating systems such as MVS/SP, MVS/XA, MVS/ESA, DOS/VSE, VM, etc. Although the capabilities of VTAM may differ from one operating system support to another, the underlying characteristics remain the same. Various components of a network have to be identified and defined to VTAM. Such definitions are done through the use of VTAM macros. Part 3 of this book covers these definitions and macros in detail.

8.3.1   SNA and VTAM services

SNA provides the architecture and protocols for a communications network. One of the SNA-defined network components, called System Services Control Point (SSCP), provides the facilities to control the network. The most important objective of VTAM is to provide the services of SSCP. There is one and only one SSCP in each VTAM address space. In general, we can say that there is only one SSCP in a host node. An SSCP looks like a PU T5 in an SNA network. In APPN, VTAM functions as a Control Point (CP) PU T2.1.

A network may have multiple host processors, and each of them may have its own VTAM. Therefore, a network in general has multiple SSCPs. VTAM provides the facility to communicate across SSCPs. For example, a terminal connected to a host via a cluster controller and a communications controller may communicate with the applications of another host. The other host may be in the same physical location or may be across the continent. In this case, the SSCP that owns the terminal provides the necessary control and connectivity information to establish a cross-domain session.

VTAM helps in the establishment of a session between two LUs. A VTAM application program (e.g., CICS) is generally a primary logical unit. In an LU-to-LU session, a PLU communicates with a SLU.

Not all the functions of network control and management are performed by VTAM. Some of the network awareness and intelligence has been moved to the communications and cluster controllers. VTAM in the host processor and NCP in the communications controller each perform specific functions for the network management.

In the early days of networking, Basic Telecommunications Access Methods (BTAM) was the popular networking software. As a predecessor of VTAM, it did *not* allow sharing of network resources. Each of the terminal devices such as 3278 could be connected to only a single BTAM application. In those days, a terminal connected to one CICS could not be used to connect to another CICS, or for that matter to a TSO. VTAM provides for sharing of the network resources. Such network resources include controllers, lines, terminals, and LUs.

8.3.2   Features of VTAM

VTAM is one of the most versatile and powerful networking software. Its salient features and characteristics are summarized as follows:

1. It supports SNA as well as non-SNA devices. Non-SNA devices

include cluster controllers such as 3274-41D and RJE/RJP stations such as 2780 and 3780.

2. It supports Advanced Program-to-Program Communications (APPC) between two LU6.2 applications. APPC makes the two participating LUs look like peers.

3. VTAM provides for network operations control through a software product called Network Communications Control Facility (NCCF). NCCF is now a component of NetView.

4. It supports both channel-attached (local) and link-attached (remote) communications controllers. Examples of such controllers are 3270, 3725, and 3745.

5. It supports channel-attached (local) and link-attached (remote) cluster controllers. Examples of local cluster controllers are 3274-41A and 3274-41D, while the remote cluster controllers are 3274-41C and 3274-61C.

6. It provides for sharing of network resources among applications. Such resources may be links, terminals, and other communications facilities.

7. Since version 2, VTAM supports cross-host (technical cross-domain) communications. Prior to version 2, the same facility was provided by an add-on called Multi-System Networking Facility (MSNF).

8. In general, VTAM is functionally compatible between DOS/VSE, MVS, and VM.

9. With some recent enhancements, it allows for dynamic network configuration and recovery.

10. Network changes and enhancements require minimal changes to the host applications.

11. Host applications are independent of the data link control protocol. Such a protocol may be SDLC or BSC.

12. VTAM architecture is based upon distribution of communications functions between host-based VTAM and communications-controller-based NCP. This results in optimum utilization of the communications controller and minimal dependence on more expensive host CPU cycles.

13. It supports a wide variety of communications requirements which include batch and interactive processing needs.

14. It allows for coexistence of other access methods, such as Telecommunications Access Method (TCAM), in the same SNA network. It also provides for running TCAM Version 3 under

VTAM. This allows for using VTAM as an access method and TCAM as a message handler, thus using the best characteristics of both VTAM and TCAM.

15. VTAM supports the IBM APPN architecture for peer-to-peer networking. In APPN VTAM acts as a PU T2.1 node.

8.3.3 VTAM application programs

A VTAM application program is a communications program that uses the services of VTAM macros to communicate and to use its services and facilities. Thus the application program can transfer data to and from other LUs, such as LU6.2 and LU2. All the complexities of network awareness and networking usage are hidden from the application. Thus most of the complexities of pacing, and device-dependent characteristics are handled by VTAM rather than by the application itself. Some of the requirements and characteristics of VTAM applications are as follows:

1. A VTAM application runs in a host under the operating system. Even though the application uses the services of VTAM, both of them are subservient to the host operating system.

2. A VTAM application looks like an LU to VTAM.

3. The interface between the application and VTAM must be written in assembler language.

4. A VTAM application can communicate with other VTAM applications.

5. It uses macros to communicate with VTAM. As in the case of CICS, a command-level interface does not exist between an application and VTAM.

You do not have to write VTAM applications. As a matter of fact, the majority of the business community uses the VTAM applications already written by IBM and third-party software companies. Such applications include CICS and IMS/DC. The user writes only the interaction with those VTAM applications to create business applications. In other words, the user writes application programs to run under a VTAM application (CICS or IMS/DC) and thus creates the business applications.

8.3.4 VTAM and other access methods

Other telecommunications access methods used in the IBM mainframe environment are Basic Telecommunications Access Methods (BTAM) and Telecommunications Access Methods (TCAM). BTAM is gradually moving toward extinction. It communicates with a telecommunications control unit (TCU) (e.g., 270X) and with channel-attached and communications-controller-attached (emulation-mode) terminals. One of the major drawbacks of BTAM is that it does not provide for sharing of network resources. For example, a terminal (e.g., IBM 3278) connected to an application (e.g., CICS) can have a session only with that application. It *cannot* log off and connect to another application (e.g., TSO). The BTAM application has exclusive control of the terminal.

TCAM, although being used less often, is still an access method of choice for environments with very large terminal networks. Pooled TCAM allows for control of very large terminal networks without having to define every terminal in the CICS Terminal Control Table (TCT). This saves considerable amounts of virtual storage in a system. VTAM does not provide for any such comparable feature. TCAM Version 3 runs as a VTAM application program, thus using the features of VTAM as a superior access method. At the same time, it retains the features of TCAM as a message handling subsystem for which there is no VTAM equivalent. Thus it provides the best of both access methods for large terminal network installations.

TCAM, VTAM, and BTAM can run and operate concurrently in an MVS system.

8.3.5 Diagnostic and recovery aids

In conjunction with the operating system, VTAM provides a number of diagnostic and recovery aids. They are briefly described as follows:

Diagnostic Aids

Trace facility.   Provides trace information on flow control, buffer contents, Transmission Group (TG), SDLC line, I/O activity, storage management services, and many other facilities. Under MVS, trace data is collected using Generalized Trace Facility (GTF). It can be printed using the Trace Analysis Program (TAP) or print dump (PRDMP) utility. In a DOS/VSE environment, trace records are collected in a trace file. They can be printed using TAP or VTAM trace print utility.

TOLTEP.   The Teleprocessing On-line Test Executive Program (TOLTEP) provides for on-line testing of communication devices. These tests are conducted while such equipment is performing normal operations. TOLTEP is not supported for releases subsequent to VTAM 2.1.

Formatted dumps (MVS only).   These are available through the services of the MVS operating system. ABDUMP is provided by the

ABEND or SNAP macro. Such dumps can be formatted using the VTAM formatted-dump program. This program prints VTAM control blocks and also provides information on broken control block chains or invalid addresses. SDUMP is provided if an error occurs during the processing of a VTAM request. However, SYS1.DUMP must exist beforehand. MVS provides for a service aid AMDSADMP to dump the contents of main storage to a printer or a tape. This dump, called the "Service-Aid Dump," does not produce formatted control blocks.

NCP dump.   The contents of the NCP can be dumped at the operator's request or automatically if an NCP failure is detected.

In addition, VTAM provides other diagnostic aids, such as hardware error recording, SDLC link test, printing error records using Error Recording Edit and Print (EREP) program, route verification, and LU connection test. More on this in *Advanced SNA Networking.*

Recovery aids.   Recovery aids facilities provide for restart of a communications controller and network configuration establishment. They try to restore a domain as soon as a failure is detected. This can be done automatically or can be at the operator's intervention. In addition, multiple SDLC links, called a Transmission Group (TG), can operate concurrently between communications controllers. Failure of one or more links in a TG does not interrupt the data flow because the data traffic is diverted to other links.

8.4 NCP

The Network Control Program (NCP) is the operating system of the communications controller. It supports various host environments, such as MVS/SP, MVS/XA, MVS/ESA, DOS/VSE, VM, etc. The NCP is generated on the host system and then loaded in the communications controller from the host NCP load libraries. Various parameters specified to generate NCP fall in the following two categories:

- NCP-only parameters
- NCP- and VTAM-specific parameters

Such parameters are discussed in detail in Part 4 of this book. Just as the SSCP component of VTAM looks like a PU Type 5 in SNA networks, an NCP looks like a PU Type 4.

8.4.1 Functions of NCP

One of the major functions of NCP is to off-load some of the communications-related routine tasks from the host and run them on a dedicated processor such as a communications controller. NCP interacts with the host access method such as VTAM or TCAM. It controls the cluster controllers (e.g., 3174) and terminal nodes attached to its various ports. It routes data to adjacent NCPs running on other communications controllers if the destination LU is not within its own subarea. It does the polling for stations and dialing and answering for switched lines. It maintains records of errors occurring in the links, NCP, and the controller itself. It allocates buffers for the data coming from and going to the host and other stations.

8.5 Related Software

VTAM and NCP alone are not enough to perform all the functions required for network support, operations, and management. IBM supports a host of related software to support various functions.

8.5.1 NTO

The Network Terminal Option (NTO) supports certain non-SNA terminals in an SNA network. It supports starts and stop terminals such as IBM 2741 and IBM 3767. NTO translates SNA session protocols and commands into a format which can be interpreted and used by such terminals. It runs in the communications controller along with NCP.

8.5.2 NPSI

The NCP Packet Switching Interface (NPSI) is a program that runs in the communications controller with NCP. It provides an interface to networks that conform to the X.25 standard. Examples of packet switching networks in the United States are Telenet and Tymnet; in Canada such services are provided by Datapac. NPSI allows application programs (e.g., CICS) and link-attached devices (e.g., IBM PC) to have sessions over an X.25 packet switching network.

8.5.3 EP/PEP

An Emulation Program (EP) runs in a communications controller and provides the functional capabilities of the Transmission Control Units (TCUs). Examples of TCUs are 2701, 2702, and 2703. They were the predecessors of communications controllers. EP runs as a separate entity and has no interaction with NCP. In order to run EP and NCP in different partitions as two mutually exclusive programs, you need the services of Partitioned Emulation Program (PEP).

8.5.4 NPM

NetView Performance Monitor (NPM) helps in managing the growth and performance of a network. It provides for analysis of historical as well as real-time data. It dynamically updates network data and identifies network bottlenecks. It comes in handy as tool for capacity planning.

8.5.5 SSP

System Support Program (SSP) consists of utilities and support programs that help in the generation, load, and dump of NCP. It runs in the host. It consists of a Configuration Report Program (CRP), NCP/EP Definition Facility (NDF), dump utility, Trace Analysis Program (TAP), loader utility, and Configuration Control Program (CCP).

8.5.6 NetView

NetView is a program that helps perform operator, information, and hardware problem analysis functions. Previously, the same functions were provided by a set of different products, such as NCCF, NPDA, NLDM, and others.

NCCF. The Network Communications Control Facility (NCCF) component is a VTAM application program such as CICS or IMS/DC. It provides a set of services for the network operators and other network management programs such as NPDA and NLDM. An operator may execute VTAM commands, receive network messages, and perform network-related functions from the NCCF terminal. In a nutshell, NCCF provides a window to the network. Under NetView, NCCF is called the Command Facility.

NPDA. The Network Problem Determination Aid (NPDA) is used to collect and display error data about different communications components. It collects data and error counts on communications controllers, SDLC lines, BSC lines, modems, cluster controllers, terminals, and start and stop lines. It works in conjunction with the NCCF component of NetView. It also provides suggested operator action for each error occurrence. NPDA is called the Hardware Monitor under NetView.

NLDM. The Network Logical Data Manager (NLDM) collects data pertaining to SNA sessions. While NCCF runs as a VTAM application, NLDM runs as a program under NCCF. The collected session data may be stored in the disk files for subsequent analysis. NLDM can collect data for single-domain as well as multiple-domain networks. Data collected by NLDM can be displayed on the NCCF terminal. NetView has incorporated NLDM, and it is now called the Session Monitor.

8.6 VTAM Applications

The end purpose of communications networks and networking software and hardware is to support applications and end users. VTAM supports a number of applications, such as CICS, IMS/DC, JES2/JES3, and TSO. All such communications subsystems are VTAM applications.

8.6.1 CICS

The Customer Information Control System (CICS) is a general-purpose teleprocessing monitor and database/data communications system. It supports a wide range of terminals and SNA logical units. It supports database management systems such as IMS/DB, DOS DLI, DB2, and SQL/DS and access methods such as VSAM and BDAM. Business-related end-user applications can be created using high-level languages such as COBOL and PL/1. CICS is supported under MVS, DOS/VSE, and VM.

8.6.2 IMS/DC

Information Management System/Data Communications (IMS/DC) is a transaction-oriented teleprocessing monitor. It supports database management systems such as IMS/DB and DB2. IMS/DC runs only under MVS and is not supported by DOS/VSE and VM. It also supports wide range of terminal devices. User applications can be written in high-level languages such as COBOL and PL/1.

8.6.3 DB2

DB2 is a Relational Database Management System (RDMS) used on an SNA mainframe running the MVS operating system. DB2 uses a nonprocedural language called Structured Query Language (SQL). SQL allows end users to create ad hoc applications and powerful transaction processing applications. VTAM applications in CICS or under TSO or even end-user non-subsystem-based applications can use SQL to access the relational database. DB2 is increasingly becoming the mainframe data access vehicle for peer-to-peer communications, whether they are using APPN LU6.2 protocols or TCP/IP protocols.

8.6.4   TSO

Time Sharing Option (TSO) is an MVS-only subsystem and primarily gives programming personnel a general-purpose time-sharing capability. It supports SNA and non-SNA terminals. Although some computer installations use TSO as a production-oriented subsystem, it is better to use it only in an application development environment. Unlike in CICS, each TSO user is a separate address space under MVS.

8.6.5   JES2/JES3

Job Entry Subsystems (JES) provide various operating system controls, batch processing facilities, and spooling capabilities for print devices and card readers. JES2/JES3 also provide Remote Job Entry (RJE) facilities where a remote workstation such as an IBM 3777 or an IBM 3780 can be used for submitting jobs and printing reports, as well as for remote console operations. JES2 is generally used in a single-host environment. JES3 can perform job scheduling for up to eight channel-attached MVS hosts.

8.7   Summary

In this chapter, we discussed various communications software to support SNA networks. VTAM running in the host and NCP running in the communications controller provide the basis for SNA networking. TCAM and BTAM are other communications access methods. Some of the sundry but specialized functions are performed by software such as NTO, NPSI, EP/PEP, and NPM. Network control and management is supported by NetView, an IBM product previously consisting of software such as NCCF, NPDA, NLDM, and others. SSP provides the basis for generating, loading, and dumping NCP. Various VTAM applications such as CICS, IMS/DC, and TSO provide the platform for end-user application development. JES2 and JES3 perform the essential functions of spooling, job scheduling, and management and support for RJE/RJP stations.

Now that we have laid the foundation by learning about SNA hardware, software, protocols, telecommunications, and SNA layers, we are ready to learn about SNA networks. This chapter completes Part 1 of this book. In Part 2, we will concentrate on SNA domains and networks, and on defining network topology.

SNA Networks

SNA Domains and Networks

In SNA there are four basic networking environments that use the various components of SNA. The simplest of these is the single-domain network. A more complex configuration is created when several single-domain networks are combined into one. This type of network is called a "multiple-domain" network. Within the multiple-domain network, the Job Entry Subsystems (JES) can have their own networking facility to direct job output and job execution. Both the single- and multiple-domain networks can participate in a multinetwork configuration using SNA Network Interconnection (SNI).

SNA's evolution from a hierarchical to a peer-to-peer network is facilitated by Advanced Peer-to-Peer Network (APPN). APPN is basically a flat networking architecture, but it can provide temporary network connections using casual connection. APPN also implements a function similar to SNI that connects autonomous APPN networks by using a border node configuration. Finally, APPN implemented on VTAM allows VTAM to act as the central directory of all resources in an APPN network through the central directory services function of APPN.

9.1 Domains

Under SNA there are single and multiple domains. A domain can be defined as the SNA resources that are controlled by an SSCP. This ability to control the resources is accomplished by the SSCP issuing an activate and/or deactivate request to the resource. These resources include the application programs, communications control units, lines, cluster controllers, terminals, and printers; in essence, the access method SSCP, PUs, LUs, links, and link stations. Once a resource issues a positive response to an activate request by an SSCP, that resource is owned and controlled by that SSCP only. However, a

PU Type 4 can be shared by up to eight SSCPs, but the PU Type 4 peripheral resources can have only one SSCP controlling them.

9.1.1 Single domain

The single-domain network consists of one SSCP and all the SNA resources in this network. Theoretically, a single-domain network can have 255 subareas (65,535 subareas with VTAM 3.2 and NCP V4.3.1/V5.2.1), of which 254 can be PU Type 4 subarea nodes that contain the Network Control Program (NCP). However, reality will dictate the actual number of PU Type 4 subareas controlled by this SSCP. Since there is only one SSCP in a single-domain network, all the resources defined in the VTAM resource definition list (VTAMLST) are owned and controlled by this SSCP. Figure 9.1 is an example of a single-domain network with three PU Type 4 subarea nodes. Note that each subarea node has two SDLC links with each adjacent subarea node. These links are associated with routes that are defined in VTAM and NCP. The routes detail the possible paths that can be taken during session establishment. Chapters 13 and 18

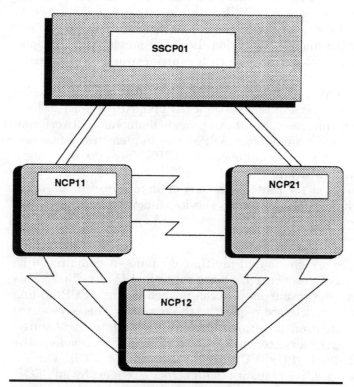

Figure 9.1 A single-domain network.

contain more information on defining routes and paths to VTAM and NCP.

9.1.2 Multiple domain

A configuration that consists of one or more SSCPs is considered to be a multiple-domain network when resources of one SSCP can communicate with resources of another SSCP via LU-LU sessions or when an NCP is being shared among SSCPs.

Figure 9.2 illustrates a simple multidomain environment between two SSCPs. In Figure 9.1, SSCP01 owns and controls NCP11, NCP21, and NCP12. In Figure 9.2, SSCP02 also activates these NCPs. NCP12 is channel-attached, and NCP11 and NCP21 are remotely activated

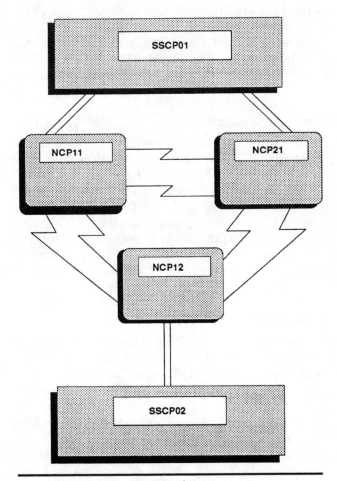

Figure 9.2 A multidomain network.

by SSCP02. All three NCPs are shared by SSCP01 and SSCP02. Both SSCPs have the same NCP major node defined in their respective VTAMLST. However, the ISTATUS operand of the resource definition statements in the NCPs may be defined differently to designate the controlling SSCP.

The NCP resources determine their ownership by responding to the first activate request received. Hence, by defining the ISTATUS operand conversely to the opposing SSCP, the individual domains can be obtained. Another option to this is the use of the OWNER operand. *Advanced SNA Networking* contains a discussion on using the ISTATUS and OWNER operands in NCP.

9.2   Multi-System Networking

To take full advantage of a multiple-domain network, you must implement the Multi-System Networking Facility (MSNF) of VTAM. This facility was incorporated into VTAM with VTAM V2 but can still be purchased as a separate product for VTAM V1 and TCAM. This facility provides the mechanism for cross-domain communications between SSCPs and LUs.

9.2.1   Cross-Domain Resource Manager

In order for an SSCP to control initiate and termination requests for resources out of its domain, a Cross-Domain Resource Manager (CDRM) function is provided. The CDRM communicates with CDRMs of other SSCPs in a multiple-domain network. The CDRM is defined to VTAM in a major node of VTAMLST by the CDRM definition statement. When a resource of this CDRM requests a session with a target resource of another CDRM, the CDRM sends a cross-domain initiate request to the target resource CDRM on behalf of the originating resource, and likewise for termination. The CDRMs communicate over an SSCP-SSCP session.

9.2.2   Cross-Domain Resources

The resources owned by CDRMs can be explicitly or dynamically defined to other CDRMs. These resources are known as Cross-Domain Resources (CDRSC). A CDRM knows a cross-domain resource by its network name and associated CDRM. In large multidomain networks it is best to use dynamic definitions for the CDRSCs. This simplifies synchronizing network updates and naming convention modifications. For cross-domain acquires, however, the resource to be acquired must be explicitly defined to the CDRM. To define cross-domain resources to VTAM, the CDRSC definition statement is used in a CDRSC major node in VTAMLST.

9.2.3   Cross-domain sessions

There are two types of cross-domain sessions, SSCP-SSCP and LU-LU. The SSCP-SSCP sessions are established by the CDRMs of the respective VTAMs when the CDRM major node that defines another PU Type 5 subarea CDRM is activated. VTAM knows that a CDRM major node being activated is not the CDRM for this domain because the SUBAREA operand value found on the CDRM definition statement does not match the subarea number specified in VTAMLST for this VTAM. A cross-domain LU-LU session is established when an LU of one domain requests a session with a resource that this VTAM does not recognize. At this point the SSCP calls upon the CDRM function to locate the owning CDRM of the target resource to satisfy the session request.

9.3   SNA Network Interconnection

The accelerated growth of SNA networks in the late 1970s and early 1980s prompted the need to connect two or more totally independent networks. Multiple-domain SNA networks were quickly reaching their subarea address limits but needed to expand to provide access for a growing number of users. SNA Network Interconnection (SNI), incorporated into VTAM V2R2, provides the capability to have two or more independent SNA networks communicate with each other. Each gateway NCP can support up to 255 interconnections.

SNI should be considered when (1) your network has reached its limits for providing adequate service to end users, (2) your network does not implement Extended Network Addressing (ENA) and the number of subareas has drastically reduced the number of elements per subarea, (3) you want to combine ENA functionality with releases of VTAM and NCP that do not support ENA, and (4) a need has arisen to connect two or more independent SNA networks (e.g., corporate mergers). Figure 9.3 lists SNI functionality with releases of VTAM and NCP.

9.3.1   Single gateway

SNI provides access to other SNA networks through gateways. A single-gateway configuration comprises a gateway SSCP and a gateway NCP. This configuration is also sometimes known as a "simple" gateway. The gateway VTAM must have the following start options defined in order for it to be a gateway VTAM: SSCPID, SSCPNAME, NETID, and GWSSCP. The presence of these four options denotes that a VTAM is at least gateway capable. A gateway NCP is defined by the following NCP definition statements: GWNAU and NETWORK.

	MVS	VM	NOTES FOR BACK-LEVEL
VTAM V3	ALL	R1.1	
VTAM V2R2	ALL	N/A	
NCP V3	ALL	N/A	
NCP V4/V5	ALL	R2	
NCP V1R2.1			Not supported by VTAM V2R2 or higher. Must be owned by VTAM V2R1 or lower.
NCP V1R3			Must be supported by VTAM V2R2 or lower.
NCP V2			Back-level NCP does not support SNI but supports all VTAMs.
TCAM V2R4			Needs MSNF for SNI support.
VTAM V1R3			Same as NCP V1R3 and TCAM V2R4. Needs a PTF for autologon with VTAM V3, V2R2 resources.
VTAM V2R1			Needs a PTF for autologon with VTAM V3, V2R2 resources.
NCP V4 SUBSET			No SNI support.

Figure 9.3   SNI functionality of VTAM and NCP.

In Figure 9.4 a simple gateway is configured. NETA01 is the sole gateway SSCP. As a gateway SSCP, VTAM can initiate, terminate, take down, and provide session outage notification for cross-network sessions. The gateway SSCP provides network name translation and assists the gateway NCP in assigning alias network addresses for cross-network resources. Note that the box depicting the gateway NCP has two labels. When NETA01 resources communicate with NETB01 resources, they are actually addressing that function of the NCP that is acting as the gateway (GW-NCP11). Likewise, when NETB01 resources communicate with NETA01 resources, they address the GW-NCP11 function of the NCP. This gateway function of the NCP acts like a pseudo-CDRM.

9.3.2   Multiple gateway

Several gateways can be configured to connect the same two independent SNA networks for cross-network sessions. In Figure 9.5 we see that NCP21 is also a gateway NCP. The rationale for defining NCP21 as a gateway NCP in this scenario is twofold: (1) backup and recovery for the gateway NCPs can easily be implemented and executed and (2) distribution of the cross-network session flows and volume. Networks are considered adjacent networks when the session path flows through only one gateway NCP. These networks are said to be

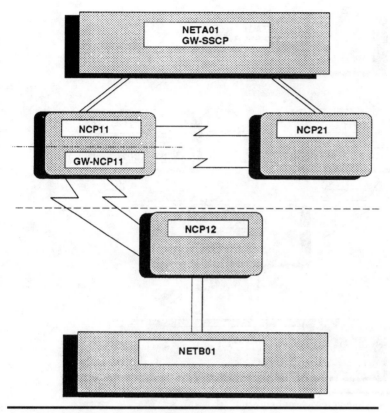

Figure 9.4 Single-gateway configuration.

nonadjacent when the session flow is through multiple gateway NCPs.

9.3.3 Back-to-back gateway

When two or more gateway NCPs are connected, an intermediate network is configured. This is called a "back-to-back" gateway (Figure 9.6). In NETX there is no SSCP involved. The gateway functions of the three NCPs can actually be considered to be a null network; that is, a network without an SSCP.

Actually, this type of SNI configuration is most favorable. The exchange of network configuration data is reduced because only the subarea numbers used by the NCPs to identify themselves to the intermediate network are required. The numbers that define the routes within the intermediate network are the only ones that need to be changed. All other routes can remain in effect. For a more detailed description and analysis of SNI, consult *Advanced SNA Networking*.

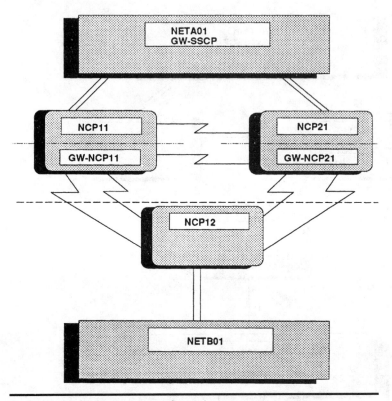

Figure 9.5   Multiple-gateway configuration.

9.4   Extended Networking

Although SNI supplied an answer to growing network resource requirements, it is not the solution for every network. For some networks the task of migrating to SNI may be too disruptive to normal operations. To allow growth in these types of networks, Extended Network Addressing (ENA) was introduced with VTAM V3 and NCP V4. ENA provides an effective means for network resource growth on the existing network configuration.

With ENA the network addressing scheme allows from 1 to 255 subarea nodes, each subarea with a maximum of 32,768 Network Addressable Units (NAUs), also known as "network elements." Before ENA, the network address was 16 bits long. The subarea-element split varied according to the MAXSUBA value specified in the VTAM start options. For instance, if the MAXSUBA value is set to 63, the maximum number of elements addressed in any subarea in this network is reduced to 1023. This is because the subarea-element split allows for 6 bits for the subarea number and 10 bits for the element

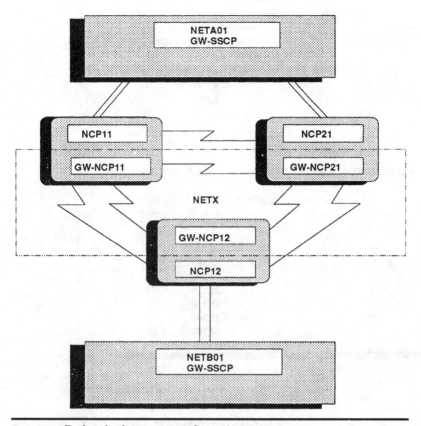

Figure 9.6   Back-to-back gateway configuration.

addresses. In ENA the network addressing scheme provides 23 bits for the network addresses. The subarea field is always 8 bits, and the element address field is always 15 bits.

In the most recent releases of VTAM and NCP, this network addressing scheme has been expanded to a maximum of 65,535 subarea nodes. Each subarea node can have a maximum of 32,768 elements. In VTAM V3.2, NCP V4.3.1, and NCP V5.2.1, the 23-bit network address has been expanded to 31-bit addressing. This allows for using 16-bit addressing for the subarea address and 15-bit addressing for the element address. VTAM V3.2 needs a Program Temporary Fix (PTF) to accommodate this new addressing scheme. Figure 9.7 illustrates the various SNA addressing schemes.

Not all levels of VTAM and NCP can participate in an ENA network. Releases of VTAM prior to VTAM V3 and NCP V4 cannot have full participation in an ENA network. A PTF must be installed on VTAM V2s and NCP V3s to allow ENA participation. If these pre-

Pre-VTAM V3 and NCP V4 (NON-ENA)

# of Subareas	Max. # of Elements/Subarea
3	16,384
7	8,192
15	4,096
31	2,048
63	1,024
127	512
255	256

VTAM V3 and NCP V4 ENA

# of Subareas	Max. # of Elements/Subarea
3	32,768
7	32,768
15	32,768
31	32,768
63	32,768
127	32,768
255	32,768

VTAM V3.2 and NCP V4.3.1/V5.2.1 ENA

# of Subareas	Max. # of Elements/Subarea
65,535	32,768

Figure 9.7 Non-ENA and ENA element tables' effect on MAXSU-BA operand.

ENA nodes communicate with ENA nodes in the same network, the ENA nodes must specify the MAXSUBA values in the VTAM and NCP. This MAXSUBA value must match the value specified for the back-level VTAM and NCPs that reside in the same network as the ENA nodes. Although MAXSUBA is not required for ENA, it is required when ENA nodes communicate to pre-ENA nodes. The ENA nodes need this information to translate the pre-ENA 16-bit addresses when communicating with pre-ENA nodes.

Problems do occur when ENA node elements try to access pre-ENA nodes. If an ENA element address exceeds the maximum element address for a pre-ENA node, the ENA element cannot establish a session with any pre-ENA element. These restrictions can be overcome by implementing SNI to separate the network into back-level nodes and ENA nodes.

9.5   LEN and APPC/LU6.2 Networks

Low-Entry Networking (LEN) is an advanced SNA network that allows connection of any resource of any network to participate in an LU-LU session through an SNA network using APPC/LU6.2 session protocols. Figure 9.8 details a sample LEN network configuration.

The implementation of LEN requires VTAM V3.2, announced June 1987, and NCP V4.3 for an IBM 3725 communications controller and NCP V5.2 for the IBM 3720 or IBM 3745 communications controllers. VTAM V3.2 incorporated APPC/LU6.2 basic functions. This allows for LU6.2 applications under VTAM to take full advantage of the APPC protocols. This release also includes the Boundary Function (BF) SNA request and response commands from NCP for establishing and tearing down peer-to-peer sessions through the network. NCP V4.3 and V5.2 provide the PU Type 2.1 boundary function necessary to allow peer-to-peer sessions through the SNA backbone network. The VTAM

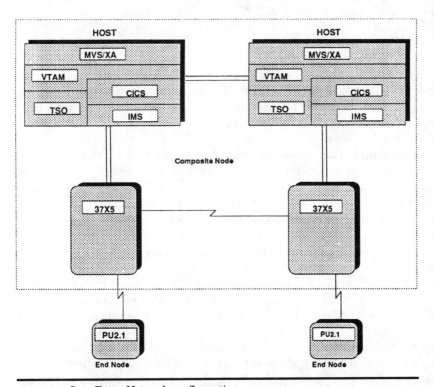

Figure 9.8   Low-Entry Network configuration.

V3.2 and NCP V4.3 or V5.2 are seen from the End Node T2.1 (EN T2.1) as a Composite Node Type 2.1 (CN T2.1).

9.6   APPN

Advanced Peer-to-Peer Network (APPN) is IBM's plan for evolving an SNA traditional hierarchical network to a peer network. APPN is implemented on VTAM V4R1 and NCP V6R2. NCP V6R2 can reside only on IBM 3745 Communication Controllers, thus requiring a sizable investment for a company that has not kept up with communications controller technology. APPN is based on Low Entry Networking.

APPN provides three functions that are similar to functions provided by SNA. One function is the ability for temporary network connections. This allows networks that are autonomous to connect and communicate on a temporary basis. This is performed by the casual connect function of APPN. A second function that provides a permanent connection between autonomous APPN networks allows full APPN support between the two networks. This is provided by the border node capability of VTAM V4R3 and AS/400s. The third function, central directory service, is provided with VTAM V4R1. This feature allows VTAM to be the primary repository of all APPN network resources.

9.6.1   Casual connection

In large APPN networks with many network node servers, it became important to break up the large APPN network into smaller networks for manageability. The earlier releases of APPN provided connection between these new networks using a function called "casual connection."

Casual connection requires that the connection be between network nodes and end nodes or between two end nodes. Connection between two network nodes is not possible with casual connection. This is illustrated in Figure 9.9. The problem with casual connection is the inability of end nodes to act as intermediate routing nodes. As shown in the figure, sessions between two end nodes or two LEN nodes are possible. Intermediate routing nodes in NET_A can be used to establish a session between a nonadjacent end node EN_A LU_A and NN_4 LU_2. To illustrate the restraints on casual connection, LU_5 in NET_B cannot establish a session with LU_X in NET_A because the end node EN_A cannot act as an intermediate routing node.

In casual connection configurations, no CP-CP sessions can be established; therefore directory services of one network cannot exchange directory information with the connected network. Hence,

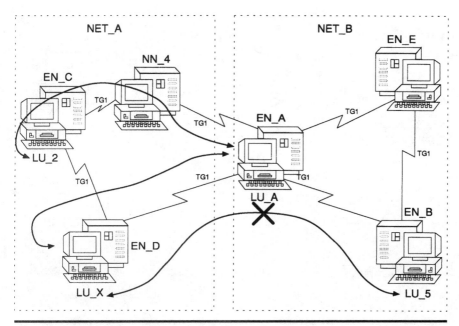

Figure 9.9 Diagram illustrating the inability of APPN End Nodes to act as intermediate routing nodes.

all resources used in a casual connection must be manually defined in each node that will participate in LU-LU sessions.

9.6.2 Border Node

Casual connection has some severe drawbacks and therefore does not provide a solution for connecting adjacent networks. APPN now implements a new function similar to SNI called border node (BN). BN allows a network designer to break up a large APPN network into smaller networks and connect them through BNs.

Each network contains its own topology database and directory database. A BN configuration is shown in Figure 9.10. In the diagram, a BN can be either an AS/400 or an IBM mainframe executing VTAM V4R3. The BN provides a service similar to that provided by an NCP gateway in SNI. In the example, to the native network NET_A, the BN is seen as an APPN NN and as part of NET_A's topology database. The nonnative network NET_B views the BN as an APPN EN and as part of the topology database of NET_B. The BN_4 in Figure 9.10 belongs to network node servers of both NET_A and NET_B. A CP-CP session is established between BN_4's EN image and the NN_1 in NET_B. The NN image on BN_4 is participating in a CP-CP session with an NN in NET_A. The CP-CP session

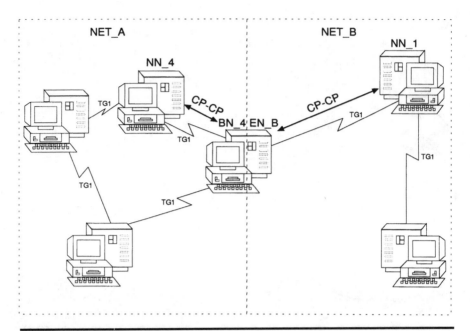

Figure 9.10 Diagram depicting the use of APPN Border Nodes.

between BN_4 EN and NN_1 allows unregistered resource searches to occur. This allows Locate(broadcast) and Locate(directed) searches to pass between the two networks. When an AS/400 is used as the BN, only adjacent networks will allow the Locate to function properly. AS/400s will allow a Locate to cross only a single BN. VTAM V4R3 as a BN will provide for nonadjacent network connections and searches to occur through it.

A single optimal route for the cross-network connection is not possible, since the topology databases of each network are isolated and not exchanged. Two routes for each network must be selected if the cross-network session is to occur. The optimal route is determined by the COS name passed on the Locate request. BNs have the ability to map COS names either to a default value or to a name found in the COS table. This requires some effort in coordinating COS names between VTAM, NNs, and BNs.

In summary, BNs support the isolation of topology databases between APPN networks but mesh the directory services of the two networks.

9.6.3 Central Directory Server

A central directory for LU resources in an APPN is supported by Central Directory Servers (CDS). A single CDS or multiple

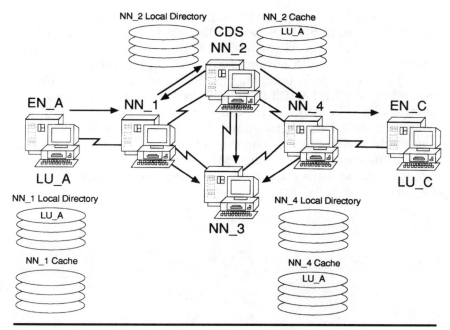

Figure 9.11 Diagram depicting a Central Directory Server configuration.

CDSs are possible in an APPN network. Figure 9.11 depicts such a configuration.

Nodes are identified as being able to act as a CDS through a properties exchange that occurs as part of the topology services. NNs providing Locate requests for LUs determine the closest CDS and send a single Locate to the CDS rather than broadcasting a Locate to all known NNs. If the CDS does not have the resource requested in the Locate request, the CDS will then send a Locate broadcast request to discover its location.

APPN ENs and NNs can avoid broadcasts by registering their resources with the CDS. When ENs register their resources with their NNs, they can also request that the resources be registered with a known CDS. Using this method, the CDS becomes populated quickly, thus avoiding unnecessary broadcasts.

9.7 Summary

In this chapter we discussed SNA domains and networks. SNA provides for establishing a single-domain or a multiple-domain network. In order to have a cross-domain session, the MSNF must be implemented. CDRM plays a major role in cross-domain sessions. SNI provides connectivity between two or more separate SNA networks. We discussed single-gateway, multiple-gateway, and back-to-back gateway possibilities in SNI. We briefly discussed LEN and APPC/LU6.2, with an emphasis on APPN and its ability to provide connectivity between adjacent APPN networks using casual connection and border node. The border node function was discussed in detail to illustrate the ability of VTAM V4R3 to provide intermediate BN connectivity functionality. The APPN central directory server function was also discussed to illustrate the need for centralizing APPN Locate requests to avoid broadcasts on the network. Now that we are familiar with the concepts of domains, networks, and network interconnection, it is time to learn about defining network topography in the next chapter.

Chapter
10

Defining Network Topography

The task of designing a viable and useful communications network is not easy. There are many actors that will dictate and direct the topography of a network. These factors should be identified by the communications systems programmer and the network designer. It is their responsibility to translate management's business needs, objectives, and requirements for end-user access to the network. This could mean restricting end-user access to specific applications or entire domains or restricting access to all applications in all domains of the network. Evaluation and documentation of the current network should be performed. How large is the network? Does management anticipate steady or rapid growth? Is the communications control unit storage size adequate for current and future needs?

Figure 10.1 outlines the major issues of concern for the communications systems programmer. The first issue listed, service level agreement, is a proviso between the communications systems programming staff and the end-user community. This accord should describe an acceptable level of service for the end users to meet their objec-

---

DEFINING NETWORK TOPOGRAPHY

o Service Level Agreement
o Host Processor Configuration
o Communications Controller Configuration
o Telecommunications Line Configuration
o Software Configuration

Figure 10.1 Network definition issues.

tives. The second issue, host processor configuration, is a network concern in that multiple hosts within the same network may need network accessibility. This can be accomplished by using MSNF. Another possibility is access to applications executing on host processors that reside in another network. We may choose to use the SNI facility for this connectivity. The third issue, communications controller configuration, is of great importance because it is directly related to the network topography. The size of the network, the end-user population, and the number of telecommunications lines influence the hardware requirements for the communications controller. The fourth issue, telecommunications line configuration, is a major factor in complying with the service level agreement. The speed of the lines, the number of end users associated with the lines, and the use of leased and switched lines must all be considered. The final issue listed, software configuration, plays a significant role in defining the network. The various levels of software must all be synchronized. Some of the end-user requirements may necessitate the acquisition of new software to support their needs.

We must remember to view the network as a vehicle that delivers information to applications and end users. Any modifications or enhancements suggested by the communications systems programmer and network designer must be reviewed with great scrutiny. Changes designed for end-user accessibility may prove to be detrimental for the delivery of information. For example, providing additional workstations on a link may allow more end users to access an application simultaneously, but the increase in traffic can diminish end-user performance by increasing response time. This change in response time may negate the original objective of allowing greater accessibility for the end user. An important factor in the final decision of this type of tradeoff is the end users' overall requirements.

10.1   End-User Requirements

Many organizations develop network requirements and objectives from the end-user community. This community is usually regarded as the recipients of the services provided by the applications that execute on the host processor. They can be the accounting, inventory, or sales departments or in some instances the public at large.

The matrix in Figure 10.2 presents the relationship between the end-user requirements and their effects on the network configuration. The volume of the end-users' applications affects all the configurations. The processing capacity of the host processor may have to be increased or even off-loaded to another host processor for effective throughput. The communications controller storage size may have to

END USER REQUIREMENTS	HOST PROCESSOR	COMMUNICATIONS CONTROLLER	LINE CONFIG	SOFTWARE CONFIGURATION
Peak Application Volume	*	*	*	*
Application Types	*		*	*
User Grouping & Dispersion		*	*	
Response Time	*	*	*	*
Recovery	*	*	*	*
Number of Users	*	*	*	*

Figure 10.2   User requirements and affected network configurations.

be increased to support the abundance of traffic to the remote network. Telecommunications lines may need a broader bandwidth to support the peak volume periods. This, in turn, can affect the hardware configuration of the communications controller by necessitating a comparable line interface coupler or line set. The operating system software may need more storage to accommodate a larger application address space to sustain the volume of applications. The types of applications that use the network have a measured effect on the network configuration. For some applications, a dedicated host processor may be necessary to achieve acceptable performance, in particular for on-line inquiry and updating applications. The use of dedicated or switched telecommunications lines should be addressed. Remote data entry using a store and forward type of operation may be used with switched lines, whereas on-line scientific or accounting applications can use leased lines. The operating systems' Job Entry Subsystem (JES) may need modification specifically for remote job processing. The degrees of software levels should be scrutinized for incompatibilities and functionality to meet the end-user requirement. VTAM's routes may need reconfiguration to balance the application mixture along with the application load. The grouping of the users can again affect the communications controller hardware configuration. Higher-capacity LICs may need to be installed, which may dictate a line attachment base expansion or installation. The telecommunications line may be set up for point-to-point or multipoint. A configuration like this can determine if full-duplex or half-duplex transmission can be used. Response time and the user population are both heavily weighed factors for all the configurations. The ability of the host processor, communications controller, telecommunications line, and software to provide acceptable response time and to support the user population is proportional to the other end-user requirements. Recovery of the networks is dependent upon the corporation's ability to function during a down period. The importance of recovery in most institutions comes down to the bottom line. Recovery can be quite expensive. Mirror-image data centers, duplication of telecommunication lines, software license fees, and added personnel can be an extremely costly insurance policy. The methodology of network recovery can be a book in itself.

The representatives of the end-user communities must supply present and future objectives to the communications systems programmer and the network designer. Several institutions relay this information by use of a service level agreement.

10.1.1   Service level agreement

A service level agreement between user communities and the network programmers and network designers outlines the objectives and requirements for network service. It should cover some if not all of the following information:

1. Recoverability of network resources. State the minimum time of outage and identify backup procedures.

2. Availability of network resources. An example would be an agreement to a specified uptime percentage for lines, workstations, communications controllers, and cluster controllers.

3. Accessibility to network resources. A proviso stating path availability to host applications.

4. Serviceability of network resources.

Here is where we define end-user response time and throughput. This agreement is probably the most visible of all the previous points. The end user is always at the mercy of network service to execute his

NETWORK SERVICE LEVEL AGREEMENT

Application Name: CICSP01

Estimated Volume: 10,000 TRANS/DAY

Estimated Number of Users: 1000

MINIMUM OUTAGE TIME

    Lines: 30 min

    FEP: 15 min

    Clusters: 30 min

RECOVERY PROCEDURES

    Lines: 1) Modem testing
                  2) FEP port swapping
                  3) Matrix switching
                  4) Dial back-up services

    FEP: 1) Matrix switching to direct lines to back-up FEP

    Clusters: 1) IML of failed cluster
                  2) Inactivate/Activate of failed resource
                  3) Contact field support

AVAILABILITY: 92-98%

ACCESSIBILITY: 98.2%

SERVICEABILITY:

    Average Response Time @ Peak Periods: 4 seconds
    Transmission Volume @ Peak Periods: 4000 transactions

Figure 10.3   A network service level agreement.

or her transaction as quickly as possible. End-user performance is determined by network performance. Figure 10.3 is an example of a service level agreement.

In reality, all the above-mentioned points have their roots in network performance. The performance is based on several factors, some of which are device types in the network, the applications using the network, and the number of domains and/or networks connected.

10.1.2   Device types

The data terminal equipment used in networks varies greatly. Usually based upon application or end-user requirements, the device types in a network determine other network factors. Most devices are developed with specific protocol usage, such as binary synchronous, SDLC, or start/stop and their ability to operate in half-duplex or full-duplex mode. Line speed may be a restriction of the device because of its configuration. For example, IBM 3274 model 61C operates in half-duplex mode using SDLC protocol and can support 56 kbps but in most cases is configured for 9.6 bps. This automatically dictates how a point-to-point line is defined for this device in the NCP. By knowing the device types connected to the network, the communications systems programmer knows the basic VTAM or NCP definition. In our first network example (Figure 10.4), we will be using both BSC and SDLC remote IBM 3274 cluster controllers, a channel-attached non-SNA 3274, and an SNA/SDLC 3274, also channel-attached.

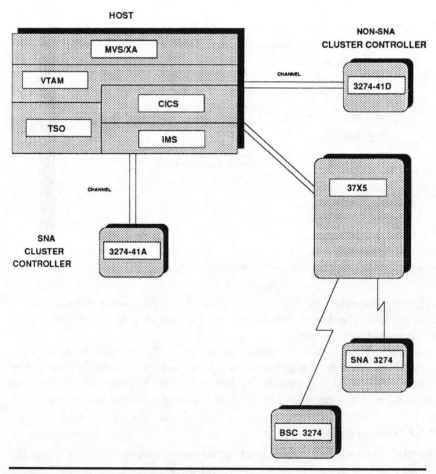

Figure 10.4 A sample single-domain network.

10.1.3   Multi-System Networking Facility (MSNF)

Requirements for cross-domain sessions concerning end users and their applications must be determined. Does the network use MSNF now? How will management objectives affect its use? Should a dynamic cross-domain resource definition be in effect, or should specific cross-domain resource definitions be used? Are there any facilities provided by MSNF that need closer review and possible consideration, such as the adjacent SSCP table? These are just some questions that should be asked by the communications systems programmer on planning the use of MSNF.

10.1.4   SNA Network Interconnect (SNI)

Some SNA networks have tens of thousands of terminals with perhaps 50 domains, all connected in the same network. To meet management's objectives and provide desirable service to the end user, the communications systems programmer and network designer may suggest the use of SNA Network Interconnect (SNI). Your corporation may have recently acquired several companies, each with its own SNA network. Accessibility to all the networks may be a requirement. However, to incorporate several networks into one large network would be an incredibly long-drawn-out process, taking several months to possibly several years to coordinate and implement. SNI can alleviate this problem, allowing for network autonomy and a rapid implementation with little end-user involvement.

10.1.5   Local area network connectivity

The number of end users that utilize today's corporate network has expanded tremendously. The ease with which end users can be added to a network has been enhanced greatly by local area networks. LANs provide end users with shared resources and media that empower the end-user community for right-sized processing. A network designer must take into account the type of applications flowing over the network. Each application must be analyzed for its characteristics. These characteristics include the type of protocol used by the application, the flow of the application data through the network, and the ability for the application to send data in bursts. All of these contribute to providing a proper network design.

10.1.6   RJE, NJE, and RJP JES3 networking

Many end-user applications don't require fast information retrieval. These applications usually execute in the background or batch processing modes. Yes, the information is important, but it does not have a high processing priority. Usually, this information is in the form of a printed report for management or customer use. The communications systems programmer must provide a path for submission and transmission of this data.

In our scenario, several remote end users have off-line processing. That is, there is no need for computer center facilities to compile data. However, applications on the host processor use this data as input for management reports. Figure 10.5 depicts an RJE scenario configuration. The data is forwarded via Remote Job Entry (RJE) in a JES2 environment or by Remote Job Processing (RJP) in a JES3 environ-

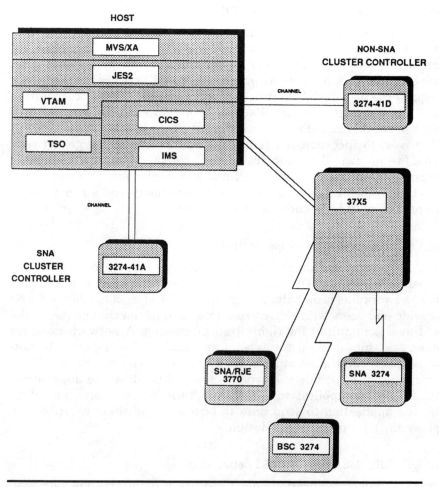

Figure 10.5   A sample single-domain network with an RJE workstation attached.

ment. Basically, these facilities allow end users to start a session with the appropriate Job Entry Subsystem (JES) to execute a batch program in the host processor that uses the remote's forwarded data. The report from this batch program can be routed to other networks for printing by using JES2/Network Job Entry (NJE) or JES3 Networking. Both these facilities use application-to-application sessions to route the report to one or more destined network printers. The device types used for RJE and RJP vary. In our primary network, JES2 will be the job entry subsystem, and the users have a need for an RJE station at a remote site. This RJE workstation will be an SNA 3770 device, and it will have two printers and an operator's console attached to it.

10.1.7 Switched lines

One more end-user requirement that needs to be considered is dial-in or switched line support. This support is accomplished by the end user or the network operator making a telephone call which establishes a communications line for the transmission of data. We place switched line support under end-user requirements for two reasons. One is the amount of time the end user needs access to the host processor facility. Many end users, as in the previous RJE/RJP scenario, need very little line time to accomplish their functions. These groups are prime candidates for switched line service. The second reason is for use as a backup procedure. Some networks are all point-to-point. An example would be one cluster control unit per line. Since cluster controllers are mostly used for interactive applications, the importance of availability is great. If the line to the cluster controller should fail, switched line support will allow access to the host processor. The communications systems programmer and/or network designer should determine if the network has the proper configuration to support switched lines. Does the communications controller have switched line support? Do the modems support switched lines? If not, what hardware features are required? Is the proper software in place for switched line support? If not, determine the software required to support the hardware.

One more note of special interest to the network designer. VTAM V3.1.2 and NCP V4.3 or V5.1.2 include support for switched subarea links. This is important specifically for providing backup communications links for subarea connectivity.

In our sample single-domain network (Figure 10.6), the corporation has a remote site that uses an IBM S/36 for off-line processing. The data entered during the workday is to be transmitted to the host at night over a switched line using SDLC protocol.

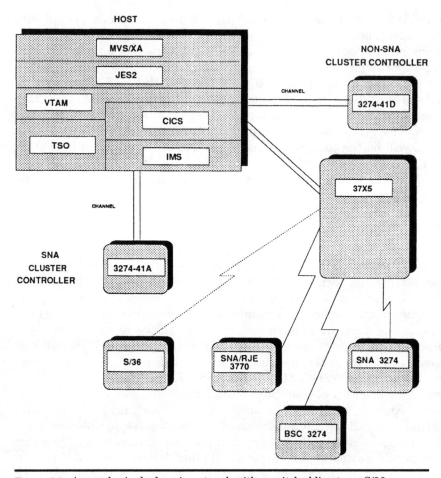

Figure 10.6   A sample single-domain network with a switched line to an S/36.

10.2   Telecommunications Line Requirements

The requirements of end users, device types, and the applications can directly determine the speed and protocol used for the network telecommunications lines. Not only is the type of communications facility of concern, but also the positioning of end-user cluster controllers and terminals on each telecommunications line. An accounting of the amount and type of data that traverses these lines must be gathered for use by the network designer to determine line saturation thresholds. This data can be found by the use of several different types of monitoring software that are used at the application, host, and network levels. Once the data is compiled, it can be used to determine such factors as line speed and actual line configurations.

10.2.1 Line speed

Many communications systems programmers and network designers believe that line speed may prove to be the quickest and easiest solution to increasing end-user response time and throughput. However, several points must be considered when deciding on line speed for your network.

Remember that links are composed of physical connections. Communications controllers, cluster controllers, modems, and telecommunications lines are all part of the link. As a communications systems programmer or network designer, you should research each of these components and determine whether the communication facility support is available for each component. For example, Figure 10.7 shows links connected to an IBM 3725 communications controller. One of these links connects a 3725 communications controller, and the other connects a 3274 cluster controller. As you can see, the link speed between the 3725 communications controllers is designated at 56 kbps, while the 3274 cluster controller link is 9.6 bps. If line speed is believed to increase response time, why not increase the 3274 cluster controller link to 56 kbps? This is a sound and logical question which deserves a logical but simple answer. This 3274 cluster controller is using BSC protocol, and there is a restriction of line speed to 9.6 bps for this device. If, however, this device were a remote 3274 SNA cluster controller with the proper hardware features, it would be able to handle a line speed of 56 kbps. Here is just one example where research about the hardware configuration of the device and the line configuration must be compared and assessed carefully.

10.2.2 Multiplexing

A different approach to line configurations allows the use of high-speed lines to devices that have a speed restriction. This is called "multiplexing." Multiplexing is a means of breaking down a high-speed line into several lower-speed lines. For example, in Figure 10.8,

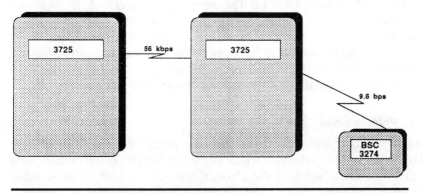

Figure 10.7 Line speeds and device restrictions.

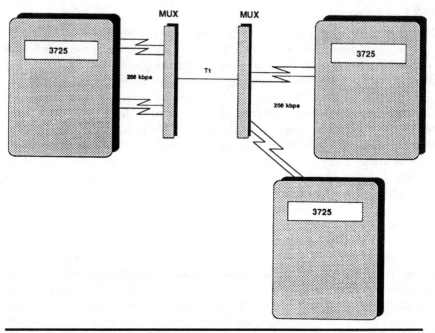

Figure 10.8 The use of a multiplexer device.

the backbone of the network is a T1 carrier. The communications control unit has defined to it four 256-kbps dedicated links to remote 3725s. To use the large bandwidth of the backbone link, a multiplexer unit can be used to concentrate and then expand the bandwidth to the appropriate speeds.

Take a look at Figure 10.8. The front-end processor has four 256-kbps lines connected to a multiplexer. The multiplexer concentrates these lines to a bandwidth of 1.44 million bits per second (Mbps). At the receiving end of the link, another multiplexer reduces the incoming data to 256 kbps and directs the data to the designated destination. The same holds true for the reverse route. Thus, high-speed transmission is accomplished by multiplexing down the speed to lower speeds for several devices attached to the front-end processor. This brings us to another factor of line requirements, the line configuration for one device attachment or several device attachments to the line. These are point-to-point and multipoint configurations, respectively.

10.2.3   Point-to-point

In the previous discussions we described a configuration of one cluster controller attached to the communications controller. We also discussed the use of a switched line for cluster controllers and other devices which use dial-up access to the network. Both scenarios illustrate a point-to-point line configuration. However, switched lines may never be used for more than one remote controller at a time. In an SNA network, it is safe to view a point-to-point configuration as having one PU attached to each link. Reasons for the use of point-to-point configurations are many. Response time, control of the type of data, security, and ease of recovery are all viable objectives, to name a few.

In Figure 10.9, we can clearly see the point-to-point configuration. Remember, when we use the term "point-to-point," we are describing the physical connection of the link and its end points, the local end point and the remote end point. There is no intervening point between them. Links with more than one remote end point are multipoint links.

10.2.4 Multipoint

Like a point-to-point line configuration, a multipoint line describes the physical connection of the remote end points. In this configuration, more than one remote point shares a single telecommunications line.

The concept of a multipoint line is not difficult if you can see the configuration. Figure 10.10 draws the picture clearly. What we see is one dedicated link with four remote points or drops attached to the link in a perpendicular arrangement. This design will allow any of the remote links that tap into the dedicated link to fail without affecting the other drops. It can also be cost-effective to the company because it decreases the number of dedicated links necessary to support the end users. A careful study of the application mix and volume will assist you in determining if a multidrop or multipoint line is sufficient. There are two factors of interest in multipointing. The first is

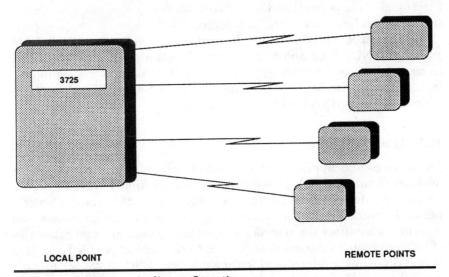

LOCAL POINT   REMOTE POINTS

Figure 10.9 A point-to-point line configuration.

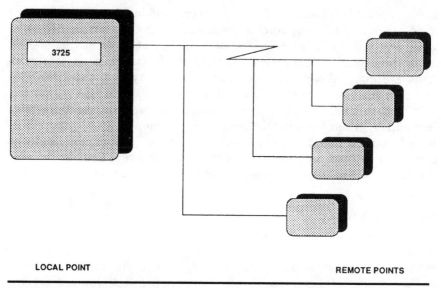

LOCAL POINT   REMOTE POINTS

Figure 10.10   A multipoint configuration.

the use of full-duplex transmission. Even if the remote device cannot function in full-duplex mode, the line can be defined to NCP as having a send and receive address. This will allow one remote point to send data while another remote point on that multipoint line is receiving data. An example of this is multipointing 3274 cluster controllers. These cluster controllers operate in half-duplex transmission protocol only. The second factor is that the link protocol must be the same for all points on the line. No mixture of BSC, SDLC, or S/S is functional. This is because the network control program's line configuration must be aware of the link protocol.

An advanced multipoint configuration is available with VTAM V3.1.2 and NCP V4.2 and higher. These levels allow a multipoint link to be used in conjunction with subarea SDLC links. This configuration will allow the network designer to use cost-effective link configurations for large SNA backbone networks.

10.3   Link Protocols

The devices and applications that use a telecommunications line or medium dictate the link protocol. Again, we see that hardware dependencies of a device for a specific link protocol can affect the design of a network. Some applications are designed for specific protocols and therefore also affect the network topography. There are two categories of link protocols to discuss that pertain to the subject of this book. One is non-SNA link protocols, which include S/S, BSC, X.25, token-ring, and Ethernet. The second category is SNA link protocol, which is SDLC.

10.3.1   Start/stop link protocol

Asynchronous protocols in today's networks are used for such devices as an IBM 3767, IBM 3101, TWX model 33/35, or WTTY-type devices. In Chapter 5 we discussed VTAM and how it supports SNA. You should also note from that discussion that VTAM does not support asynchronous link protocol. Therefore, the asynchronous data being transmitted must be converted to a format that is supported by VTAM. This can be accomplished by the use of an IBM program called Network Terminal Option (NTO). NTO resides in the communications controller with the network control program (Figure 10.11). A main function of NTO is to allow asynchronous devices access to the VTAM host by converting the start/stop commands to an SNA format. Another and more recent means of allowing asynchronous devices to access VTAM is by using protocol converters. One such protocol converter has already been previewed, the IBM 3710 network controller. This device can have several start/stop devices attached to it. In turn, the 3710 provides the protocol conversion function and passes the data to the network control program in SDLC format, eliminating the need for protocol conversion software residing in the communications controller. The 3710 can also be used for BSC protocol conversion. It is up to you, the communications systems programmer or network designer, to determine if either of these is feasible in the design of your network.

10.3.2   Binary synchronous link protocol

BSC has been by far the most widely used link protocol since 1966. Because of this large user base, VTAM supports only one type of non-

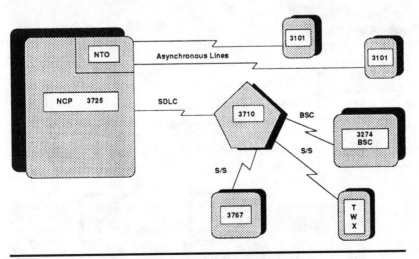

Figure 10.11   A network protocol conversion configuration.

SNA device: IBM 3270 BSC. All other non-SNA devices must either emulate an IBM 3270-type device or have a software package of some type to provide the emulation for the device. We have already discussed the NTO program product in the previous section for S/S devices. NTO can also be used for BSC devices as well.

More often than not, another IBM software package, the Emulation Program (EP) (Figure 10.12), is used for non-SNA BSC devices. This emulation program executes with the NCP in the communications control unit. It provides the proper interface for BSC devices to their respective applications that are executing on the main processor. When both NCP and EP are loaded into the communications controller, it is a Partitioned Emulation Program (PEP). An example of a non-SNA BSC device is the IBM 2770 remote workstation. This workstation is used for RJE processing. As with all non-SNA devices, NCP and VTAM have no knowledge of these devices. This includes data transmission and session setup. It is all provided by EP.

One other program that is worth mentioning here is the Non-SNA Interconnect (NSI) program offered by IBM (Figure 10.13). This program allows BSC RJE networks and BSC NJE networks to use SNA by enveloping the BSC data with SNA headers. This is done in the communications controller where NSI executes with the NCP.

10.3.3   X.25 NPSI

Packet switch data networks allow several telecommunications users to transmit their specific data over public telephone circuits. This idea came about as a way to reduce a corporation's line costs by using

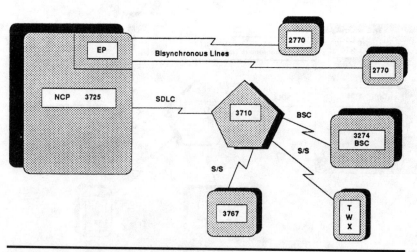

Figure 10.12   A PEP configuration.

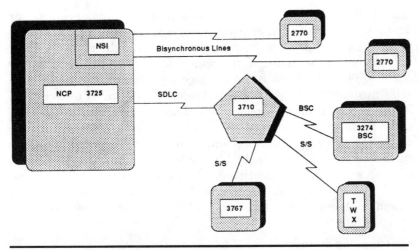

Figure 10.13 An NCP/NSI network configuration.

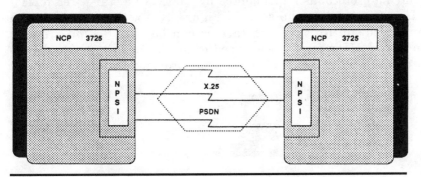

Figure 10.14 An X.25/NPSI network configuration using a Public Switched Data Network (PSDN).

existing communications lines rather than having to pay large installation and maintenance fees for private circuits.

NPSI, executing with the NCP in the communications controller, provides the interface and link procedures needed for transmission of data through the public switched data network (Figure 10.14). It does this by creating a packet and then affixing the X.25 linkage headers. If your company is either planning to use a PSDN or looking at other cost-effective communications, serious consideration should be given to X.25 NPSI.

10.3.4 SDLC

Of all the communications link protocols being used today, SDLC for the greatest part has become the industry standard for network link protocols. In part because of the emergence of SNA, SDLC gives the user improved problem determination capabilities over the previous protocols, BSC and S/S. Also, SDLC is a more efficient protocol for data transmission and transmission recovery. Its ability to transmit data with great integrity at high speeds has brought it to the forefront of telecommunications. If the opportunity arises to redesign or create a new network, indicate to management the uses of SDLC and how it is incorporated into SNA. Remember, any SNA device uses SDLC link protocol, and therefore no emulation program is required for the device to access the network.

We have seen how the various telecommunications line requirements are affected by the devices attached to the lines and how these different device types may require additional software in order for them to be supported. But also to be considered is the operating system software, the communications system software, and the subsystem software.

10.3.5 Token-ring protocol

Token-ring protocol as implemented by IBM uses a star-wired topology. This topology can be seen in Figure 10.15. Depending on the wiring used, up to 255 workstations can be installed on a single token-ring LAN segment. The speed of the token-ring is also dependent on the wiring used and the support provided by the adapters on

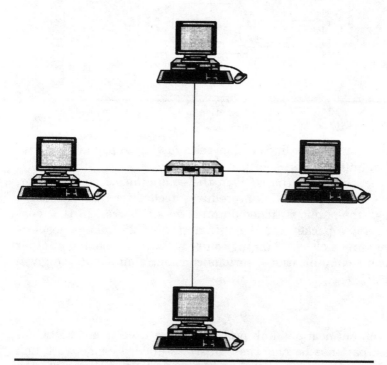

Figure 10.15 Star-wired topology for token-ring networks.

the token-ring LAN segment. Token-ring carries some of the inherent single-threadedness found with SDLC. In token-ring networks, only one station on the ring at any given time has the right to deliver data to a destination. This feature provides for a structured management approach to managing a token-ring network.

The bridging and/or routing of token-ring networks can connect several isolated LAN segments and create a single large token-ring network. Source route bridging is used for many of the protocols that run over a token-ring network. Such a restraint on the protocol allows a session between workstations on token-ring to span only eight token-ring LAN segments. This constraint must be thought out as your token-ring network grows.

10.3.6 Ethernet protocol

Ethernet LANs use a bus topology architecture. A bus topology is shown in Figure 10.16. In a bus topology, any workstation can send data to any other workstation at any time. Hence, there are instances when many stations send data on the bus at the same time. This creates collisions on the bus and leads to lost data. Ethernet uses CSMA/CD to identify and circumvent collisions on the bus.

Ethernet is not constrained by the number of LAN segments between workstations as token-ring is. Ethernet, however, is more distance-constrained. The specific wiring used and the speed at which the Ethernet is driven will determine the number of workstations on a segment.

Network designers need to understand Ethernet LAN in an enterprise network because of the need for communications between all entities of a company. Ethernet LANs are mostly found at installations where the main application is CAD/CAM/CAE. However,

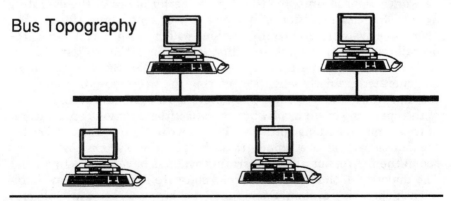

Bus Topography

Figure 10.16 Bus topology for Ethernet LANs.

Ethernet has gained ground on token-ring as the LAN of choice for corporate America because of its low cost for wiring and adapter cards.

10.4 Software

All software in an effective system must be carefully selected and evaluated. The different levels of software available may not provide the support needed for you to implement your network. The pieces to the puzzle must fit in such a way that proper support is maintained. As you will see later on, many different levels of software may or may not provide the support you design for your network.

10.4.1 Operating system software

Much consideration has to be given to the operating system that controls the main processor. Any channel-attached communications controller or cluster controller or any EP devices must be defined to the operating system. If an EP device is used in the network, as a requirement, the communications controller that defines that EP device must be attached to a byte multiplexer channel. A byte multiplexer channel sends data to the main processor 1 byte or multiple bytes at a time. This yields a maximum transfer rate of 1.5 MB/s. The importance of the transfer rate is related to the processing speed of your channel-attached front-end processor. The newest communications controllers can execute 1 million instructions/second (mips), and to let the controller lie idle at that instruction speed is not cost-effective. However, if your network configuration will allow for the front-end processor to be channel-attached via a block multiplexer channel, the data transfer rate may reach a maximum of 4.5 MB/s. The size of the block that is transferred is dependent on application device buffer sizes. This high volume of data transfer is important not only for the network response time, but also for the performance of the operating system. So, consideration of network device types can influence the operating system's I/O configuration, which in turn can affect the overall response time and throughput of the operating system.

The three major operating systems, MVS, DOS/VSE, and VM, must be generated properly when considering the telecommunications system hardware and software. The different versions and release levels of the operating system software can affect the functions and features of the communications software. The chart in Figure 10.17 describes the functions that are unique to each operating system. As you can see in the figure, not all the operating systems have unique functions. The majority of the communications software functions are available to each operating system. Some functions, however, are available only

FUNCTION	MVS	MVS/XA	VSE	VM
31-Bit Addressing		*		
Direct Link-Attached Devices, Hosts, NCPs				*
VSCS Display Capability FORCE Command Internal Trace				*
Extended Recovery Facility (XRF)		*		
Programmable Operator Message Exchange (PMX)				*
Symptom String Subset		*		

Figure 10.17   Unique function support for operating systems.

if the proper version and release of related communications software is installed.

10.4.2   Communications system software

During the development of VTAM, each new version and release provides an increase in function of different VTAM components as well as the addition of new ones. Associated with this growth is the natural creation of communications network management (CNM) software. As VTAM's capabilities increased, so too did the abilities of management software. We should point out to you that although VTAM executes under the MVS, VM, and VSE operating systems, some functions of VTAM do not cross operating systems. You as the communications systems programmer must be aware of the differences between the versions of VTAM, NCP, and CNM products and how they differ between operating systems. Figure 10.18 illustrates the various components of VTAM.

FUNCTION	VTAM					NCP						
	V2.2	V3.1	V3.2	V4.1	V4.3	V4.1	V4.2	V4.3.1	V5.2	V5.2.1	V6.2	V7.1
CTC attached hosts	●	●	●	●	●							
ENA (MVS,MVS/XA)			●	●	●	●	●	●	●	●	●	●
ENA (VSE,VM)			●	●	●	●	●	●	●	●	●	●
SNI (MVS, MVS/XA)	●	●	●	●	●	●	●	●	●	●	●	●
SNI (VM)			●	●	●	●	●	●	●	●	●	●
APPN Border Node					●						●	●
XRF (MVS,MVS/XA)		●	●	●	●	●	●	●	●	●	●	●
Dynamic TABLE reconfiguration			●	●	●		●	●	●	●	●	●
Dynamic PATH reconfiguration			●	●	●		●	●	●	●	●	●
Dynamic LU definition				●	●		●	●	●	●	●	●
PORT SWAP (37X5,3720)								●	●	●	●	●
Dial-up SDLC subarea links								●	●	●	●	●
Mixed Multipoint line support								●	●	●	●	●
Token-ring subarea links								●		●	●	●
Frame relay terminal equipment											●	●
Frame relay switch equipment												●
TCP/IP Support											●	●

Figure 10.18   Some functional components of VTAM/NCP in relation to version and release levels.

For instance, there are differences between VTAM versions executing on an MVS system as compared to a VM system. An example is SNI, which was supported in an MVS system from version 2, release 2 of VTAM but was not incorporated for a VM operating system until VTAM version 3, release 2. This, however, is dependent upon the level of NCP executing in the communications controller. In order for SNI to function, the NCP must be, at a minimum, version 3. So you can see how this software-level dependency can cascade. This software compatibility not only affects the communications software functions but can have a great effect on the subsystem applications that request these functions.

10.4.3   Subsystem software

The final piece of the software puzzle is the application software. The different types of application software can and will affect the performance and makeup of the network. One of the most popular interactive subsystems being used today is IBM's Customer Information and Control System (CICS). This software package has become the industry standard for use in information retrieval and updating. Under CICS, a company may execute several different applications at the same time, all sharing the same address space in the host computer. Accounts receivables, name and address applications, inventory, and order entry applications may all execute under the same CICS subsystem. Two other very popular database software products may also execute under CICS. One is IBM's Information Management System (IMS), the other is Cullinet's Information Database Management System (IDMS). A coordinated effort by both the communications systems programmer and the CICS and/or the database systems programmers will result in smooth network implementation and a satisfactory performance curve for the end user.

One other major software package that is a primary development tool for most computer shops is IBM's Time Sharing Option (TSO). This software facilitates the development and systems programmers' throughput by providing increased productivity for maintenance and development by using TSO's powerful commands and command list facility. A long with its own features, TSO has an extensive offering of programming tools that can execute in the TSO subsystem. Some of these are ISPF/PDF, a menu-driven development facility; on-line compiler program products for COBOL, FORTRAN, and PL/I; a graphics display product called GDDM; and many more.

10.5 Summary

This chapter has covered the basis for network design. This included the analysis of user requirements and its effect on network requirements, the research needed on device types and how they will interface with the network, and the interdependency of the software levels and the effect on available functions to complete the network design. All these points are necessary in order to evaluate and implement your network. There are, of course, many other variables that need to be discussed. We will bring them to your attention as we proceed to the next chapters, which concern the actual coding of the VTAM and NCP parameters that are needed to map a network.

VTAM

Defining a
Single-Domain Network

In this chapter we will lay the foundation for all VTAM network definitions, including multiple-domain and SNI networks. No matter what type of configuration your network may have, all networks consist of a base design. That basis is the definition of a single-domain network. But, before we start coding the VTAM definition statements, we must introduce you to the format of definition statements used in this book and the "sift-down" effect.

11.1 Definition Statement Format and Sift-down Effect

All definition statement formats in this book will follow standard assembler language rules (Figure 11.1). The NAME field assigns a symbolic name that identifies the definition statement or minor node.

The NAME field has the following format:

1. The field must be one to eight characters.

1 - 8	10 - 17	19 - 71
NAME	DEFINITION OPERANDS	STATEMENT

Figure 11.1 Definition statement format.

2. The first character must be uppercase alphabetic (A to Z) or the national characters @, #, or $.

3. The second to eighth characters must be uppercase alphabetic (A to Z), numeric (0 to 9), or the national characters @, #, or $.

4. The field must start in column 1 of the definition statement and must be followed by at least one blank.

The DEFINITION STATEMENT field identifies the definition statement. It must be preceded and followed by one or more blanks.

The OPERANDS field contains the required and optional definition statement operands. Some of these operands are called keyword operands. They are denoted by being followed by an equal (=) sign which is followed by the keyword value or values. Required keywords will be identified in **bold type,** the default operands will be underlined, and the optional keywords will be surrounded by brackets ([ ]). If more than one keyword is possible, the "or" bar (|) will be used.

All definition statements are coded in columns 1 through 71 of a statement. The continuation of a statement is accomplished by placing a nonblank character in column 72. The continued part of the statement must begin in column 16 of the following card. Comments can appear after the last operand field or by placing an asterisk (*) in column 1 of the statement.

The sift-down effect (Figure 11.2) relieves the communications systems programmer from repetitive operand coding. This is done by coding the operand on a higher-level node, and the following lower-level nodes will use this value. However, the lower-level node may override the sifted value by coding the same operand, and this value will be used instead.

```
GROUP      USSTAB=USSSNA
 LINE
 PU
    LU     (uses USSSNA)
 PU
    LU     USSTAB=USSRMT      uses USSRMT
    LU     (uses USSSNA)
 PU        USSTAB=,
    LU     (uses IBM supplied default)
    LU     USSTAB=USSRMT      uses USSRMT
    LU     USSTAB=USSSNA      uses USSSNA
```

Figure 11.2   Example of sift-down effect.

11.2 VTAM Start Options List ATCSTR00

The start options list in VTAM is used at initialization time to define to the SSCP the initial topology of the network. The start list is found in a member of the MVS library named SYS1.VTAMLST. This default member name is ATCSTR00. This member is always read at initialization time. It may, however, be overridden by an operator command during start-up. Figure 11.3 lists the start options that should be assessed for a single-domain network.

```
SSCPID=n,
NETID=name,
SSCPNAME=name,
GWSSCP=YES|NO,
[,bufpoolname=(baseno,bufsize,slowpt,F,xpanno,xpanpt)]
[,COLD/WARM]
[,CONFIG=xx|00|name]
[,CSALIMIT=0|n|nK|nM]
[,CSA24=0|n|nK|nM]
[,DLRTCB=n|32]
[,HOSTPU=name of VTAM PU|ISTPUS]
[,HOSTSA=n|1]
[,IOINT=n|180]
[,ITLIM=n|0]
[,LIST=xx]
[,MAXSUBA=n|15]
[,MSGMOD=YES|NO]
[,NODELST=name]
[,PPLOG=YES|NO]
[,PROMPT|NOPROMPT]
[,SONLIM=([m|60][,t|30])]
[,SUPP=NOSUP|INFO|WARN|NORM|SER]
[,TNSTAT[,CNSL|NOCNSL][,TIME=N|60]|NOTNSTAT]
[,TRACE|NOTRACE,ID=nodename,TYPE=BUF|IO|[,EVERY]]
[,TRACE|NOTRACE,ID=linename,TYPE=LINE[,COUNT=n|ALL]]
[,TRACE|NOTRACE,ID=nodename,TYPE=SIT[,COUNT=n|ALL]]
[,TRACE|NOTRACE,ID=VTAMBUF,TYPE=SMS]
[,TRACE,TYPE=VTAM[,MODE=INT|EXT][,SIZE=n|2]
      [,OPTION=ALL|option|(option,option,...,option)]]
[,NOTRACE,TYPE=VTAM]
[,USSTAB=tablename]
```

Figure 11.3 VTAM start options for VTAM.

11.2.1   Required start parameters

The SSCPID=n parameter is used when VTAM constructs a 48-bit identification sequence that is sent to a PU during session establishment using the SNA command ACTPU. The n value is a decimal integer from 0 to 65,535 and must be unique within its own network and interconnected networks.

For our sample network we will use the decimal value 2. Therefore we will code in SYS1.VTAMLST member ATCSTR00

```
SSCPID=2,
```

The NETID=name parameter assigns a network name to this VTAM. It is a required statement for VTAM V3.2 only. In pre-VTAM V3.2, this parameter is optional and is used for specifying gateway SSCP capabilities in an SNI network. The name contains one to eight alphanumeric characters. In VTAM V3.2 it is used as a prefix to all network resource names. This name is used for all VTAMs in a single SNA network.

The SSCPNAME=name parameter is a network-unique name for the SSCP of this VTAM. It is also a required parameter in VTAM V3.2. For pre-VTAM V3.2 networks, the parameter along with the NETID parameter specifies that this VTAM can participate in gateway SSCP responsibilities for cross-network sessions over an SNI network. The name can be one to eight alphanumeric characters in length, and it is highly recommended that the name specified here match the CDRM name assigned to this VTAM when defining a multisystem network. For more on defining a multisystem network, consult *Advanced SNA Networking*.

The final required parameter under VTAM V3.2 is the GWSSCP=YES|NO parameter. This parameter is required because unlike pre-VTAM V3.2, the NETID and SSCPNAME parameters do not denote gateway SSCP capabilities. It is used specifically for SNI networks. More information on SNI can also be found in *Advanced SNA Networking*.

11.2.2   Buffer pool start options

These options define to VTAM the buffer pool allocations to use for the holding of data and building of control blocks. The format of the buffer pool start option is

```
bufpoolname=(baseno,bufsize,slowpt,F,xpanno,xpanpt)
```

The *bufpoolname* value is one of the names listed in Figure 11.4. It identifies which buffer pool the following options apply to. The *baseno*

POOL NAME	USAGE
CRPLBUF	Request parameter list (RPL) copy pool in pageable or virtual storage
IOBUF	Input/output message pool in fixed storage
LFBUF	Large buffer pool in fixed storage (also serves as message pool)
LPBUF	Large buffer pool in pageable or virtual storage
SFBUF	Small buffer pool in fixed storage
SPBUF	Small buffer pool in pageable or virtual storage
WPBUF	Message-control buffer pool in pageable or virtual storage

Figure 11.4   VTAM buffer pools for MVS.

option defines the initial number of buffers VTAM is to allocate at start-up. The range is 1 to 32,767. The *bufsize* option defines to VTAM in bytes the size of the buffer being defined. This value is applicable only to the fixed-storage message pools IOBUF and LFBUF. It is ignored for all other buffer pools. It is a requirement that this *bufsize* value match the UNITSZ operand of the HOST statement defined in the resource definition list for a channel-attached NCP definition. If these two values do not match, the NCP will not be activated and VTAM will issue an error message. The *slowpt* value will determine when VTAM is to perform slowdown mode. Slowdown mode occurs when the number of available buffers in this buffer pool is equal to or less than the value specified for *slowpt*. VTAM will then honor only priority requests, such as reads from a channel-attached device, until the available buffer number rises above the *slowpt* value. Note that if you choose a *slowpt* value that is too high, VTAM may never come out of slowdown mode, thus affecting network performance tremendously.

The *F* stands for fixed. It defines to VTAM that this buffer pool is to be fixed in storage on our MVS system. In other words, the buffers are always resident and in the CPU memory rather than being paged from storage every time a buffer is requested. This greatly increases your performance specifically for the IOBUF pool. The *xpanno* value is the amount of buffers VTAM will expand this buffer pool if the initial allocation from the *baseno* value has been exhausted and all buffers are allocated. This is performed by VTAM dynamically. If, however, a value of 0 is coded, VTAM will not perform dynamic buffering, which may very well lead to slowdown mode. The range for this value is 0 to 32,767. The value you pick is rounded upward to the nearest whole page of storage. A page of storage is 4096 bytes. If your value for expansion causes the number of buffers times the *bufsize* to go over one page by only a few bytes, VTAM will allocate two pages. Therefore, care should be taken when estimating the expansion number. The *xpanpt* value is directly related to *slowpt* and *xpanno*. This value defines to VTAM at what point of free buffers for this pool should the buffer pool be expanded to avoid slowdown mode. This value should be larger than *slowpt*.

The buffer pool of most concern is the IOBUF. We will code an appropriate size for any VTAM system before capacity analysis has been determined. The remaining pools will be defaulted. Figure 11.5 shows ATCSTR00 with the buffer pool defined.

11.2.3 CONFIG start option

This option identifies to VTAM the location of the list of major nodes to be activated at initialization time. The format is

```
CONFIG=xx|00|name
```

The xx value is a one- to two-alphanumeric-character value that is appended to the SYS1.VTAMLST member named ATCCONxx. VTAM upon start-up will access ATCCONxx for a list of major nodes that should be made active to VTAM. If xx is coded, this value will override the default appendage of 00. The member ATCCON00 is always accessed by VTAM unless xx is coded or the operator overrides the list with the CONFIG option during VTAM start-up. The name value specifies the three- to eight-character file name (DD name) of the VSAM configuration restart file. This file contains a list of all the major nodes that were active and the dynamic recovery data set files and the PATH statements that were applicable at the time of failure or deactivation of VTAM. The name value should be coded on the NODELST option also to include updating of major node activation after VTAM is active. The file must have been used prior to failure or deactivation for the configuration restart facility to execute. If the file is empty, VTAM does not activate any major nodes during initialization.

```
SSCPID=2,                                                    *
IOBUF=(32,256,4,F,16,8),                                     *
```

Figure 11.5 ATCSTR00 after defining buffer pools.

```
SSCPID=2,                                           *
IOBUF=(32,256,4,F,16,8),                            *
COLD,CONFIG=01,                                     *
```

Figure 11.6 ATCSTR00 for MVS after defining CONFIG start option.

In our sample start list we are going to append the value 01 to the ATCCONxx member name. Figure 11.6 now shows the start list after adding the CONFIG start option.

11.2.4 CSALIMIT start option

This option defines to VTAM the maximum amount of the MVS operating system Common Service Area (CSA) that VTAM will use. The format for this option is

```
CSALIMIT=0|n|nK|nM
```

The 0 value is the default value and tells VTAM that there is no limit to the amount of CSA it can use. In an MVS/370 system the maximum is 16 million bytes (16 MB), and in an MVS/XA system the maximum is 2 billion bytes, or 2 gigabytes (2 GB). Taking the default could seriously impede the operating system if VTAM were to take advantage of being nonconstrained from CSA usage. The $n \mid nK$ value denotes the number of 1000-byte (1 kB) areas of storage in decimal that VTAM will use as the CSA maximum. This value is rounded to the next multiple of 4. The nM value is the number of 1-million-byte (1 MB) areas of storage VTAM will use as the CSA limit. If just the n value is coded, kilobytes are assumed. If you code a value that is greater than the real available CSA storage, no limit is used. If a CSALIMIT is coded, you should carefully define the LPBUF values so there is no buffer expansion. If LPBUF must be expanded but the CSALIMIT is reached, VTAM may enter a locked condition and cause messages to be lost or session initiation and termination failures.

After conferring with the MVS systems programmer for our sample network, we are going to code a CSALIMIT of 512 kB. Figure 11.7 shows the start list after adding the CSALIMIT start option.

11.2.5 CSA24 start option

Prior to the MVS/XA operating system, a 24-bit addressing scheme was in use with the MVS operating system, allowing the highest address of storage to be 16 MB. With the 31-bit extended storage

```
SSCPID=2,                                                    *
IOBUF=(32,256,4,F,16,8),                                     *
CONFIG=01,CSALIMIT=512,                                      *
```

Figure 11.7 ATCSTR00 after defining CSALIMIT start option.

addressing scheme provided by MVS/XA, the highest available storage address was moved up to 2 GB. Because of the VTAM architecture for such VTAM elements as IOBUF, some VTAM code must execute in the 24-bit addressing mode. MVS/XA has provided an area of CSA storage below the 16-MB line for programs to use the CSA in 24-bit addressing mode. This start option determines the amount of CSA storage VTAM may use below the 16-MB line. The format for this option is

```
CSA24=0|n|nK|nM
```

The 0 value is the default value and tells VTAM that there is no limit to the amount of CSA it can use. The maximum is 16 million bytes (16 MB). The default for CSA24 is usually taken. This is caused by the new structure of VTAM's buffers. In VTAM V3, only the IOBUF pool is below the 16-MB line. All the other pools are above the line. The n I nK value denotes the number of 1000-byte, or 1 kB, areas of storage in decimal that VTAM will use as the CSA maximum. This value is rounded to the next multiple of 4. The nM value is the number of 1-million-byte or 1 MB areas of storage VTAM will use as the CSA limit. If just the n value is coded, kilobytes are assumed.

In our sample network, the operating system is MVS/XA, and we will be taking the default value of 0. It is not necessary to code the option, but for documentation purposes we will enter it into the start list. Figure 11.8 shows the start list after the addition of the CSA24 option.

11.2.6 HOSTPU start option

Remember that VTAM's SSCP is a PU Type 5 and like other PUs in the network is given a name. This option allows the communications

```
SSCPID=2,                                                    *
IOBUF=(32,256,4,F,16,8),                                     *
CONFIG=01,CSALIMIT=512,CSA24=0,                              *
```

Figure 11.8 ATCSTR00 for MVS/XA after defining CSA24 start option.

systems programmer to assign a name to VTAM's PU that should be network unique. It is highly recommended that you use this option to specify a unique PU name for ease of debugging purposes or if you are using the IBM program products NetView or NLDM. This name should not be the same as this VTAM's cross-domain manager name or the SSCP name assigned to this VTAM if you have a multidomain or SNI network. The format for this start option is

```
HOSTPU=VTAM PU name|ISTPUS
```

If the VTAM PU name or this option is not coded, the default name ISTPUS will be used. For our network we will assign the name VTAM01 to the VTAM PU.

11.2.7   HOSTSA start option

This option assigns the subarea number to this VTAM. This number is between 1 and 255 decimal. In VTAM V3.2+, with the help of a Program Temporary Fix (PTF), this value can range from 1 to 65,535. It is a good idea to always code the subarea number for VTAM. This number is the unique address for this VTAM in a network and is important for multidomain or SNI networks. Any duplication of sub-area numbers in a multidomain network will result in one of the sub-areas not being activated. The format is

```
HOSTSA=n|1
```

The default value is always 1 if the option is not coded. The n is the number you assign to the VTAM subarea.

For our network, we will use the default value of 1 as the subarea number for VTAM. Again, for documentation purposes, we will code the option in the start list as is shown in Figure 11.9.

11.2.8   MAXSUBA start option

It is important for us to keep in mind the direct relationship of the different version levels of VTAM and NCP and how they relate func-

```
SSCPID=2,                                                        *
IOBUF=(32,256,4,F,16,8),                                         *
CONFIG=01,CSALIMIT=512,CSA24=0,HOSTPU=VTAM01,                    *
HOSTSA=01,                                                       *
```

Figure 11.9 ATCSTR00 for MVS/XA after defining HOSTPU and HOSTSA start options.

tionally to each other. For this option, if we are using VTAM version 3 or higher, we have the capability of using the Extended Network Addressing facility (ENA). However, ENA can be used only if (1) all other VTAMs in the network are version 3 or higher and (2) the version of NCP is version 4 or higher. If either one of these factors is not found in your network, you should code the MAXSUBA option. The one requirement when coding MAXSUBA is that all VTAMs in the network must specify the same MAXSUBA value. See the discussion on ENA in Chapter 5 for more information on ENA and non-ENA addressing. Remember, in a non-ENA network this value dictates the number of resources addressable in the network. The format is

```
MAXSUBA=n|15
```

where n is the value of the maximum subarea defined in the network. The range for this value is 3 to 255. The default value is 15, and for our sample network this value will give us more than enough addressable resources. But, since our single-domain network consists of VTAM V3 and NCP V4 or V5, the MAXSUBA option is not needed, since VTAM will be using extended network addressing.

11.2.9   NODELST start option

This option defines the file or DD name VTAM is to open to maintain a list of currently active major nodes. This list is accessed if the CONFIG option is coded with the name operand. If the names on the CONFIG and NODELST options match, VTAM will activate all the major nodes that were active prior to failure or deactivation of the network. This NODELST data set is updated whenever a major node is activated or deactivated. This assists in recovery of major nodes that need to be active after start-up from a failure. If the name operands of the CONFIG and NODELST options differ, the NODELST data set is erased before any major nodes are activated. Since we have already chosen the CONFIG=01 option, we need not use the NODELST option for our sample network start list.

11.2.10   PPOLOG start option

This option is valid only for VTAM V3R1.1 and higher. It refers to the primary program operator log that interfaces with VTAM to capture VTAM commands and messages. The format is

```
PPOLOG=YES|NO
```

The default NO means that VTAM commands, messages, and responses entered on the systems' console will not be recorded on the primary program operator's log. The value of YES means that recording will be in effect and also that NetView is executing on the system, since this option is also only valid if NetView is installed on the system. For our sample network in its early stages, we will use the default value of NO.

11.2.11   TNSTAT start option

In the beginning of this chapter we briefly discussed the different buffer pools VTAM uses. This option tells VTAM to keep records of the buffer pools and the amount of CSA storage that is being used during VTAM's execution. This is the only way you can track the buffer pool and CSA usage. If TNSTAT is not specified at start-up time, it is not possible to keep statistics on these storage uses. The format is

```
TNSTAT[,CNSL|NOCNSL][,TIME=n|60]|NOTNSTAT
```

The default value is NOTNSTAT and is not recommended. If you take this default, you will not be able to start tuning statistics at a later time by using the operator MODIFY command. However, if TNSTAT is coded, you may stop the tuning statistics at a later time and restart them if you wish. The CNSL operand denotes recording of tuning statistic records to the system console as well as the MVS Systems Management Facility (SMF). NOCNSL specifies recording only to the SMF data set. The TIME operand specifies the time interval for recording the tuning statistics records. The default is 60 minutes, which can be modified using the MODIFY command if greater determination of storage usage is desired. For our network we will use the TNSTAT,CNSL,TIME=60 option.

11.2.12   TRACE start option

It is our recommendation that the TRACE options not be used and allowed to default to NOTRACE. This is because of the heavy burden placed upon VTAM to trace. The fact is that trace should only be turned on if it is needed to perform problem diagnosis. This can be accomplished by using the operator MODIFY command. In our start list, all traces will be allowed to default to NOTRACE. Figure 11.10 is the completed ATCSTR00 start list for our sample VTAM single domain.

11.3   APPN Start Options

VTAM V4R1 is the first version of VTAM to have full APPN function support. The new start options can be grouped into the various

```
SSCPID=2,                                                       *
IOBUF=(32,256,4,F,16,8),                                        *
CONFIG=01,CSALIMIT=512,CSA24=0,HOSTPU=VTAM01,                   *
HOSTSA=01,ITLIM=0,PPOLOG=YES,PROMPT,                            *
TNSTAT,CNSL,TIME=60
```

Figure 11.10 ATCSTR00 for MVS/XA pre-VTAM V3.2 after defining all the appropriate start options. Note the absence of the continuation character in the last line, denoting the end of the start list.

VTAM V4R1 APPN Specific Start Options		
NODETYPE	CPCP	SORDER
SSEARCH	CDSERV	DIRSIZE
DIRTIME	APPNCOS	RESUSAGE
RESWGHT	ROUTERES	CONNTYPE
DYNADJCP	VERIFYCP	NQNMODE

Figure 11.11 ATCSTR00 VTAM start options needed to define APPN support in VTAM.

requirements for VTAM to support an SNA subarea network with an APPN network and to provide the functions of directory services, central directory server, connectivity, and topology and route selection services. Figure 11.11 lists the APPN VTAM start options.

11.3.1 NODETYPE start option

The NODETYPE start option indicates whether VTAM will take on the role of an APPN NN or an APPN EN. The NODETYPE start option in conjunction with the HOSTSA start option will dictate the type of migration functions VTAM will provide. The values available for this start option are NN for network node and EN for end node. If the NODETYPE is set to NN and the HOSTSA start option is specified, then the VTAM will function as an interchange node. When the NODETYPE start option is defined with a value of EN and the HOSTSA start option is defined, then the VTAM will operate as a migration data host. The VTAM will operate as a pure NN or EN when the HOSTSA start option is not defined.

11.3.2 CPCP start option

The CPCP start option indicates whether the VTAM node and its attached APPN nodes will support CP-CP sessions. The CPCP start option sets a networkwide default for all adjacent APPN nodes attached to this VTAM. The CPCP start option can be defined with the value YES, NO, SWITCHED, or LEASED. CPCP=YES is the default when VTAM V4R1 is defined as an EN, NN, ICN, or MDH node. CPCP=LEASED is always the default for all other configurations. The value coded in the start options is used for all node connections. Switched connections default to CPCP=NO. The CP-CP session capabilities of APPN nodes can also be controlled by the CPCP parameter definition found on the resource's GROUP, LINE, or PU definition statement. If the CPCP parameter is not coded here, then the CPCP value defined in the start options is used.

11.3.3   SORDER start option

In an APPN VTAM node, resource searches must be conducted for both APPN and SNA subarea resources. The SORDER start option determines which search method is to be used first when searching for a resource. The SORDER start option can be defined with the APPN, SUBAREA, or ADJSSCP value. SORDER=APPN is the default definition and indicates to VTAM that the APPN Locate request should be used and exhausted prior to the traditional SNA subarea CDINIT/DSRLIST commands. Recall from Chapter 6 that this dual functionality is found only in an interchange node. If SORDER=SUBAREA is defined, then VTAM will issue CDINIT/DSRLIST commands prior to trying APPN Locate requests. The last option, ADJSSCP, is used when the ICN is part of a multidomain SNA network. In the VTAM's adjacent SSCP table that is created for a multidomain configuration, the generic name of ISTAPNCP can be coded to represent the APPN CDRM. If SORDER=ADJSSCP is coded and ISTAPNCP is found in the search, the ICN will issue an APPN Locate request. If SORDER=ADJSSCP is defined, then ISTAPNCP must be defined in the appropriate places in the ADJSSCP table. When SORDER=APPN is defined, VTAM will put ISTAPNCP at the top of each ADJSSCP table. Finally, VTAM will place ISTAPNCP at the end of each ADJSSCP table when the SUBAREA value is defined. By coding SORDER=ADJSSCP and omitting ISTAPNCP from the ADJSSCP tables, you can prevent VTAM from forwarding subarea searches to the APPN part of the network.

11.3.4   SSEARCH start option

The SSEARCH start option provides greater control of APPN-to-subarea routing. A VTAM ICN uses this option to control how the search from APPN to the subarea is to be done. A value of YES indicates that the search is to use cross-domain subarea routing. YES is the default value for SSEARCH. The default indicates that the search will take place as indicated by the SORDER value. A value of NO tells the ICN that it is not to use the subarea network, and the CACHE value indicates that the ICN can route to a cross-domain subarea only if VTAM knows from the cache that the target resource is an SNA subarea resource.

11.3.5   CDSERVR start option

VTAM will initialize as a Central Directory Server (CDS) if the CDSERVR start option is coded with a value of YES and the NODE-TYPE is defined as NN. You can define VTAM as a NN but not have it defined as a CDS by leaving the CDSERVR option out of the start list or by specifying the default of CDSERVR=NO.

11.3.6   DIRTIME, DIRSIZE, and INITDB start options

VTAM V4R1 requires the ability to dynamically manage the directory of resources for the APPN network. To do this, VTAM V4R1 includes three new data sets. The SYS1.DSDBCTRL file contains information that identifies the APPN directory database to load at initialization. The other two file names SYS1.DSDB1 and SYS1.DSDB2, are data files of APPN directory information used for checkpointing APPN directory resources. The DIRTIME and DIRSIZE start options dictate the deletion of entries and the size of these files. DIRTIME is the number of days a resource has remained in the files without being referenced. The default for DIRTIME is 8, and it can range from 1 to 15. The DIRSIZE value specifies the maximum number of resources that can be contained in the files. The default value is zero, indicating that there is no limit to the number of resources in the files.

Entries in the two directory files are dynamically updated through normal network searches. But when the DIRSIZE value is reached, the directory files reduce their size by deleting the least recently used resources. The files are also dynamically updated when the DIRTIME value is reached and the entries remaining in the directory files are searched for a time stamp that indicates that the entry is equal to or older than the DIRTIME value.

The INITDB start option specifies whether VTAM at initialization should automatically reload the topology and directory databases. The topology database is a new file for VTAM V4R1 and is generically named SYS1.TRSDB. This topology database is used to hold network topology data whenever a VTAM MODIFY, CHKPT, HALT, or HALT QUICK operator command is issued. The default for INITDB is ALL, indicating that the three dynamic databases discussed here for directory services and topology and routing services are to be reloaded into

VTAM data areas at VTAM initialization. A value of INITDB=TOPO means that only the topology database is to be reloaded, and a value of DIR means that only the directory services databases are to be reloaded.

11.3.7 APPNCOS start option

With the value found on the APPNCOS start option, VTAM can substitute a COS for unknown APPN COS names specified on a session request. APPNCOS=NONE is the default value, indicating that no substitution is provided. If the APPNCOS start option identifies a COS name, then this is the name of the COS to be used in place of the unknown COS. Figure 11.12 gives a table that indicates the COS that will be used for APPN and subarea COS.

11.3.8 RESUSAGE, RESWGHT, and NUMTREES start options

Routes that have been calculated are stored in cache as routing trees. When these routing trees are determined to be the best route for many sessions, the route and its resources may be overused. The RESUSAGE, RESWGHT, and NUMTREES start options are used to evenly distribute the sessions over equally weighted paths.

The RESUSAGE option is used to indicate the number of times a resource may be used for routing before it is considered to be overused. The default value is 100. If the distribution of sessions over the equally weighted paths proves to be ineffective, then the RESWGHT start option may be used. The RESWGHT option is defined as a percentage that indicates the amount by which an overused resource's weight should be increased every time the resource reaches the RESUSAGE limit. The value for RESWGHT can range from 1 to 100. The NUMTREES start option limits the number of routing trees found in the TRS cache. The default value is 100. When the NUMTREES value is reached, the oldest unused routing trees are deleted. The value for NUMTREES can range between 2 and 10,000.

11.3.9 ROUTERES start option

This start option is used by VTAM when it is defined as an NN. It adds a route resistance value to the characteristics of the VTAM APPN NN. The value specified here identifies VTAM's desire to perform intermediate session routing services. This will be used during CP-CP session initiation to indicate to partner CPs that VTAM is at a certain level of congestion. This characteristic will be taken into account when NNs calculate routes that may pass through VTAM

Mode	APPNCOS	COS	Description
A	X	Y	APPN COS X is used in the APPN part of the network. Subarea COS Y is used in the subarea part of the network.
B	X	None	APPN COS X is used in the APPN part of the network. The default COS (8 blanks) is used in the subarea part of the network.
C	None	None	The default COS (#CONNECT) is used in the APPN part of the network. The default COS (8 blanks) is used on the subarea part of the network.
D	None	Y	COS Y is used in both the subarea and APPN parts of the network.

Figure 11.12 The various combinations and usage by VTAM of APPNCOS and COS specified values.

APPN NNs. The lower the value, the more desirable the node is for intermediate session routing. A value of 128 is the default, with a possible range of 0 to 255. This start option is valid only when NODE-TYPE=NN is specified.

11.3.10 CONNTYPE start option

The CONNTYPE start option specifies whether the Node Type 2.1 connections established by VTAM are to be APPN connections or LEN connections. Again, this start option is valid only when the NODETYPE start option is defined. The APPN value is the default and indicates that the T2.1 connection supports parallel transmission groups, CP-CP sessions, and CP name change support. The APPN value also means that the links established are reported to the topology database. The VTAM start option value for CONNTYPE can be overridden by the CONNTYPE parameter on the GROUP, LINE, or PU definition that defines the T2.1 node being connected to VTAM.

11.3.11   CACHETI start option

The value for the CACHETI start option specifies the amount of time that VTAM will keep procedure correlation identifications. The value is specified in minutes. The default is 8 minutes.

11.4   Summary

In this chapter we discussed the fundamentals of single-domain definitions to VTAM. We learned about the sift-down effect when coding the VTAM macros for network definition. The VTAM start options list, coded in the ATCSTR00 member of the SYS1.VTAMLST library, helps SSCP in defining the initial network topology at initialization time. We also learned how to define VTAM as an APPN NN, an EN, and an interchange node and migration data host. In the next chapter, we will learn about VTAM major nodes.

12

VTAM Major Nodes—
Application, Non-SNA, and SNA

12.1   Application Program Major Node

During our research of the topography of the network, we discussed
some of the applications and subsystems that may be employed on
the host computer. These interactive programs use VTAM functions
to communicate with the end user. The interface that the application
will use to VTAM and consequently to the end user is defined in the
application program major node. This major node differs from our
previous discussion on nodes in Chapter 5. In that discussion we were
concerned with the nodes of an SNA network. In this discussion and
several to follow, the nodes are VTAM symbolic names that represent
resources to VTAM. The major node name is actually the member
name found in VTAM's definition library called SYS1.VTAMLST. The
VTAM major node consists of a set of resources that can be activated
or deactivated by VTAM as a group. These major nodes contain a set
of related minor nodes. The minor node is a uniquely defined resource
within the major node and is also assigned a symbolic name.

   The major node may contain one or more or all of the applications
in your network. It is recommended that an application major node be
coded for each application. This gives you greater control and flexibility in your network when you have to activate or inactivate major
nodes because of problems with the application. This way only the
users on the application experiencing the problem will be affected and
not the entire user community.

12.1.1   VBUILD statement

For every major node defined, a VBUILD statement must appear as
the first definition statement. This statement defines to VTAM the

```
name    VBUILD    TYPE=major node type
                  [,CONFGDS=ddname]
                  [,CONFGPW=password]
                  [,MAXGRP=n]
                  [,MAXNO=n]
```

Figure 12.1 VBUILD statement and applicable operands.

OPERANDS	VBUILD TYPE=							
	APPL	LOCAL	CDRM	CDRSC	CA	DR	SWNET	ADJSSCP
CONFGDS=		•	•	•	•		•	
CONFGPW=		•	•	•	•		•	
MAXGRP=							•	
MAXNO=							•	

Figure 12.2 Some of the VBUILD types and applicable operands.

type of major node VTAM is to build from the minor nodes that follow. Figure 12.1 shows the possible operands in VTAM for a VBUILD statement. The major node type field is any of the column names from Figure 12.2. The purpose of each operand is as follows:

TYPE=APPL defines to VTAM that this is an application major node definition. There are no other optional operands applicable when defining an application major node. APPL is also the default type if the TYPE operand is not coded. However, we have denoted the format with the TYPE operand as a required operand.

TYPE=LOCAL defines to VTAM that this is a local SNA device major node. The applicable optional operands are CONFGDS and CONFGPW.

The CONFGDS operand is the Data Definition statement name (DDname) that is assigned to the configuration restart data set from the VTAM start procedure that is to be used for this major node.

CONFGPW is the password for the VSAM configuration restart data set referred to by the CONFGDS operand. This operand is coded only if CONFGDS is coded. If a password is not supplied here, the VTAM operator is prompted for the correct password whenever VTAM accesses the configuration restart data set. In our example we will not be using the configuration restart data set.

TYPE=CDRM tells VTAM that this major node contains the cross-domain manager definitions for this VTAM. CDRM is discussed in greater detail in *Advanced SNA Networking*. The valid operational operands are CONFGDS and CONFGPW.

TYPE=CDRSC tells VTAM that this major node contains the list of cross-domain resources for this VTAM. The CDRSC is discussed in *Advanced SNA Networking*. The valid operational operands are CONFGDS and CONFGPW.

TYPE=CA tells VTAM that this major node contains the definitions for a channel-attached major node. Examples include a channel-to-channel attachment between VTAMs and a channel-attached NCP major node. Again, the optional operands are CONFGDS and CONFGPW.

TYPE=DR tells VTAM that this major node contains the definitions for dynamic reconfiguration. There are no optional operands for this VBUILD type.

TYPE=SWNET tells VTAM that this major node contains the definitions for a switched communications resource. Besides the CONFGDS and CONFGPW optional operands, the MAXGRP and MAXNO optional operands are also applicable. This major node provides a list of switched resources that can gain access to the network over switched lines.

The MAXGRP operand is coded only if a switched PATH statement is coded. The maximum value is 32,767. This value is the number of unique path group names defined by the GRPNM operand of the PATH statement for this switched major node.

MAXNO is coded only if a switched PATH statement is coded. Its maximum value is 32,767. This value is the number of unique telephone numbers defined by the DIALNO operand of the PATH statement in this switched major node.

TYPE=ADJSSCP tells VTAM that this major node contains the adjacent SSCP table. This table is used by VTAM to search for unknown resources controlling SSCP. We will discuss this major node in greater detail in *Advanced SNA Networking*.

12.1.2 APPL definition statement

One APPL definition statement is required for each application that is to be identified to VTAM (Figure 12.3). The name field is used to assign the minor node name to the application. The assigned name must be unique within a network. This name is also referred to as the applications network name. An example would be CICSP01 or IMSP01.

The ACBNAME=acbname operand is also the minor node name for this application. ACB stands for application control block. This is a VTAM control block that contains all the information VTAM needs to know about the applications characteristics. Again, this acbname

```
name       APPL        [ACBNAME=acbname]

                       [,APPC=YES I NO]

                       [,REGISTER=CDSERVR I NETSRVR I NO]

                        [,AUTH=([ACQ I NOACQ] [,CNM I NOCNM]

                        [,PASS I NOPASS] [,PPO I SPO I NOPO]

                        [,TSO I NOTSO]

                        [,VPACE I NVPACE])]

                       [,DLOGMOD=default logmode name]

                       [,EAS=n I 491 I 509]

                       [,ENCR=SEL I REQD I OPT I NONE]

                       [,HAVAIL=YES I NO]

                       [,MAXPVT=0 I n I nK I nM]

                       [,MODETAB=logon mode table name]

                       [,PARSESS=YES I NO]

                       [,PRTCT=password]

                       [,SONSCIP=YES I NO]

                       [,SPAN=(NCCF or NetView spanname)]

                       [,SRBEXIT=YES I NO]

                       [,SSCPFM=USSNOP I USSPOI]

                       [,USSTAB=name]

                       [,VPACING=n I 0]

                       [,VTAMFRR=YES I NO]
```

Figure 12.3 APPL definition statement format.

must be network unique. If the ACBNAME operand is not coded, the name of the APPL definition statement is used as the ACB-NAME. In our network configuration, we will assign the name to the acbname. Having the APPL statement name and the ACB-NAME the same helps to avoid confusion as to what the application's acbname really is.

The APPC=YES|NO operand is specific to VTAM V3.2. This operand tells VTAM V3.2 that the application defined here may wish to use the basic functions of LU6.2 that VTAM V3.2 supplies. This is denoted by specifying APPC=YES. The default is APPC=NO.

VTAM applications and dependent and independent LUs that are defined to VTAM can be either registered or nonregistered. VTAM determines how the systems programmer wants the resource to be registered through the use of the REGISTER parameter of the definition statement. The REGISTER parameter has three possible values and is valid only with VTAM V4R1 or higher. These are CDSERVR, NETSRVR, and NO. The default for applications is CDSERVR, which indicates to VTAM that the application name is to be registered at VTAM's NN and CDS node servers. If the resource is a dependent LU, then the default is NETSRVR, which indicates that the resource is to be registered at the NN server only. The default for independent LUs is NO and indicates that the resource is not to be registered at the NN or CDS server. Currently only a VTAM V4R1 EN will register resources to a CDS. The REGISTER parameter is valid for the APPL, LU, LOCAL, and CDRSC definition statements.

Let's take a look at Figure 12.4, which illustrates how VTAM activates and uses the various operands defined for an application major node. In Figure 12.4 we see how VTAM correlates the major node name in VTAMLST to an operator activate command for the major node A01TSO. Actually, upon initialization, VTAM has created a Major Node Table (MNT) that contains pointers to the specific ACBs for each application in its domain. Let's look at the minor node definition of each of the listed major nodes in SYS1.VTAMLST.

For each CICS subsystem, only one application statement is necessary, even though the subsystem can have several different applications under its address space (Figure 12.5). Each application task shares CICS resources with the other application tasks executing in

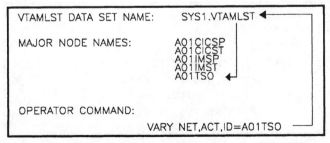

Figure 12.4 Relationship of VTAM operator command to VTAMLST and major nodes.

```
Major Node Name: A01CICSP
Minor Node Name:
                  VBUILD      TYPE=APPL
        CICSP01   APPL        ACBNAME=CICSP01,                    X
                              AUTH=(ACQ,PASS,VPACE),              X
                              MODETAB=MTLU62,EAS=100,             X
                              PARSESS=YES,VPACING=4,SONSCIP=YES
```

Figure 12.5 Application major and minor nodes for a production CICS address space.

the CICS address space. Two of the coded operands indicate this. The EAS operand has defined 100 concurrent estimated active sessions. The PARSESS operand tells us that this subsystem can have multiple sessions with other applications on an LU-LU session. The AUTH operand defines this subsystem to have the ability to acquire an LU (ACQ) and to pass session establishment requests to other applications (PASS) and specifies that this subsystem will adhere to the methodology of VPACING to LUs (VPACE). The SONSCIP operand tells VTAM that it can terminate sessions with the SLU on behalf of the application.

The test CICS example in Figure 12.6 is quite similar to the production CICS example. The only differences are in the naming convention for the application and the estimated number of active sessions. One operand not discussed for Figure 12.6 is the MODETAB operand. This operand defines to VTAM the logon (logmode) mode table VTAM is to search for locating the LOGMODE that is to be used by the application when it participates as an SLU. This mode table contains entries that define session parameters for SLUs. The default

```
Major Node Name:  A01CICST
Minor Node Name:
                  VBUILD      TYPE=APPL
        CICST01   APPL        ACBNAME=CICST01,                    X
                              AUTH=(ACQ,PASS,VPACE),              X
                              MODETAB=MTLU62,EAS=25,              X
                              PARSESS=YES,VPACING=4,SONSCIP=YES
```

Figure 12.6 Application major and minor nodes for a test CICS address space.

LOGMODE operand (DLOGMOD) is where you can specify the default session parameter entry name that is to be used with this application. The session parameters are coded in a LOGMODE entry of the logon mode table in VTAM. These session parameters pertain to the application only when the application is acting as the secondary logical unit, for instance, during an application-to-application session where this application is the SLU. If DLOGMOD is specified, VTAM will search the value given from the MODETAB operand. If the LOGMODE entry is not found in the specified MODETAB, VTAM will select the first entry from the default logon mode table supplied by IBM. If MODETAB is specified and DLOGMOD is not coded, the first LOGMODE entry in the specified MODETAB table will be used. If either DLOGMOD or MODETAB is not coded, the first entry in VTAM's default logon mode table will be used. For more information on session parameters and LOGMODES see Chapter 15.

In both the CICS minor nodes a logon mode table name of MTLU62 has been specified. This tells VTAM to search in the logon mode table named MTLU62 for a LOGMODE entry to use for application-to-application sessions. However, a default LOGMODE (DLOGMOD) was not specified. VTAM will select the first LOGMODE entry in the MTLU62 mode table.

Just like the CICS examples, the IMS production address space can acquire LUs, pass session establishment requests to other applications, and follow the rules for VPACING. However, the IMS subsystem is not expected to be used by as many users as the CICS subsystem. This is apparent from the EAS operand value of 20 (Figure 12.7). However, the application program must handle failure termination, since the operand SONSCIP was not coded and the default is SONSCIP=NO.

We see in Figure 12.8 that the number of estimated active sessions is reduced to 10. Can you determine what the name of the ACB for

```
Major Node Name:    A01IMSP
Minor Node Name:

                    VBUILD      TYPE=APPL
        IMSP01      APPL        ACBNAME=IMSP01,              X
                                AUTH=(ACQ,PASS,VPACE),       X
                                MODETAB=MTLU62,EAS=20,       X
                                PARSESS=YES,VPACING=4
```

Figure 12.7 Application major and minor nodes for a production IMS address space.

```
Major Node Name:   A01IMST
Minor Node Name:

                   VBUILD    TYPE=APPL
       IMST01      APPL      AUTH=(ACQ,PASS,VPACE),        X
                             MODETAB=MTLU62,EAS=10,        X
                             PARSESS=YES,VPACING=4
```

Figure 12.8 Application major and minor nodes for a test IMS address space.

```
Major Node Name:   A01TSO
Minor Node Name:
                   VBUILD  TYPE=APPL
       TSO01       APPL  AUTH=(PASS,NVPACE,TSO),           X
                         EAS=1,ACBNAME=TSO
       TSO01001    APPL  AUTH=(PASS,NVPACE,TSO),           X
                         EAS=1,ACBNAME=TSO0001
       TSO01002    APPL  AUTH=(PASS,NVPACE,TSO),           X
                         EAS=1,ACBNAME=TSO0002
       TSO01003    APPL  AUTH=(PASS,NVPACE,TSO),           X
                         EAS=1,ACBNAME=TSO0003
       TSO01004    APPL  AUTH=(PASS,NVPACE,TSO),           X
                         EAS=1,ACBNAME=TSO0004
       TSO01005    APPL  AUTH=(PASS,NVPACE,TSO),           X
                         EAS=1,ACBNAME=TSO0005
```

Figure 12.9 Application major and minor nodes for the TSO subsystem.

this application is? If you remember our discussion on the ACBNAME operand, the acbname value will be taken from the network unique name (the name of the APPL statement) for this application minor node. The name for the ACB is IMST01.

As you can see from Figure 12.9, the TSO application definition is quite different from the previous examples. This is because TSO is actually part of the operating system control program. TSO primarily consists of two components, the Terminal Control Address Space (TCAS), which is a component of the operating system, and the VTAM Terminal I/O Coordinator (VTIOC), which is a component of VTAM. The TCAS component accepts the end-user's logon and creates an address space for each user that requests a TSO/VTAM session; hence the extra APPL definition statements, one for each user that logs on to TSO. TCAS is represented by the first application definition statement. Its ACBNAME must always be TSO. No substitutes are allowed. The VTIOC component coordinates the interface between TSO and VTAM.

Two operands coded on the application definition statement should stand out to you. One is the coding of the TSO parameter for the AUTH operand. This should tell you immediately that this APPL statement is for the definition of a TSO application. The other major difference is the EAS operand. Because TSO users are allocated their own address spaces, coding an EAS greater than 1 would waste valuable storage in the host's Common Storage Area (CSA).

There is just one more requirement for coding TSO application definition statements. The APPL statements that define the user's TSO application must have an ACBNAME in the format of TSOnnnn. The nnnn value must be a decimal integer, and the first value coded must start with 0001, with the remaining APPL ACBnames ascending sequentially as depicted in Figure 12.9.

When a user issues a logon to TSO, the next available user application will be attached to that user by TCAS. Suppose one user is logged onto TSO and the name of the application statement (applid) assigned to the user is TSO01002. Then the next user logging on will be assigned TSO01003, even if TSO01001 is available.

This concludes our discussion on defining applications to VTAM. Remember that it is important for you as the communications systems programmer to converse with the systems programmer responsible for the applications to obtain any necessary information on how the application will interface with VTAM.

Now that we have defined the applications, we need to describe to VTAM the locally attached terminals and logical units the end user will use to access the applications.

12.2 Local Non-SNA Device Major Node

There are two types of locally attached (channel-attached) terminal devices the user can use to access the network. One is the non-SNA device, and the other is an SNA device. The non-SNA device terminal is defined to the operating system as an addressable unit unto itself. The cluster controller the terminal is attached to is not defined to VTAM. Therefore each terminal, although attached to different cluster controllers, may be grouped into a major node based on location or perhaps by application. This may be an approach you may wish to pursue in your shop. An advantage of this is that not all users attached to a local non-SNA cluster controller will be affected if the major node is inactivated. Figure 12.10 diagrams the attachment of local cluster controllers.

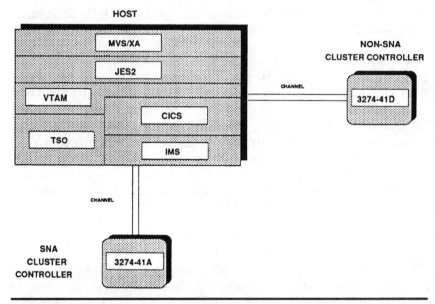

Figure 12.10 Local attached cluster controllers.

12.2.1 I/O GEN considerations

For all locally attached devices, the host control program for input/output services must be aware of the device channel and unit address. This is accomplished by the operating systems programmer when he or she performs an I/O generation (GEN) process. For most local non-SNA channel-attached terminals, the I/O GEN parameters should resemble the example in Figure 12.11.

In Figure 12.11, we can see that the cluster controller being defined is attached to channel 2. The 3274 cluster controller is defined to

```
CONSOLE   MCONS=221,ROUTCDE=ALL
CHAN1     CHANNEL ADDRESS=2,TYPE=BLKMPXR
CNTLUNIT  CUNUMBR=210,UNIT=3274,PATH=(02),              X
          UNITADD=((20,32)),SHARED=N,PROTOCL=D
IODEVICE  UNIT=3278,MODEL=2,ADDRESS=(220,30),           X
          FEATURE=(EBKY3277,DOCHAR,KB78KEY),            X
          CUNUMBR=210
IODEVICE  UNIT=3286,MODEL=2,ADDRESS=(23E,2),            X
          CUNUMBR=210
```

Figure 12.11 MVS I/O GEN parameters for local non-SNA cluster controller.

have 30 3278-2 type terminals and 2 3286-2 type printers. From the IODEVICE macro we can determine that the terminals will be attached to Channel Unit Address (CUA) 220 to 23D, which correlates to ports 0 to 30 of the cluster controller. The second IODEVICE defines two printers for the cluster controller on ports 31 and 32, which correlates to channel unit addresses 23E and 23F, respectively.

We added the CONSOLE macro here to make you aware of an important factor when considering non-SNA major and minor node definitions. Assume, for a moment, that the host is an IBM 3090-180 mainframe. On these host computers and all the 30xx mainframe computers IBM makes, the main operator console and all secondary operator consoles will take a port on the non-SNA locally attached cluster controller. In this example, the main console will be attached to channel unit address 221, which will correlate to port 1 on the cluster controller. Figure 12.12 gives you a pictorial representation of this addressing.

There are two important distinctions between local non-SNA and SNA cluster controllers. The first point is the protocol. The local non-SNA cluster controller uses channel protocol. The local SNA cluster controller uses SDLC protocol. The second distinctive characteristic is the addressing scheme. Each local non-SNA terminal and/or printer is assigned a unique channel unit address. Therefore, each device is polled individually by VTAM. The operating I/O subsystem has com-

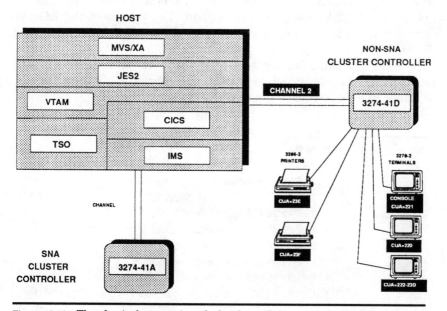

Figure 12.12   The physical connection of a local non-SNA cluster controller.

plete control of each device. On a local SNA cluster controller, there is only one polling address for all the attached devices. Consequently, the I/O subsystem has control over this one address. Unfortunately, if this one address is brought off-line by the I/O subsystem, all the end users connected to the local SNA cluster controller are brought off-line. The use of this one channel unit address is discussed in the following sections.

12.2.2 LBUILD statement

Now that we know the non-SNA device addresses, we can begin to code the local non-SNA major node. To tell VTAM that this major node is for local non-SNA devices, we must code the LBUILD definition statement. The format of this statement is

```
[name] LBUILD [CONFGDS = name][,CONFGPW=password]
```

As you can see from the format, the name parameter is optional. If it is coded, it is purely a symbolic name that VTAM ignores. The CON-FGDS and CONFGPW operands have the same function here as they do in the VBUILD statement

12.2.3 LOCAL definition statement

For each local non-SNA terminal that is to be in VTAM's domain, we must code a LOCAL definition statement. The format of the LOCAL definition statement is outlined in Figure 12.13. An explanation of it follows.

The name operand defines to VTAM the network unique symbolic name for this locally attached non-SNA terminal. This name is also the minor node name assigned to this network resource.

```
name     LOCAL   CUADDR=channel unit address
                 ,TERM=3277|3284|3286,
                 [,DLOGMOD=default logmode entry name]
                 [,FEATUR2=([EDATS|NOEDATS][,MODEL1|MODEL2]
                 [,ISTATUS=ACTIVE|INACTIVE]
                 [,LOGAPPL=application program name]
                 [,LOGTAB=interpret table name]
                 [,MODETAB=logon mode table name]
                 [,SPAN=(NCCF or NetView spanname)]
                 [,USSTAB=USS definition table  name]
```

Figure 12.13 Format of the LOCAL definition statement.

The CUADDR is the actual channel unit address for this device. The CUAs defined in Figure 12.12 are the values that should be coded for the local non-SNA device on the CUADDR operand.

The TERM operand tells VTAM what type of station is being defined for this CUA. Only those values specified on the TERM operand can be coded. Any other value will cause an error.

The DLOGMOD operand is optional and defines the default LOG-MODE entry name to be used for this device. This name will refer VTAM to what session parameters this device will use for session establishment. See Chapter 15 for more information on the LOG-MODE entry.

The FEATUR2 operand identifies the specific features for this device. The EDATS | NOEDATS parameter tells VTAM if this device supports the extended data streaming feature. The default is NOE-DATS. The MODEL1 | MODEL2 parameter tells VTAM the default screen and buffer size. MODEL1 denotes a size of 480 bytes, and MODEL2 denotes a size of 1920 bytes. The default is MODEL1, despite the fact that the majority of 3270 family type displays the printers have screen and buffer sizes of 1920 bytes and greater.

The ISTATUS operand tells VTAM if this device should be activated when the major node is activated. The default is ACTIVE. If INACTIVE is coded, it can be overridden with an operator command, which is

```
VARY NET,ACT,ID=major node name,SCOPE=ALL
```

The SCOPE=ALL operand tells VTAM to activate all the minor nodes associated with this major node. The LOGAPPL operand tells VTAM that when this device is activated, automatically begin a LOGON to the specified application program coded on this operand. This application becomes the controlling primary logical unit for this device. You can verify this by displaying the LUname for this device and searching for the CONTROLLING PLU=field in the displayed output.

The MODETAB operand defines the name of the LOGON mode table to be used for this device when searching for session parameters.

The SPAN operand is coded only if NCCF or NetView is to be used to manage the network.

The value coded in the USSTAB operand tells VTAM to look in the USS definition table named by this operand. Usually, this name is the name assigned to a user-created Unformatted System Services (USS) table. If you omit this operand, VTAM will search a default USS table. See Chapter 15 for more information on the USS table and how to create your own USS definition table.

```
MAJOR NODE NAME:   L01TERMS
MINOR NODE NAMES:
               LBUILD
   L01T220  LOCAL   CUADDR=220,TERM=3277,DLOGMOD=L3270,          X
                    MODETAB=MT01,FEATUR2=(NOEDATS,MODEL2),       X
                    USSTAB=USSNSNA
   L01T222  LOCAL   CUADDR=222,TERM=3277,DLOGMOD=L3270,          X
                    MODETAB=MT01,FEATUR2=(NOEDATS,MODEL2),       X
                    USSTAB=USSNSNA
                    "
                    "
                    "

   L01T23D  LOCAL   CUADDR=23D,TERM=3277,DLOGMOD=L3270,          X
                    MODETAB=MT01,FEATUR2=(NOEDATS,MODEL2),       X
                    USSTAB=USSNSNA
   L01P23E  LOCAL   CUADDR=23E,TERM=3286,DLOGMOD=L3270,          X
                    MODETAB=MT01,FEATUR2=(NOEDATS,MODEL2),       X
                    USSTAB=USSNSNA
   L01P23F  LOCAL   CUADDR=23F,TERM=3286,DLOGMOD=L3270,          X
                    MODETAB=MT01,FEATUR2=(NOEDATS,MODEL2),       X
                    USSTAB=USSNSNA
```

Figure 12.14 Coding of the non-SNA cluster controller from Section 12.2.2.

All of the above-mentioned optional operands have the same function for an SNA device. The exception to this is the FEATUR2 operand. It is applicable to non-SNA devices only.

Figure 12.14 shows how we coded the example non-SNA channel-attached cluster controller from the example in Section 12.2.2. Notice that we did not code CUA 221 for the console in this major node. Since the operating system owns that device, VTAM will not be able to activate it.

In summary, remember for non-SNA local devices that each device is assigned a specific channel unit address. Each address is associated with a unique minor node name. The non-SNA terminals and printers need not be grouped into the same major node. Although they are physically attached to the same cluster controller, they can be logically defined in different VTAM major nodes that describe local non-SNA devices.

12.3 Local SNA Device Major Node

Locally attached SNA devices are defined to VTAM using a local PU statement for each physical unit SNA cluster controller and a local

LU statement for each logical unit that is associated with the physical unit. There can be more than one PU statement in a local SNA major node, but each LU associated with a specific PU must be defined after the associated PU in the same major node. This is in contrast to the non-SNA terminal definitions, in which the cluster controller is not even defined and each terminal definition associated with a cluster controller can be coded in a different non-SNA major node for the locally attached cluster controllers. This difference between the two cluster controllers exists because the SNA terminal definitions are logically connected to the cluster controller. There is no physical connection to the cluster controller by means of a channel unit address for each terminal. Instead, there is only one channel unit address, and that address represents the cluster controller address.

12.3.1 I/O generation considerations for local SNA

From our discussion in Section 12.2.1, we must define all channel-attached devices to the operating system. This is done by assigning a channel unit address. Figure 12.15 outlines a sample coding for a local SNA cluster controller attached to an MVS operating system.

In this example, we can see that the cluster controller being defined is attached to channel 3. The unit being described is a 3791L cluster controller, though the actual device is a 3274-41A. The cluster controller's channel unit address is 320. The individual terminal addresses are assigned during configuration of the cluster controller. Notice that there is no operating system console support on a locally attached SNA cluster controller. Figure 12.16 depicts both non-SNA and SNA channel-attached cluster controllers.

There are two important distinctions between local non-SNA and SNA cluster controllers. The first point is the protocol. The local non-SNA cluster controller uses channel protocol. The local SNA cluster controller uses SDLC protocol. The second distinctive characteristic is the addressing scheme. Each local non-SNA terminal and/or printer is assigned a unique channel unit address. Therefore, each device is polled individually by VTAM. The operating I/O subsystem has complete control of each device. On a local SNA cluster controller, there is

```
CHAN3        CHANNEL ADDRESS=3,TYPE=BLKMPXR
CNTLUNIT     CUNUMBER=310,UNIT=3791L,PATH=(03), X
             UNITADD=((20,1))
IODEVICE     UNIT=3791L,ADDRESS=(320,1)
```

Figure 12.15 MVS I/O GEN parameters for a local channel-attached SNA cluster controller.

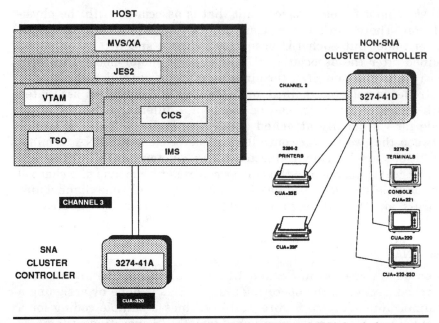

Figure 12.16 The physical connection of both non-SNA and SNA local cluster controllers.

only one polling address for all the attached devices. Consequently, the I/O subsystem has control over this one address. Unfortunately, if this one address is brought off-line by the I/O subsystem, all the end users connected to the local SNA cluster controller are brought off-line. The use of this one channel unit address is discussed in the following sections.

12.3.2 PU definition statement for local SNA major node

Prior to coding the PU definition statement, a VBUILD statement must be coded specifying TYPE=LOCAL to notify VTAM that the following statements are defining local SNA devices. More than one PU can be defined under this VBUILD statement, but this reduces the flexibility of operation.

A PU definition statement must be coded for each physical unit that is to be included in this local SNA major node definition. Figure 12.17 has the complete format of the PU definition statement.

The name parameter is required and assigns the minor node name to this physical unit. This name must be network unique. The operand CUADDR is the hexadecimal value of the channel unit address for this PU. Remember that in Section 12.3.1 we assigned a

```
name       PU    [CUADDR=channel unit address]
                 [,DISCNT=([YES|NO][,F|NF)]
                 [,ISTATUS=ACTIVE|INACTIVE]
                 [,MAXBFRU=n|1]
                 [,PUTYPE=2]
                 [,SECNET=YES|NO]
                 [,SPAN=(spanname)]
```

Figure 12.17 Complete format of the PU definition statement for local SNA major node.

channel unit address of 320 to this PU. Code the hexadecimal value for this operand. If you do not code CUADDR, the ISTATUS value must be INACTIVE and the operator must supply the channel unit address in the VARY active command. The DISCNT operand defines how to terminate the SSCP-PU and SSCP-LU sessions when the last LU-LU session on this PU is terminated. The YES|NO tells VTAM which option to use. If YES is coded, VTAM terminates all SSCP-LU sessions and the SSCP-PU session when the last LU-LU session is terminated. IF NO, the default, is coded, then VTAM terminates the SSCP sessions after normal session terminations from the PU and all the associated LU-LU sessions on that PU. The F|NF tells VTAM to flag this PU in final-use status when deactivating the PU if DISCNT =YES has been coded. This operand is not used if DISCNT=NO is coded. The F is the default and indicates "final-use" status. The NF indicates "not-final-use" status. You should consult the appropriate device publication on the use of the DISCNT operand. The ISTATUS tells VTAM if this PU should be activated when the associated major node is activated. The default is ACTIVATE. This operand can also be used to sift down to the PU's associated LUs. The MAXBFRU value defines to VTAM the maximum number of buffer units from the IOBUF buffer pool that can be used to receive or send data from or to the physical unit. The n value is a decimal integer with a range of 1 to 65,535. The PUTYPE operand describes to VTAM the type of PU attached. The default value of 2 is the only valid PU type that can be coded in a local SNA major node. The SECNET operand should be coded only if you have the IBM 3710 network controller in your network. It tells VTAM that this PU is associated with a secondary network. The resources in this secondary network are not defined to VTAM. If YES is coded, the PU data requires special problem determination considerations when a communications management application receives the data. The SPAN operand value is the span name from NCCF or NetView that is associated with this PU.

```
LS320      PU      CUADDR=320,                              X
                   DISCNT=NO,                               X
                   ISTATUS=ACTIVE,                          X
                   MAXBFRU=1,                               X
                   PUTYPE=2,                                X
                   SECNET=NO
```

Figure 12.18 Sample coding for PU definition statement for local SNA major node.

The PU definition statement in Figure 12.18 tells us that the PU will be made active when the SNA major node is activated. The definition also tells us that this is not a 3710 resource. Now that the PU has been defined, we must define the LUs associated with this PU.

12.3.3 LU definition statement for local SNA major node

The LU definition statements must follow the associated PU definition statement within this local SNA major node. There must be one LU definition statement for each LU defined for this physical unit. Figure 12.19 has the complete format of the local LU definition statement.

The name parameter is required and assigns the minor node name to this logical unit. This name must be network unique. The operand LOCADDR is also required. This value is the logical unit's local address that is assigned at this physical unit. The n value is a decimal integer from 1 to 255. This is the address assigned to the LU being

```
name       LU      LOCADDR=n
                   [,DLOGMOD=default logon mode entry name]
                   [,ISTATUS=ACTIVE|INACTIVE]
                   [,LOGAPPL=autologon application name]
                   [,LOGTAB=interpret table name]
                   [,MODETAB=logon mode table name]
                   [,PACING=n|0|1]
                   [,SPAN=(spanname)]
                   [,SSCPFM=FSS|USSSCS]
                   [,USSTAB=USS table name]
                   [,VPACING=n|0|1]
```

Figure 12.19 Complete format of the LU definition statement for local SNA major node.

defined by this statement. The LOCADDR need not be consecutive, and an LU statement is not required for every possible local address.

According to the IBM 3274 architecture for a 3274 cluster controller, 128 logical units can be defined, even though the address range is from 1 to 255. However, the actual maximum number of LUs a local SNA IBM 3274 PU TYPE 2 device can have assigned to it is 125 logical units. This is a restriction of the 3274, not of SNA. The restriction is because of port 0 of the 3274. Port 0 may not have more than one LOCADDR value assigned to it because this port is used for the 3274 configuration process. Therefore, instead of having 128 (32 ports times 4 LUs per port), the maximum is 3 LUs fewer, or 125 logical units. There is one more requirement for the IBM 3274 SNA cluster controller. The LOCADDR value must start at 2. Therefore, the maximum LOCADDR value that can be coded for the SNA 3274 is 127. The LOCADDR values are assigned to the ports during the customization process for SNA cluster controllers.

The SSCPFM operand tells VTAM whether this logical unit supports the SNA character string coded messages (USSSCS) or formatted messages (FSS) during SSCP communications. For LU statements defining terminals, the default FSS is used. For LU statements defining SNA printers, the USSSCS parameter is used.

The PACING operand defines how VTAM will pace data flowing to the logical unit during a LU-LU session from the channel-attached host for this LU's physical unit. The default value of 1 means that VTAM will send one message to the LU and not send any more messages until VTAM receives a pacing response from the LU. This default value or any other n value is ignored if the LU-LU session is in the same domain. The range for n is 1 to 63. The value 0 means that no pacing is to be performed.

The VPACING value determines the number of messages the primary logical units owning VTAM can send to the secondary logical units owning VTAM before waiting for a pacing response. This type of pacing is called two-stage and is between VTAM hosts. If the LU-LU session is in the same domain, the value specified for the PACING operand is ignored and the value of the VPACING operand is used. This is called one-stage pacing.

Usually, the pacing values are set to 0 for locally attached SNA devices unless the device receives large amounts of data, such as printers or graphic terminals. In this case, a PACING value of 1 and a VPACING value of 2 should be adequate for initial trials.

The remaining operands not discussed here have been reviewed in Section 12.2.3. Turn to this section if you need to review their meaning and purpose. All the operands for an LU definition statement may be coded on the PU definition statement to use the sift-down effect.

```
MAJOR NODE NAME:   L01SNA20
MINOR NODE NAMES:
                    VBUILD   TYPE=LOCAL
    LS320     PU      CUADDR=320,DISCNT=NO,ISTATUS=ACTIVE,            X
                      DLOGMOD=D4A32782,MAXBFRU=10,USSTAB=USSSNA,      X
                      PACING=0,VPACING=0
    LS320T02 LU      LOCADDR=02
    LS320T03 LU      LOCADDR=03
    LS320T04 LU      LOCADDR=04
                  "

                  "

    LS320P32 LU      LOCADDR=32,LOGMODE=DSC4K,USSTAB=ISTICNDT,       X
                      PACING=1,VPACING=2,SSCPFM=USSSCS
```

Figure 12.20   Coding of the local SNA major node with the PU and LU definition statements.

The exceptions are the LOCADDR and SPAN operands. These must be coded for each LU definition statement.

From the sample local SNA major node definition in Figure 12.20, we can see that we are using the sift-down effect. All the LUs defined will use the same default LOGMODE entry name of D4A32782, the same USS table name of USSSNA, the formatted system services (FSS), and a PACING and VPACING value of 0. This is true for all the LUs except the last one that is defined. This LU was created to define a SNA printer. The LU name tells us that from its format LS320P32, where the P denotes a printer for this LU definition. The pacing values of PACING=1 and VPACING=2 were coded so that the printer does not monopolize the PU during large volumes of print destined for this LU. Finally, the DLOGMOD was overridden to DSC4K and the SSCPFM=USSSCS tells us that this LU uses USS SNA character string mode when communicating with the SSCP.

12.4   Defining the IBM 3172 Interconnect Controller Major Node

The definition of the IBM 3172 to the mainframe is dependent on the mainframe operating system in use. IBM mainframe computers offer two widely used operating systems, the Virtual Machine (VM) and Multiple Virtual Storage (MVS) operating systems. Each operating system is available with various features. For our purposes we will refer to all the various releases of the operating systems as VM and MVS.

Channel attachment of the IBM 3172 to the mainframe is provided by parallel and serial channels. Attachment to an IBM System/370 (S/370) mainframe is accomplished using a block multiplexer channel. This channel can transmit data at up to 4.5 MB/s. Channel attachment to IBM Enterprise System/390 (ES/390) and Enterprise System/9000 (ES/9000) mainframe computers may be accomplished by utilizing IBM's Enterprise Systems Connection (ESCON), providing data rates up to 20 MB/s. In both channel configurations the IBM 3172 is defined to the operating system as an IBM 3088 Channel-To-Channel (CTC) control unit.

The IBM 3172 is assigned a Unit Control Word (UCW) or Input/Output Configuration Program (IOCP) address by the VM or MVS systems programmer through a system generation (SYSGEN) or IOCP generation (IOCPGEN) process. The addresses assigned to the IBM 3172 are dependent on the use of the adapters within the IBM 3172. The assigned channel must be defined as nonshareable between operating systems. An address range of 32 or 64 must be defined in the VM operating system environment to compile an IOCP definition for an IBM 3088 control unit. The MVS operating system environment does not have this constraint. The addresses assigned to the channel can be used by the IBM 3172 to connect to access methods on the mainframe. Connecting to TCP/IP requires a contiguous even-odd pair of addresses. The even address is used by the IBM 3172 to send LAN traffic to the mainframe. The odd address is used to send data from the mainframe to the IBM 3172 and out to the LAN. SNA VTAM connectivity to the IBM 3172 requires only one address. The address used to connect to VTAM can be either odd or even. These channel configurations are shown in Figure 12.21.

12.4.1   I/O GEN definitions for VM and MVS

In a non-ESCON environment, there are three types of channel configurations that can be chosen, and each type can be logically connected to only one channel adapter on the mainframe host operating system. The definitions for these are printed in Figure 12.22. The CNTLUNIT macro identifies the type of control being defined along with the channel and address requirements. The difference between the three definitions specified in Figure 12.22 is the PROTOCL keyword. The value specified for PROTOCL defines the type of channel that will be attached to the device. A value of S4 indicates that the device is attached to a 4.5-MB/s data-streaming channel as shown in the first example. A value of S indicates that the control unit is attached to a data-streaming channel other than 4.5 MB/s. Usually this data-streaming channel data rate is up to 3 MB/s. Finally, a value of D for the PROTOCL keyword defines a Data Channel Interlock (DCI) chan-

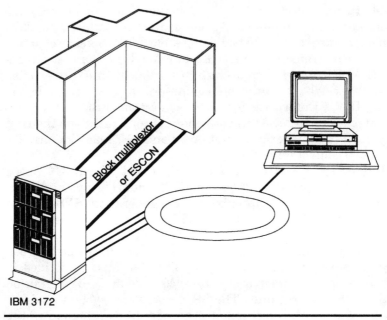

IBM 3172

Figure 12.21   Channel configuration of an IBM 3172.

4.5MB data streaming channel		
C3172	CNTLUNIT	CUNUMBR=040,PATH=(05), PROTCL=S4,SHARED=N,UNIT=3088, UNITADD=((04,8))
D3172	IODEVICE	UNIT=CTC,ADDRESS=((040,8)), STADET=N,CUNUMBR=040
Other data streaming channel		
C3172	CNTLUNIT	CUNUMBR=040,PATH=(05), PROTCL=S,SHARED=N,UNIT=3088, UNITADD=((04,8))
D3172	IODEVICE	UNIT=CTC,ADDRESS=((040,8)), STADET=N,CUNUMBR=040
Data Channel Interlock Channel		
C3172	CNTLUNIT	CUNUMBR=040,PATH=(05), PROTCL=D,SHARED=N,UNIT=3088, UNITADD=((04,8))
D3172	IODEVICE	UNIT=CTC,ADDRESS=((040,8)), STADET=N,CUNUMBR=040

Figure 12.22   Channel definitions used for IBM 3172 in the operating system I/O GEN process.

nel as the connection to the control unit. The SHARED keyword of the CNTLUNIT macro has a value of N, indicating that the channel is not to be shared by other operating systems. The CUNUMBR value is an internal number for the operating system control program representing the control unit being defined. This value is usually set to match the starting address for the device. The UNITADD keyword identifies the starting unit address and the range of addresses allowed on this channel for the device. The UNIT keyword specifies the type of device being defined. As discussed previously, the IBM 3172 is defined as an IBM 3088 channel-to-channel control unit.

In an ESCON environment, the ESCON adapter for the IBM 3172 can support up to 16 logical hosts by using an ESCON Director. The ESCON Director is basically an intelligent device that maps the connection from the IBM 3172 to other operating system channels. Since the IBM 3172 can support up to two ESCON channel adapters, a logical total of 32 host operating systems may be supported. A typical ESCON connection and definition are depicted in Figure 12.23. Note that in this configuration the IBM 3172 is not directly connected to the mainframe. Instead, the IBM 3172 is attached to an ESCON Director, which is actually a channel I/O switch. The CHPID macro defines the channel path from the mainframe to the ESCON Director. It also defines the type of channel being defined. In this case the TYPE keyword with the value of S indicates that it is a serial channel. The SWITCH keyword defines the ESCON Director address the CHPID is connecting.

The CNTLUNIT macro points back to the CHPID definition through the PATH keyword. The LINK keyword assigns a link address to the channel. The UNITADD keyword defines the starting address of the device being defined. Note that when ESCON channel addresses are defined, a range of addresses is not specified. Finally, a major difference between non-ESCON and ESCON channel and control unit definitions is the UNIT keyword on the CNTLUNIT macro. Under ESCON the Interconnect Controller is actually defined to the CNTLUNIT macro as a 3172. This is in contrast to the non-ESCON definition just discussed, where the UNIT keyword for an Interconnect Controller is specified as 3088. Note: The IBM 3172

```
ESCON IOCP Definition
P3172              CHPID       PATH=((34)),TYPE=S,SWITCH=00
C3172              CNTLUNIT    CUNUMBR=100,PATH=(34),
                               LINK=(C1),CUADD=1,UNITADD=((B0)),
                               UNIT=3172
D3172              IODEVICE    UNIT=CTC,ADDRESS=(040),
                               UNITADD=B0,CUNUMBR=100
```

Figure 12.23 ESCON channel definition for the IBM 3172.

Interconnect Controller must be defined as the last device on a channel for both non-ESCON and ESCON channel definitions.

12.4.2 Device definitions for VM and MVS

The type of device being defined is referenced by VM in the RDEVICE macro. The VM operating system releases have different requirements and must be broken out. In a Virtual Machine/System Product (VM/SP) operating system environment, the RDEVICE macro and the RCTLUNIT macro must specify an address range of 32 or 64 for the 3088 and a minimum of 8 for CTC as the address range. The definition for the IBM 3172 on the RDEVICE macro for VM/SP looks like

```
RDEVICE ADDRESS=(640,32),DEVTYPE=3088
RCTLUNIT ADDRESS=640,CUTYPE=3088,FEATURE=32-DEVICE
```

These macros are added to the real I/O configuration file DMKRIO in VM/SP operating system environments.

The first variable of the ADDRESS keyword on the RDEVICE macro specifies the base address of the IBM 3172 control unit. The second variable indicates the address range that was defined in the IOCP generation. The DEVTYPE keyword of the RDEVICE macro identifies the type of device associated with the address. For IBM 3172 devices, this keyword value is 3088.

The RCTLUNIT macro specifies the base unit address for the control unit being defined on the ADDRESS keyword. The CUTYPE keyword indicates the type of control unit being defined, in this case a 3088, and the FEATURE keyword indicates the number of device addresses that are assigned to the control unit starting with the control unit base address. In VM/SP, this value must be 32 or 64.

In a Virtual Machine/Extended Architecture (VM/XA) or Virtual Machine/Enterprise System Architecture (VM/ESA) operating system environment, only the RDEVICE macro is required. The IBM 3172 in these environments is defined as a Channel-To-Channel adapter (CTCA) device. The RDEVICE macro for VM/XA and VM/ESA is defined in the real I/O configuration file named HCPRIO. The macro definition looks like

```
RDEVICE ADDRESS=(640,8),DEVTYPE=CTCA
```

In this definition, the second variable, denoting the address range, is valued at 8, since the device being defined is a channel-to-channel adapter. The MVS environment defines the IBM 3172 as a channel-to-channel device on the IODEVICE macro in the MVS control program. The definition of the macro looks like

```
IODEVICE UNIT=CTC,ADDRESS(640,8)
```

The UNIT keyword is defined with the value of CTC, indicating that the I/O device acts like a channel-to-channel device. The ADDRESS keyword indicates that the base address of the IBM 3172 is 640 and that up to 8 addresses (i.e., 640–647) starting with the base address are reserved for the device.

12.4.3 Defining the External Communications Adapter (XCA) major node

Connectivity to IBM's Virtual Telecommunication Access Method (VTAM) is used by IBM 3172 to access mainframe applications written to IBM's SNA that execute under VTAM. This SNA application access over token-ring through an IBM 3172 requires IBM VTAM V3R4 and higher. This release of VTAM includes a new definition called an External Communication Adapter (XCA). It is the XCA definition that allows VTAM to define token-ring resources off of the IBM 3172 as switched devices. The importance of this connectivity configuration is the absence of a front-end processor as the SNA VTAM token-ring gateway.

The XCA definition defines to VTAM all downstream SNA resources attached to the token-ring through the IBM 3172. VTAM still provides the SNA routing, session, and data flows as it would under standard SNA environments. The four major types of SNA nodes are supported on the IBM 3172 using the XCA definition. Each adapter installed on the IBM 3172 and configured for VTAM support, without OF/2 support, has a maximum of 255 active SNA physical units. This totals 1020 active SNA physical units per IBM 3172 without OF/2 support. OF/2 on a VTAM-supported token-ring adapter takes up 10 of the possible 255, hence limiting the full capacity support to 1010 active SNA physical units.

The IBM 3172 Interconnect Controller provides VTAM SNA LAN gateway support with VTAM V3R4. The added functionality is provided by the new VTAM XCA major node. An XCA major node must be coded in VTAM's SYS1.VTAMLST data set for each token-ring LAN adapter that is to be supported by VTAM on the IBM 3172. In our example, only one XCA major node needs to be defined to support the SNA LAN gateway. A second XCA major node may be defined to allow the IBM 3172 to be managed by VTAM and a communication network management application on the mainframe. The XCA major node contains three distinct sections: VBUILD, PORT, and GROUP.

An example of an XCA major node is found in Figure 12.24. The VBUILD definition statement identifies this major node to VTAM as an XCA major node by having the TYPE operand value equal to XCA.

```
ICLAN1              VBUILD      TYPE=XCA
ICLAN1P             PORT        CUADDR=042,ADAPNO=1,
                                MEDIUM=RING,SAPADDR=4
ICPUGRP             GROUP       DIAL=YES,CALL=IN,ANSWER=ON
ICPULNE1            LINE        ISTATUS=ACTIVE
ICPU01              PU          ISTATUS=ACTIVE
ICPULNE2            LINE        ISTATUS=ACTIVE
ICPU02              PU          ISTATUS=ACTIVE
```

Figure 12.24   The XCA definition for an IBM 3172 to support downstream SNA devices.

A single PORT definition statement must be coded for each XCA major node.

The PORT definition statement defines the VTAM host connection to the token-ring network attached to the IBM 3172. The CUADDR keyword defines the subchannel address assigned to the LAN adapter during the IBM 3172 adapter configuration process. Recall that in our scenario the VTAM-supported token-ring adapter is defined with subchannel address 042 as the SNA LAN gateway subchannel address. The ADAPNO keyword identifies the relative adapter number generated by the IBM 3172 during the VTAM-supported token-ring adapter definition. In our example, the relative adapter number generated for this adapter is 1. If a new profile for the IBM 3172 is created and the VTAM-supported token-ring adapter is defined second, then this value will change to a 1 and must be changed here in the XCA major node to reflect the new IBM 3172 configuration. The MEDIUM keyword of the PORT definition statement identifies the type of LAN medium attached to the IBM 3172. In our scenario, the value will be RING, since we are concerned only with token-ring connectivity. The SAPADDR keyword specifies the Service Access Point (SAP) address for the LAN connection through the IBM 3172. The SAPADDR is used in conjunction with the CUADDR and ADAPNO keywords to route information between the LAN and VTAM.

The keywords and values discussed are required for every XCA major node defining SNA Node Types 2.0, 2.1, 4, and 5. Each SNA node type is defined within the XCA major node.

12.4.4   Defining SNA Node Type 2.0/2.1 in the XCA major node

SNA Node Types 2.0 and 2.1 provide IBM logical unit access to VTAM applications. Typically, these nodes provide IBM 3270 terminal connectivity to the mainframe. The platforms executing these node types are varied. Some platforms that can support these SNA node types are IBM 3174 Establishment Controllers, IBM AS/400, DOS- and OS/2-based personal computers, and UNIX-based workstations. The

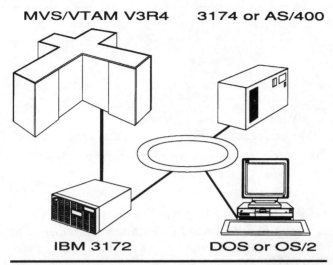

MVS/VTAM V3R4   3174 or AS/400

IBM 3172   DOS or OS/2

Figure 12.25 Network diagram depicting IBM 3172 as the SNA gateway to VTAM applications.

hardware platform is not what matters in this type of connectivity. It is the software available on the various platforms that gives them SNA connectivity using SNA Node Types 2.0 and 2.1. SNA Node Types 2.0 and 2.1 are also referred to as peripheral nodes.

In Figure 12.25, a typical configuration is diagramed showing the connectivity of an SNA Node Type 2.0 and 2.1 device to VTAM on the mainframe using an IBM 3172 as the gateway. These devices must be defined to the XCA major node. Following the PORT definition in the XCA major node will be a GROUP definition statement defining the line groups associated with the SNA Node Types 2.0 and 2.1 as shown in Figure 12.24.

The SNA Node Types 2.0 and 2.1 are defined as switched resources to VTAM in the XCA major node. This switched connectivity is specified by the DIAL keyword being equal to the value YES. The CALL keyword value of IN indicates to VTAM that the resources will always initiate the connectivity process. A value of INOUT is also possible, but this indicates that VTAM, on behalf of an application, may initiate the session to the SNA Node Type 2.0 and 2.1 device. In current networking environments the CALL=IN operand will suffice. The ANSWER keyword, with a value of ON, specifies that the switched resource may initiate communications when the line associated with the node is active. The ANSWER and CALL keywords actually apply to the LINE definition statement following the GROUP definition statement but are coded here to reduce repetitive coding of these parameters. The values coded on the GROUP definition statement are sifted down to the LINE definition statements that follow.

SWICLAN1	VBUILD	TYPE=SWNET,MAXGRP=1,MAXNO=2
SWICPU01	PU	ADDR=C1,CPNAME=ARDEPT,PUTYPE=2,
		MAXDATA=1033
ICPATH01	PATH	DIALNO=010440000000005D,
		GRPNM=ICLAN1G
SWICLU01	LU	LOCADDR=0
SWICPU02	PU	ADDR=C2,IDBLK=017,IDNUM=31741,
		PUTYPE=2,MAXDATA=265
ICPATH02	PATH	DIALNO=0104400000000017,
		GRPNM=ICLAN1G
SWICLU02	LU	LOCADDR=02

Figure 12.26   The switched major node definition used in conjunction with the IBM 3172 XCA definition.

The values may be overridden by coding the keywords on the LINE definition statement. One LINE and one PU definition statement are required for each SNA Node Type 2.0 and 2.1 resource having connectivity through the IBM 3172.

VTAM requires a switched major node for all switched resources. A sample switched major node for our example is listed in Figure 12.26. The switched major node, like the XCA major node, has a VTAM VBUILD statement identifying the type of major node being defined. TYPE=SWNET indicates to VTAM that this major node is defining switched resources. The MAXGRP keyword specifies the number of unique group names defined in this switched major node. The MAXNO keyword defines the number of unique dial numbers in the switched major node.

The PU definition statement defines the SNA switched physical unit attributes. The ADDR keyword defines the switched resource's unique SNA station address. The value is a hexadecimal number ranging between 01 and FF. An SNA Node Type 2.1 contains a control point function. This function is given a name, and the keyword that identifies the name is the CPNAME keyword. This value must match the definition assigned in the Node Type 2.1. The presence of the CPNAME keyword indicates to VTAM that the physical unit definition is that of a Node Type 2.1. For Node Type 2.0 devices, the IDBLK and IDNUM values must be defined. The 3-digit hexadecimal IDBLK value is dependent on the type of device being used for the SNA Node Type 2.1 connectivity. For example, IBM 3174 Establishment Controllers use a value of 017, OS/2 platforms use 05D, and DOS platforms use 017. The IDNUM value is assigned by the network administrator and provides a unique 5-digit hexadecimal number for the device. Frequently the device serial number is used for this number. The PUTYPE keyword for both peripheral types has a value of 2. The MAXDATA keyword defines the largest frame size that can be sent to the device.

The PATH definition statement follows the PU definition statement in a VTAM switched major node. The DIALNO keyword on the PATH definition statement for an IBM 3172 attached resource has the following syntax:

```
aaccccccccccc
```

The aa is a 2-digit place holder and is not used in the IBM 3172. It may be useful for documentation purposes to assign the relative adapter number to the aa position. The remaining 12 characters are the MAC address (i.e., LAA) for the station associated with this definition. The GRPNM keyword value must match the XCA GROUP name associated with the switched peripheral node definitions.

Following the PATH statement is the LU definition statement for the switched major node. This statement defines the characteristics of the SNA logical units used on the physical unit. The LOCADDR keyword of the LU definition statement defines the address of the logical unit on the physical unit. LOCADDR=0 indicates that the logical unit being defined is an SNA LU6.2 logical unit. LU6.2 is IBM's peer-to-peer protocol for cooperative processing between applications. This type of logical unit with a LOCADDR=0 can reside only on SNA Node Type 2.1. This logical unit can initiate sessions without the assistance of VTAM. Instead, it utilizes the control point function of SNA Node Type 2.1. These logical units are also called Independent Logical Units (ILU). A LOCADDR specifying anything greater than 0 is referred to as a Dependent Logical Unit (DLU). DLUs require the assistance of VTAM to establish sessions with other logical units.

SNA Node Types 2.0 and 2.1 require an XCA major node definition and a switched major node definition. The following sections illustrate the differences when defining SNA Node Types 4 and 5 connectivity through the IBM 3172 Interconnect Controller.

12.4.5 Defining SNA Node Type 4 in the XCA major node

The XCA major node can also include an SNA Node Type 4 definition along with the Node Type 2.0 and 2.1 definitions. Figure 12.27 depicts the addition of a Node Type 4 to the sample configuration. A Node Type 4 in SNA is basically synonymous with a communication controller executing the Network Control Program (NCP). Communication controllers attach to a token-ring network using Token-ring Interface Couplers (TIC). The NCP defines the token-ring interface and assigns a MAC address along with a Transmission Group Number (TGN). The TGN is used in SNA routing. The NCP as well as VTAM is also assigned an SNA subarea number. The subarea number is a network-

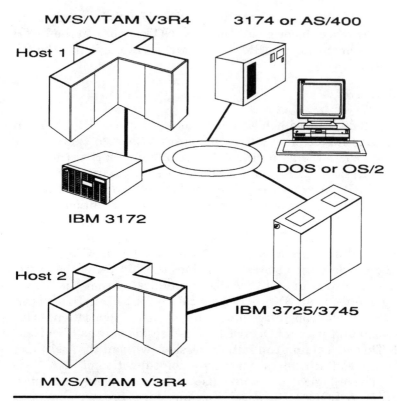

Figure 12.27   Configuration for IBM 3172 to act as the gateway for an
IBM 37x5 Communication Controller.

wide unique address for the SNA Node Type 4. Different SNA node
types may access VTAM through an IBM 3172 SNA LAN gateway
because each connection is actually viewed as a point-to-point connection. The SNA Node Type 4 must be controlled (i.e., owned) by an SNA
Node Type 5 VTAM host through either channel attachment, SDLC
attachment, or token-ring attachment from a channel-attached SNA
Node Type 4. SNA Node Type 4s cannot be activated, loaded, or
dumped when connected to VTAM through an IBM 3172.

The sample listing shown in Figure 12.28 shows the addition of the
SNA Node Types 4 and 5 definition to the current XCA major node.

A GROUP definition statement is coded. However, note that the
DIAL keyword now specifies a value of NO. This indicates to VTAM
that this is not a switched definition but a dedicated link definition.
Because of this, a switched major node is not necessary to define the
token-ring-attached Node Type 4 device when using an IBM 3172.
The LINE definition statement for a Node Type 4 definition indicates
that the line is using SNA protocols to access the IBM 3172 over the

```
ICLAN1              VBUILD      TYPE=XCA
ICLAN1P             PORT        CUADDR=042,ADAPNO=1,
                                MEDIUM=RING,SAPADDR=4
ICPUGRP             GROUP       DIAL=YES,CALL=IN,ANSWER=ON
ICPULNE1            LINE        ISTATUS=ACTIVE
ICPU01              PU          ISTATUS=ACTIVE
ICPULNE2            LINE        ISTATUS=ACTIVE
ICPU02              PU          ISTATUS=ACTIVE
ICNCPGRP            GROUP       DIAL=NO
ICNCPLNE1           LINE        USER=SNA,ISTATUS=ACTIVE
ICNPC11             PU          PUTYPE=4,MACADDR=400000374501,
                                SAPADDR=4,SUBAREA=11,TGN=1
ICVTMGRP            GROUP       DIAL=NO
ICVTMLNE            LINE        USER=SNA,ISTATUS=ACTIVE
ICVTMGPU            PU          PUTYPE=5,MACADDR=400000317201,
                                SAPADDR=4,SUBAREA=2,TGN=1
ICVTMLNE            LINE        USER=SNA,ISTATUS=ACTIVE
```

Figure 12.28 IBM 3172 XCA major node definition detailing the definition statements needed to support PU TYPE 2.0, PU TYPE 5, and PU TYPE 4.

LAN. The PU definition statement specifies the MAC address of the Node Type 4 device on the MACADDR keyword. The PUTYPE keyword identifies the device as a Node Type 4. The SUBAREA keyword defines the Node Type 4 SNA subarea value assigned to the communication controller, and the TGN keyword identifies the link as transmission group number 1. The SAPADDR keyword defines the SAP address of the LAN connection through the IBM 3172.

The NCP in the Node Type 4 defines its MAC address on the LINE definition statement of the physical GROUP definition statement. The LOCADD keyword is the MAC address associated with the TIC in the communication controller. This LOCADD keyword value must match the MACADDR keyword value in the XCA definition of this Node Type 4. The ADDR keyword on the PU definition statement of the NCP logical GROUP definition specifies the MAC address of the IBM 3172 VTAM-supported token-ring adapter. The logical line associated with the logical physical unit identifies transmission group number 1, matching the definition in the XCA major node.

In the configuration diagramed in Figure 12.29, SNA routes may be established between all the SNA subareas. SNA routes will utilize the token-ring network as the medium for transmission.

12.4.6 Defining SNA Node Type 5 in the XCA major node

IBM's mainframe, SNA Node Type 5 devices, may now be directly connected over token-ring using the IBM 3172. In Figure 12.30, a second SNA host has been added to the network using a second IBM

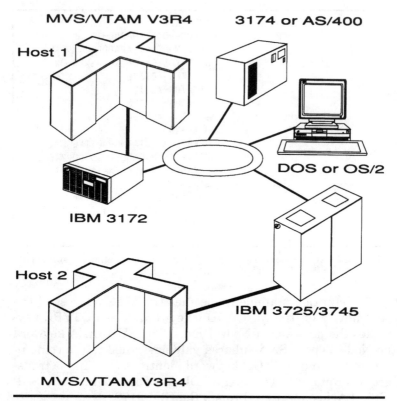

Figure 12.29 SNA subarea routing is made possible through the use of the token-ring and the 3172 PU TYPE 4 support.

3172. The XCA definition defined for Host 1 now includes the definition for Host 2. Note that the SNA Node Type 5 definition in the figure is comparable to that just discussed for Node Type 4 connectivity. The differences here are the values specified on the PU definition statement for the Node Type 5. The MACADDR value specifies the MAC address assigned to the IBM 3172 token-ring adapter defined for VTAM support on the IBM 3172 channel attached to Host 2. The PUTYPE keyword value of 5 indicates that this definition is for an SNA Node Type 5 resource, namely VTAM on an SNA host computer. The TGN keyword identifies the transmission group number to use when determining SNA routing paths, and the SAPADDR value is the service access point on the LAN connection.

12.5 APPN VTAMLST Definitions

VTAM V4R1 has added several new definition statements to define APPN services. These new definitions define to VTAM V4R1 adjacent

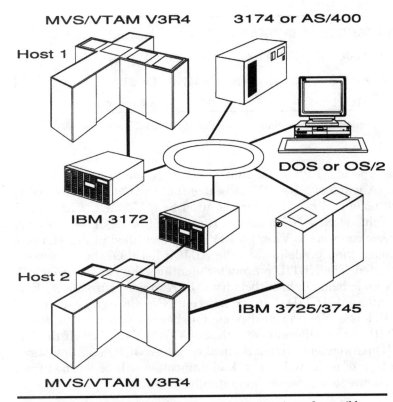

Figure 12.30 Direct VTAM to VTAM communications is made possible by adding an IBM 3172 channel attached to Host 2.

control points, the profile of the transmission groups used for communications to the adjacent control points, and the list of network node servers that this EN-functioning VTAM may use as its network node server.

12.5.1 Adjacent CP major node

VTAM creates a default adjacent CP major node called ISTADJCP. This major node is created when VTAM V4R1 encounters a definition that contains the CPNAME parameter on PU definition statements. Systems programmers can create a major node that provides for predefined adjacent CP names. This node can be used in conjunction with ISTADJCP or without it. Figure 12.31 lists a sample adjacent CP major node with the two possible parameters NN and NETID defined.

The ADJCP major node is identified to VTAM V4R1 by using the VBUILD definition statement with the TYPE=parameter set to ADJCP. Each ADJCP definition statement must have a name defined to it. The NN parameter of the ADJCP statement indicates to VTAM

ADJCP01	VBUILD	TYPE=ADJCP	Adjacent CP node.
3174A	ADJCP	NN=YES	3174 NN
OS2EN	ADJCP	NETID=NETB,NN=NO	OS/2 EN
AS400	ADJCP		AS/400 NN I EN

Figure 12.31 Format and sample definition for the APPN ADJCP major mode.

V4R1 the type of APPN node that is being defined for this ADJCP statement. A value of NN=YES indicates that the CP being defined is a network node and can therefore be used by a VTAM EN as its NN server. A value of NN=NO indicates that the adjacent CP is an APPN EN. If a node connects to VTAM V4R1 and is identified in the XID as a different node than predefined in the ADJCP list then the connection will be rejected. The NETID parameter identifies the network to which the APPN node being defined belongs. APPN architecture allows ENs to have a different NETID than the adjacent NNs. So, as an example, if VTAM V4R1 is a NN and an EN establishes a connection to it, the EN's NETID can be different from that of VTAM V4R1 NN. If the NN and NETID parameters are not defined on the ADJCP definition statement the type of node and network identification will be learned from the XID process during connection establishment.

12.5.2 Transmission Group Profile (TGP) major node

Each physical connection defined to an adjacent APPN node is called a Transmission Group (TG). The TGs are defined in VTAM V4R1 with a major node that defines the typical profiles used for the TGs. The profiles are referred to on a GROUP, LINE, or PU definition statement. The TG Profile (TGP) major node provides defaults for TG characteristics. Values defined in the TGP major node and referenced for a connection are used by TRS in the routing selection process.

Figure 12.32 lists a sample definition of a TGP major node. Note that a VBUILD definition statement is not used for this major node. The characteristic types shown are valid for all TGs used for APPN connection. More than one TGP definition statement can be defined in a single major node in SYS1.VTAMLST.

The CAPACITY value specifies the effective capacity of the TG. For example, if you are defining a TG with a data rate of 56 Kbps, it would be specified here as 56K. The value for CAPACITY can be specified as nnnnK or nnnnM. The nnnnK is kilobits, and the nnnnM is megabits. The nnnn can be a value from 1 to 1000. The default is 8K.

TGP56K	TGP	CAPACITY=56K,	Total line capacity
		COSTBYTE=128,	Arbitrary cost
		COSTTIME=128,	Arbitrary cost
		PDELAY=TERRESTR,	Prop. delay
		SECURITY=PUBLIC,	Security
		UPARM1=128,	User parameter
		UPARM2=128,	User parameter
		UPARM3=200	User parameter

Figure 12.32 Characteristic parameters and sample TG profile definition.

The COSTBYTE characteristic is a value that indicates the cost per byte associated with the TG. This value can be used to provide a heavier weight associated with a TG because of line speed or actual costs associated with a given connection. As an example, a higher value for COSTBYTE means that the TG for this connection is less desirable than a TG for a connection with a lower COSTBYTE value. The value for COSTBYTE can range from 0 to 255. A value of 0 is the least expensive, and a value of 255 is the most expensive. The default of COSTBYTE is 0.

The COSTTIME characteristic is used to provide a weight for the amount of time units needed to transmit data to the partner node over the TG. Again the range for this value is 0 to 255. A value of 0 is the least expensive, and a value of 255 is the most expensive. This value is especially useful for switched TG connections, since the time of the connection is directly related to the cost of the transmission of data.

The PDELAY characteristic specifies the type of delay that can be expected using this TG. There are four possible values for this characteristic. The NEGLIGIB value indicates to TRS that the propagation delay is negligible, meaning that it is less than 0.48 µs and hence a favorable value. The TERRESTR value identifies the TG as being a terrestrial link with a propagation delay between 0.48 and 49.152 µs. The PACKET value indicates that the TG link uses a packet type of transmission, and hence data must be packetized and depacketized through the various packet switch nodes of the TG. The propagation delay for the PACKET value is calculated between 49.152 and 245.76 µs. The fourth value allowed for the propagation delay characteristic is the LONG value. This value can be coded for such TG links as satellite or long-distance lines operating at low data rates. The propagation delay for the LONG value is determined to be 245.76 µs or greater.

The SECURITY characteristic of the TGP definition statement specifies the level of security provided for on this TG link. Security here refers not to data security but to the ability of the physical media used to handle outages. Media used for TGs can be "tapped" through the use of electronic devices. The SECURITY value allows the systems programmer to identify the possibility of data being intercepted or changed through such devices with different types of media. There are seven levels of security that can be defined for this TG. A value of UNSECURE indicates that this link is the lowest level of physical security. The PUBLIC value indicates that the TG operates over a switched public network; this may be useful when defining switched 56K connections, X.25, or ISDN switched connections. The UNDERGRO value indicates that the TG defined is actually underground cable but that the cable may be prone to outages because of public utilities maintenance or due to construction work. The SECURE value means that the physical medium is found in a conduit but is not guarded against unforeseen accidents as well as it could be. The GUARDED value indicates that the medium is found in a conduit and is physically protected from faults. The ENCRYPT value indicates that the TG connection supports encryption of the data and the SHIELDED value indicates that the media is in a guarded conduit and is physically shielded to prevent electromagnetic radiation leakage. The default for SECURITY is UNSECURE.

The last three characteristics are end-user defined. The UPARM1, UPARM2, and UPARM3 values default to 128 each and may be used by a systems programmer in further quantifying an optimal route. The values for the user parameters can range from 0 to 255. Not all three need be defined, but the use of them is taken into account by TRS when determining optimal routes.

The TG profile is selected through a definition parameter on the GROUP, LINE, or PU definition statement. The parameter TGP on these statements specifies the name of the TG profile member in VTAMLST. Any of the characteristics defined in this member can be overridden on the GROUP, LINE, and PU definition statements. The characteristics are overridden by defining the characteristics keyword along with the appropriate value as discussed above.

12.5.3 Network Node Server List Major Node

A VTAM V4R1 EN obtains network services from a NN server using the CP-CP session with the NN. Usually an EN will select its NN server based on the first NN which supports a CP-CP session with it.

NNSL	VBUILD	TYPE=NETSRVR,	NN server list
		ORDER=FIRST	Selection order
NN1	NETSRVR		Same NETID as VTAM
NN2	NETSRVR	NETID=NETB	NETB.NN2
	NETSRVR		Any NN

Figure 12.33   Network Node Server List major node format and sample definition.

VTAM V4R1 allows the systems programmer to have some control over which NN servers can be used by VTAM. This is done by defining a Network Node Server List major node in SYS1.VTAMLST.

The network node server list major node is defined using a VBUILD definition statement as shown in Figure 12.33. The VBUILD definition statement identifies this major node as an NN server list by the TYPE=NETSRVR parameter. A second parameter on the VBUILD definition statement is the ORDER parameter. This parameter can have either the value FIRST or NEXT defined. This keyword specifies the method used by VTAM when it attempts to restart a failed CP-CP session with a new NN server. If the ORDER keyword is set to FIRST, VTAM will start its SP-SP session attempts from the top of the network node server list defined in the major node. If the ORDER keyword value is set to NEXT, it will try the next CP name found in the list that follows the CP name that just failed, and then continue down the list until a CP-CP session is established.

The NETSRVR definition statement may have a CP name assigned to it at the beginning of the statement. The NETSRVR definition may also have a NETID parameter defined. If the NETID parameter is not coded and a CP name is coded then the NETID is determined to be the same as that found on the EN. If the CP name and NETID are not specified then any NN may be used as the server as long as the XID3 exchange indicates that the NN is CP-CP session capable with this EN. The example in Figure 12.33 illustrates the use of the NETSRVR definitions statement. The first definition identifies NN1 as the first NN server to use by this VTAM EN. NN1 is considered to be in the same network as the NN1 since the NETID parameter is not defined on the NETSRVR definition statement. NN2 would be the next NN to try in the network should NN1 fail. The NETID value of NETB indicates that NN2 is in network NETB which does not necessarily have to be the same as the VTAM EN. The last NETSRVR definition indicates that if any of the other NN servers in the list failed to establish a CP-CP session then VTAM EN can issue a broadcast to the network to find a NN server.

12.6 Summary

We have defined all the applications, local non-SNA devices, local SNA devices, and token-ring devices attached via an IBM 3172. We also discussed the various parameters and major nodes required to use APPN under VTAM V4R1. In the next chapter, we will define devices to VTAM that are not channel-attached but remotely attached through a communications controller executing NCP using switched lines. These resources are defined to VTAM in a switched major node.

Switched Major Nodes and Path Tables

13.1 Switched Major Node

This major node is used to define SNA resources that gain access to the network by using dial-up facilities, that is, a phone call from either the network control center operation facility or the remote end-user's facility. This dial arrangement can provide a number of remote sites with access to the network without the added expense of using a private dedicated telecommunications line. The link protocol used is SDLC. Each switched SDLC line must have a switched line definition in the NCP. For products such as NTO that use virtual physical and logical units, these NCP switched line definitions are also needed. The switched major node is defined using a VBUILD statement, a PU (switched) statement, a PATH statement for dialout operations, and an LU (switched) statement. Figure 13.1 diagrams a switched connection.

In our discussion of switched lines in Section 10.1.7, we had determined that a remote System/36 (S/36) processor will transmit data to the host over a switched line during the night production shift. This will be accomplished by the S/36 emulation of an SNA/RJE 3770. We will now define to VTAM this remote switched SNA device in a switched major node.

13.1.1 Switched PU definition statement

Prior to the switched PU definition statement, a VBUILD definition statement must be coded that defines this switched major node. See Section 12.1.1 for more on the VBUILD definition statement.

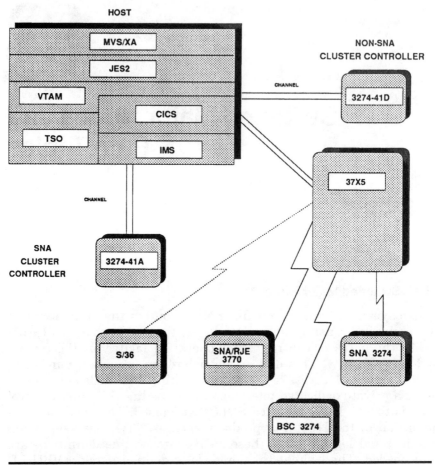

Figure 13.1   Single-domain network with a switched line to a remote System/36 processor.

A PU definition statement must be coded for each physical unit that is to be included in this switched major node. Figure 13.2 has the complete format of the switched PU definition statement.

The name parameter is required and assigns the minor node name to this physical unit. This name must be network unique. The operand ADDR is the hexadecimal 1-byte value of this PU's SDLC station address. This address must match the SDLC station address that was assigned to the PU during the configuration process of the control unit. This value does not need to be enclosed in quotes or apostrophes.

An identification structure that is unique to a switched PU is the station ID. This station ID is used during the exchange identification (XID) command during the dial procedure. The station ID must also be network unique. Figure 13.3 shows the format of the station ID.

```
name       PU          ADDR=physical unit station address

                       [,IDBLK=identification block number]

                       [,IDNUM=identification number]

                       [,CPNAME=Node Type 2.1 control point name]

                       [,TGP=TG profile name]

                       [,TGN=preferred TG number]

                       [,NETID=node network identifier]

                       [,CPCP=YES I NO I SWITCHED I LEASED]

                       [,TOPO=CONNECT I 0-255]

                       [,CONNTYPE=APPN I LEN]

                       [,DISCNT=([YES I NO] [,F I NF]}

                       [,IRETRY=YES I NO]

                       [,ISTATUS=ACTIVE I INACTIVE]

                       [,MAXDATA=size I 261 I 265]

                       [,MAXOUT=n I 1]

                       [,MAXPATH=n I 0]

                       [,PASSLIM=n I 1]

                       [,PUTYPE=n I 2]

                       [,SECNET=YES I NO]

                       [,SPAN=(spanname]
```

Figure 13.2 Complete format of the switched PU definition statement for a switched major node.

```
BITS:      0 - 3       Reserved
           4 - 7       PUTYPE value
           8 - 15      "00"
          16 - 27      IDBLK value
          28 - 47      IDNUM value
```

Figure 13.3 Station ID format for the XID command.

The PU TYPE operand defines to VTAM the type of physical unit. The default is PU TYPE 2. But notice that here, in a switched PU definition statement, you may code something other than a Type 2 device. You should consult the device's component description manual to determine the value for this operand. The IDBLK operand is required. This value is a three-digit hexadecimal number that represents the device type. Each device type has a unique identification block number. Again, consult the component description manual for the device you are defining. The final piece of the station ID is the IDNUM operand. It is also a required operand. The value coded here can be any five-digit hexadecimal value. Some shops use the serial number of the device, others devise their own number. This part of the station ID format is where the station ID uniqueness is derived. Make sure no switched PUs in your network have the same IDNUM value. Although the DISCNT operand was covered in the local PU definition statement, it acts differently for a switched PU. For a switched PU, the YES|NO value determines if the actual physical connection is to be disconnected after the last LU-LU session has terminated for that PU, literally hanging up the phone. This will save telephone costs, especially if a user's session is timed out by the system but the dial connection is not broken. However, the switched PU and LU status in VTAM remains active. This will allow a new dial connection to be reestablished immediately. The default NO will keep the switched link connected. It is advisable for you to code YES for this parameter. The IRETRY operand pertains to the switched PU's connecting NCP. If YES is coded, the NCP will retry a polling sequence immediately following an Idle Detect Timeout for the device. The MAXDATA is the largest Path Information Unit (PIU), in bytes, that can be received by the physical unit. The maximum value you can code is 65,535 bytes. If MAXDATA is not coded, the default is 261 for a PU Type 1 and 265 for a PU Type 2 device. The MAXOUT operand value supplies the maximum number of PIUs that can be sent to the PU before requesting a response from the PU. The value range for n is a decimal integer from 1 to 7. The MAXPATH operand value is the number of dial-out paths available to the PU. A value range of 0 to 256 is valid. The default value 0 signifies that only dial-in paths are available for this PU. The final operand is PASSLIM. This value will determine the number of PIUs that the NCP will send to the PU in one transmission. The valid range is 1 to the value of MAXOUT for switched PUs. The remaining operands not discussed can be reviewed in the previous section for local SNA major node definition.

Included on all PU definition statements for APPN support are six new keywords. The TG Number (TGN) assigned to the connection for an APPN node defined by this switched definition is defined with the TGN keyword. The default value for the TGN keyword is 21. This value on the TGN keyword is the preferred TG number and may not be the final number after negotiation. The CPNAME and NETID keywords are required on the PU definition statement when the TGN keyword is defined. The CPNAME is the name of the control point in the adjacent node. The NETID keyword is the network identifier of the network associated with the adjacent PU.

The profile to use for the TG associated with the CP-CP connection is specified on the TGP keyword of the PU definition statement. This value is the name assigned to a TGP definition statement in the TGP member of VTAMLST. The CPCP keyword can be used on the PU definition statement to override the default value found in the VTAM ATCSTR00 start list.

The TOPO keyword is defined on the PU definition statement to provide control of the way the link station is reported to the topology data base. The TOPO keyword can have two values. The first is the CONNECT value. This is the default for the TOPO keyword. The CONNECT value means that connections established over this link station are reported to the topology database when the connection is made and reported as inactive when the connection is dropped. A value of 0 to 255 on the TOPO keyword indicates that the switched TG is to be included in the topology database before the link is activated. The value coded here is used as the COSTTIME characteristic of the TGP. If TOPO is coded, the TGN, NN, NETID and CPNAME keywords must also be defined.

The CONNTYPE keyword on the PU definition statement determines whether the Node Type 2.1 lines are to be established as APPN connections or LEN connections. The CONNTYPE parameter is valid only when the NODETYPE parameter is defined in VTAM. A value of APPN indicates that previously defined LEN nodes are now treated as APPN connections. APPN is the default value for CONNTYPE. A value of APPN indicates support for parallel TGs, CP-CP sessions, and CP name change support. The APPN value also means that the link is reported to the topology database while LEN links are not. The LEN value for CONNTYPE means that the node and link defined by this PU definition are to be treated as LEN-type nodes. CONNTYPE may also be defined on NCP GROUP, LINE, and PU definition statements.

13.1.2 Switched LU definition statement

The switched LU definition statements must follow the associated PU definition statement within this switched major node. There must be

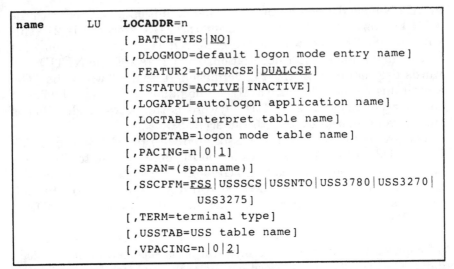

```
name        LU     LOCADDR=n
                   [,BATCH=YES|NO]
                   [,DLOGMOD=default logon mode entry name]
                   [,FEATUR2=LOWERCSE|DUALCSE]
                   [,ISTATUS=ACTIVE|INACTIVE]
                   [,LOGAPPL=autologon application name]
                   [,LOGTAB=interpret table name]
                   [,MODETAB=logon mode table name]
                   [,PACING=n|0|1]
                   [,SPAN=(spanname)]
                   [,SSCPFM=FSS|USSSCS|USSNTO|USS3780|USS3270|
                            USS3275]
                   [,TERM=terminal type]
                   [,USSTAB=USS table name]
                   [,VPACING=n|0|2]
```

Figure 13.4 Complete format of the switched LU definition statement for a switched major node.

one LU definition statement for each LU to be defined for this physical unit. Figure 13.4 shows the complete format of the switched LU definition statement.

The name parameter is required and assigns the minor node name to this logical unit. This name must be network unique. The operand LOCADDR is also required. This value is the logical unit's local address that is assigned at this physical unit. The n value is a decimal integer. Its range varies depending on the PU type for the physical unit. A PU Type 1 allows the range to be 0 to 63; PU Type 2 has a range of 1 to 255. Consult the device's component description manual for the correct value range. The LOCADDR need not be consecutive, and an LU statement is not required for every possible local address.

The BATCH operand provides the LU's processing priority for service by the NCP. If NO is coded, the LU will have a high priority. This is particularly important if the LU is used for interactive applications. If YES is coded, NCP assigns the lowest possible priority for service. The FEATUR2 operand is used for NTO devices. This operand tells VTAM how to send the data coded with the TEXT operand of the USSMSG macro coded in the USS table for non-SNA terminals during the SSCP-LU session. The LOWERCSE operand tells VTAM to send only lowercase characters to the terminal. The DUALCSE means that VTAM is to send the characters as they are coded on the USSMSG macro to non-SNA terminals. DUALCSE is the default. The values for the SSCPFM operand are self-explanatory, determining device type SSCP-LU session support. The TERM operand is for virtual LUs supported through the NTO program product. It identifies the device data stream compatible characteristics. The VPACING value on a switched LU is different from the local LU definition in that the default for a switched LU is 2. Otherwise, all the operands in Figure 13.4 are the same as those defined for a local LU definition statement.

13.1.3 Switched PATH definition statement

The switched PATH definition statements must immediately follow the associated PU definition statement within this switched major node. The PATH statement is used to define dial-out paths to the PU. Some of the operands are exclusive to X.21 switched protocol. You can code up to 256 PATH statements for each switched PU definition statement. The search for a path is performed in the order of the coded PATH statements by VTAM. Figure 13.5 outlines the format of a PATH definition statement.

The name parameter is operational and is not used by VTAM. The DIALNO operand is the telephone number used to call the PU over a switched link. Special characters can be coded in the phone number. A vertical bar can be used as a separator character or a dialing pause character. An underscore can be a separator character, and an *, %, or @ can be an end-of-number character. The length of this operand can be up to 32 characters. For autocall or address call with X.21 switched lines, a unique end-of-number character must follow the digits. However, you do not need to code this. VTAM will supply the unique end-of-number character for the X.21 service. If an NCP/Token-Ring Interconnection (NTRI) is used, the DIALNO must be in the following format: DIALNO=xxyy4000ccccccccc, where xx is the Token-Ring Interface Coupler (TIC) number of the communications controller. This value must be between 00 and 99. The yy value is the system access point address of the terminal. This value must be a multiple of 4. Finally, zzzzzzzz is the last four digits of the termi-

```
[name]      PATH  [DIALNO=telephone number|LINENM=linename]
                  [,GID=n]
                  [,GRPNM=groupname]
                  [,PID=n]
                  [,REDIAL=n|3]
                  [,SHOLD=YES|NO]
                  [,USE=YES|NO]
```

Figure 13.5 Complete format of the switched PATH definition statement for a switched major node.

nal's ring-station address. The first digit must be between 0 and 7. Each zz represents 1 byte. All the x, y, z values are decimal numbers. The LINENM option of the DIALNO operand is mutually exclusive with DIALNO. LINENM is a VSE-only operand. The line name specified must be defined as a direct call line by the common carrier facility. The GID operand identifies a group of paths that exists between all the PUs in the switched major node. The value may be 0 to 255. The GRPNM operand value is the symbolic name assigned to the NCP GROUP definition statement to which this switched line is associated. This parameter is required if you code the SHOLD=YES parameter for this PATH definition. The SHOLD parameter is applicable to VSE operating systems only. If you code YES, VTAM is aware that the path is an X.21 Short Hold Mode/Multiple Port Sharing (SHM/MPS) path. Also, the DIALNO and GRPNM operands must be coded. This feature is available on VTAM V3R1. See the GROUP statement for channel-attached major nodes in Chapter 14 for VM and VSE operating systems. The PID operand is a unique identifier for the path being defined. The REDIAL operand defines to VTAM the number of dialing retries before a dialing error is issued to VTAM. The minimum value is 0. This means there are no dial retries. The maximum is 254 retries. The final operand is the USE operand. This operand is similar to the ISTATUS operand for PUs and LUs. It tells VTAM if this path is made usable at activation of the switched major node.

In the sample switched major node, Figure 13.6, the remote SNA RJE station can be dialed by using the phone number supplied by the DIALNO parameter from the PATH statement. When the operator activates the switched major node, the logical unit will remain inac-

```
MAJOR NODE NAME:    SWS36RJE
MINOR NODE NAME:
                 VBUILD   TYPE=SWNET,MAXGRP=1,MAXNO=1
  SWS36PU    PU     ADDR=C1,IDBLK=016,IDNUM=19874,PUTYPE=2,      X
                    DLOGMOD=BATCH,MAXPATH=1,MAXDATA=265,         X
                    MAXOUT=7,SSCPFM=FSS
  SWPATH     PATH   DIALNO=15551212,GID=1,GRPNM=SWRJEGRP,        X
                    PID=1
  SWRJEL01 LU       LOCADDR=01,PACING=1,LOGAPPL=JES2RJE,         X
                    ISTATUS=INACTIVE
  SWRJEL02 LU       LOCADDR=02,PACING=1,LOGAPPL=JES2RJE,         X
                    ISTATUS=INACTIVE
```

Figure 13.6 Coding of the switched major node for dial-up S/36 emulating an SNA/3770 3777 multiple logical unit.

tive. The operator may activate the LU with a VARY NET,ACT,ID= SWRJEL01 command after the PU has been activated or by using the SCOPE=ALL operand of the VARY ACT command when activating the major node or the PU. The IDBLK value is the coded ID of the 3777 multiple logical unit that the S/36 had configured for the SNA/3770 emulation. The two other operand values that must agree between the switched major node definition and the SNA/3770 emulation configuration are the ADDR and IDNUM operands. Both must be entered into the emulation configuration. These three values will be used in the XID command to verify the device. If the values do not match, VTAM denies the session request.

13.1.4   Token-ring considerations

SNA views PU token-ring connectivity through the IBM 37X5 gateway as switched or dial connections. VTAM on the SNA host computer defines switched connections using a VTAM switched major node. The switched major node identifies the token-ring station physical units, and their associated logical units, that can "dial" to the mainframe or that the mainframe may "call."

The VTAM switched VBUILD statement defines to VTAM a switched major node definition. The format for defining token-ring resources is listed in Figure 13.7. This format is the same regardless of the token-ring station device type.

The PU definition for token-ring resources uses traditional switched SNA parameters. The PU definition statement itself identifies the station address representing this physical unit. Specific to switched connection are the IDBLK and IDNUM keywords on the PU definition statement, as shown in Figure 13.7. These fields together are the unique identifiers of the physical unit and must match the value sent in the PU's Exchange Identifier (XID). The IDBLK keyword value identifies the actual resource type. For instance, an IBM 3174 will use the value 017. The IDNUM keyword value indicates the unique identifier for the resource. The IDNUM value must match the configuration question 215 on an IBM 3174 Establishment Controller. The MAXDATA keyword specifies the largest unit of information that the physical unit can receive. This value must match the value speci-

```
TRSWNET            VBUILD    TYPE=SWNET
DSPU01             PU        IDBLK=017,IDNUM=E1101,MAXDATA=1033,
                             ADDR=C1
TRSWPATH           PATH      DIALNO=0104400031741101
DSPULU01           LU        LOCADDR=02
DSPULU02           LU        LOCADDR=03
```

Figure 13.7 Example of a token-ring switched major node definition.

fied on the physical unit definition. If the resource is a Type 2.1 node, then the CPNAME parameter may be used instead of the IDBLK IDNUM combination to uniquely identify the resource.

The PATH definition statement specifies the LAA address of the station being defined by the PU definition statement. The keyword DIALNO identifies the address of the station. The format of the DIALNO keyword value is xxyy4000abbbbbbb. The xx value specifies the actual TIC position in the IBM 37X5 Communication Controller. The yy value is the Service Access Point (SAP) for an NCP. The 4000abbbbbbb matches the MAC address defined on the station. If the station is an IBM 3174 Establishment Controller, the value here should match the value defined on the configuration question 380 of the establishment controller.

The LU definition statement defines the address of the logical unit on the physical unit. The LOCADDR keyword defines the local address of the logical unit. Typically SNA physical units can support up to 255 logical units. If the LU being defined is an independent LU 6.2 resource, the LOCADDR value is specified with a value of 0.

13.2   PATH Table

Communications between SNA subarea nodes is made possible by the defining routes between each adjacent subarea. In SNA, each subarea must have a map of subareas that can be reached. These are called the "destination" subareas. This mapping of destination subareas provides the possible routes a message may use to reach its destination. The actual mapping of the routes is accomplished by coding PATH definition statements.

The PATH definition statement consists of a destination subarea operand, DESTSA; an explicit route operand, ERn; a virtual route operand, VRn; and a virtual route pacing window size operand, VRPWSnn. Review Chapter 5 for a more detailed description of a subarea, ER, and VR.

13.2.1   PATH definition statement

More than one PATH definition statement may be coded in the PATH major node. For this table the VBUILD statement is not used. The name operand (Figure 13.8) assigns a name to the PATH statement being defined. Its value is used during a definition error or a warning message during this path's activation.

The DESTSA operand defines what subareas this VTAM can communicate with over this path. The value specified for n is the subarea number of the destination subarea. For example, in our single-domain network, all of our remote devices are attached to a single NCP/3725.

```
name      PATH          DESTSA=n|(n1,n2,n3...)
                        [ER0=(adjsub[,tgn])]
                        [ER1=(adjsub[,tgn])]
                              .
                              .
                        [ER7=(adjsub[,tgn])]
                        [ER8=(adjsub[,tgn])]
                              .
                              .
                        [ER15=(adjsub[,tgn])]
                        [,VR0=er#]
                        [,VR1=er#]
                              .
                              .
                        [,VR7=er#]
                        [,VRPWSvr#tp=(min#,max#)]
```

Figure 13.8 Format of the PATH definition statement.

This NCP has been assigned a subarea number of 2. Therefore, this operand is coded as

```
DESTSA = 2
```

Since our single-domain example (Figure 13.9) has only one NCP, we code only one destination subarea in the PATH statement. However, later on when we discuss multiple NCPs in the network, you will see that more than one destination subarea may be coded on the DESTSA operand.

The ER0...ER15 operand identifies the adjacent subarea that VTAM can use to route data to the destination subarea. ER8 to ER15 can be used with VTAM V3.2 or higher only when the adjacent NCP is NCP V4.3.1/V5.2.1. Each Explicit Route (ER) can have a transmission group number assigned to it. For channel-attached subareas, VTAM always uses transmission group number 1.

In Figure 13.10, subarea 2 is reached by VTAM via Explicit Route 0 (ER0) over Transmission Group Number 1 (TG#1). The ER statement is coded as

```
ER0 = (2,1)
```

where the adjacent subarea is 2 and the transmission group to which this subarea is attached is transmission group number 1.

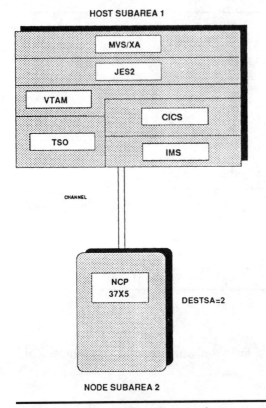

HOST SUBAREA 1

MVS/XA

JES2

VTAM

CICS

TSO

IMS

CHANNEL

NCP
37X5

DESTSA=2

NODE SUBAREA 2

Figure 13.9 Single-domain subarea nodes.

The VR0...VR7 operand associates the explicit route to an adjacent subarea to a virtual route. The virtual route in VTAM is a logical route, whereas the explicit route is a physical route. Figure 13.11 diagrams virtual routes to subarea 2. Virtual route 0 will be mapped to ER0. The VR statement is coded as

```
VR0 = 0
```

The final operand is the virtual route pacing window size. The VRPWS*vr#tp* statement defines to VTAM the number of messages that can be sent through this virtual route at any given time. The *vr#* is the virtual route number associated with this window. The *tp* is the transmission priority for this virtual route. The transmission priority ranges from a low priority of 0 to a high priority of 2. This means that we can have up to 24 VR pacing window size statements, from VRPWS00 to WRPWS72.

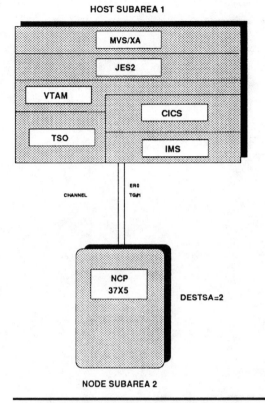

HOST SUBAREA 1

MVS/XA

JES2

VTAM

CICS

TSO

IMS

CHANNEL

ER0
TG#1

NCP
37X5

DESTSA=2

NODE SUBAREA 2

Figure 13.10   The use of explicit routes to the destination subarea 2.

The *min#* defines the minimum window size for this VR. This value must be greater than 0 but less than or equal to the maximum window size. The *max#* is the maximum window size for this VR. The value coded here must be greater than the minimum window size and less than or equal to 255. If either the *min#* or the *max#* is not coded, the VRPWS is ignored by VTAM.

If you do not code the VRPWS operand, VTAM will determine the *min#* and *max#* of the window size by:

1. Setting the *min#* to the number of subareas associated with the ER (ER length).

2. Setting the *max#* to 3 times the ER length.

3. If the VR ends in an adjacent subarea to VTAM, the *max#* is set to the larger of the following values:

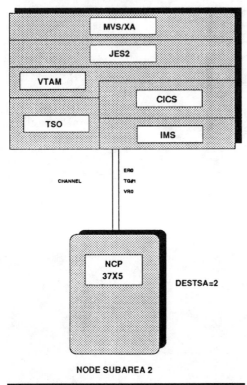

HOST SUBAREA 1

MVS/XA

JES2

VTAM

CICS

TSO

IMS

CHANNEL

ER0
TG#1
VR0

NCP
37X5

DESTSA=2

NODE SUBAREA 2

Figure 13.11 The use of virtual routes to the destination subarea 2.

```
PATH12   PATH       DESTSA=2,
                    ER0=(2,1),
                    VR0=0
```

Figure 13.12 Format of the coded PATH definition statement for our single-domain network.

A. 15

B. 255 minus $(16 \times n)$ where n is the number of ERs that originate in this VTAM host and pass through the adjacent subarea.

These options allow VTAM to increase or decrease the window size accordingly. For the most part, you are better off letting VTAM decide on the size of the window.

In conclusion, the PATH definition statement for our single domain network is coded as in Figure 13.12.

13.3 Summary

We have now reviewed and coded major and minor nodes that are configured into our single-domain network in an MVS operating system environment. All the operands discussed are valid for an MVS, MVS/XA, MVS/ESA, VM, or VSE operating system environment except as noted. The following chapter reviews the VTAM operands that are unique to both the VM and VSE environments but are not used in an MVS environment.

VM and VSE Channel-Attached Configurations

The VM and VSE operating systems need special consideration when defining a VTAM network. The channel-attached SNA devices are defined to the operating systems as switched or nonswitched lines, and the coding resembles that used in NCP. These VTAM resource definitions are used with either the Integrated Communications Adapter (ICA) of the IBM 4361 host processor or the Telecommunication Subsystem Controller (TSC) of the IBM 9370 host processor. For the VSE operating system, this is really the only consideration for you to remember. However, on the VM operating system we must point out the considerations in more detail.

As the name "VM operating system" implies, all devices addressed by VTAM are defined as virtual addresses rather than real addresses like those used in MVS. The virtual addresses you code for the VTAM devices must be used with the following VM commands:

1. DEDICATE statements coded in the VM directory

2. CP ATTACH commands coded for PROFILE GCS EXECs for the VTAM and VSCS virtual machine

3. CP ATTACH command entered from the operator console

4. CP DEFINE commands entered on the VTAM operator console or used in a profile

One other requirement on a VM operating system that affects VTAM definitions is the VM userid. All logical unit device names that will log onto VM cannot be the same as any VM userids in the system.

14.1   SDLC Nonswitched Channel-Attached Major Node

This major node defines to the VM/VTAM and VSE/VTAM all the characteristics and number of PUs and LUs that are grouped into this major node.

14.1.1   The GROUP definition statement

For SDLC nonswitched channel-attached devices, this GROUP statement defines line characteristics for the lines defined following the GROUP statement and, if you desire, for the PUs and LUs associated with each line. Figure 14.1 contains the operands used for the GROUP definition statement.

As you can see from Figure 14.1, the statement is not extensive. The name parameter assigns the minor node name for the line group. It can be one to eight alphanumeric characters but cannot begin with a $. The LNCTL operand is the only required operand. In this case, it defines the line control for this group as SDLC protocol. The DIAL=NO operand identifies this line group as a nonswitched line group. If switched network backup is provided, this operand is still coded as DIAL=NO. The last operand is the SPAN operand. It defines to NCCF or NetView an operator's span of control for this VTAM resource.

14.1.2   The LINE definition statement

The LINE statement is actually describing the channel to which the physical units are attached. A LINE statement is coded for each SDLC nonswitched line in your VM/VTAM channel-attached network. Figure 14.2 outlines the operands for the LINE definition statement.

Both the name and the SPAN operand have been discussed under the GROUP statement, and their meanings also apply to the LINE statement. The ADDRESS operand is the channel unit address for the SDLC nonswitched line. The default address is 030. You may override the address by specifying the U= parameter on the VARY

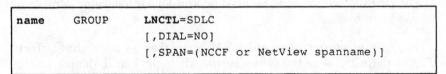

```
name      GROUP     LNCTL=SDLC
                    [,DIAL=NO]
                    [,SPAN=(NCCF or NetView spanname)]
```

Figure 14.1   GROUP definition statement format for SDLC nonswitched channel-attached devices.

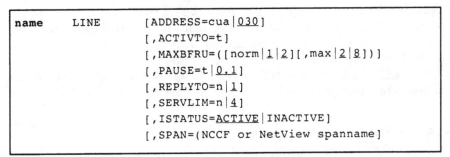

Figure 14.2 LINE definition statement format for SDLC nonswitched channel-attached devices.

NET,ACT operator command. The ACTIVTO operand defines the interval that the communications adapter is to wait without detecting an SDLC frame from another NCP or VTAM.

The MAXBFRU operand defines the number of buffers VTAM will use for normal channel program reads and writes. This value must be coordinated with the IOBUF pool for VM and the LFBUF pool for VSE. These pools define the size of VTAM's message buffers. You must determine the average or normal PIU size and the maximum or largest PIU size being transmitted on this line. Once this determination has been made, you can code an appropriate value for MAXBFRU based on the message buffer size and the average and maximum PIU sizes. The *norm* value should result in a final message PIU size that is 1 to 2 message buffers larger than the normal PIU size. The default for PU Type 1 and 2 devices is 1, and for PU Type 4 and 5 devices it is 2. The *max*imum value coded should be large enough to hold the largest PIU that is to be transmitted over this line. The default for PU Type 1 and 2 is 2, and for PU Type 4 and 5 it is 8.

The PAUSE operand controls the polling to the SDLC PU. The value specified is the time to wait before sending another poll if the PU had no data to send after receiving the first poll. The default is 0.1 s. The valid range is 0 to 25.5 s. This operand is not used with PU Type 4 (NCP) devices.

The REPLYTO is the time-out value for the line when it is the primary station. If a response to a poll has not been received in the interval specified, an idle time-out is detected. VTAM will retry the poll up to the limit specified by the RETRIES operand of the PU statement. The range is 0.1 to 25.5 seconds. The default is 0.1 s.

The SERVLIM operand sets the ratio value of data polls to contact polls for PUs on the line. This means that all PUs that have been contacted and are active will be polled as many times as specified by the

SERVLIM value before trying to contact PUs that have not responded to the contact request. This provides more service to PUs with active sessions rather than being concerned with contacting nonproductive PUs. The range is from 0 to 255. The default is 4.

The ISTATUS operand tells VTAM if this resource should be made active when the major node is activated. The default is ACTIVE.

14.1.3 The PU definition statement

In the VM or VSE environments, a PU statement is coded for each physical unit type (PU Types 1, 2, 4, and 5) that VTAM is to communicate with over this nonswitched SDLC line. We may intermix PU Types 1 and 2 on the same line, creating a multipoint channel-attached link. But PU Type 4 or 5 must be coded as point-to-point lines; any other configuration for these PU types will result in a VTAM definition error. The format of the PU statement is as shown in Figure 14.3.

The MAXDATA operand is the maximum size of a PIU or segment. The size operand is a value you code for the PU. This value can be found by consulting the documentation provided with the device. For PU Type 1 devices, the default is 261; for PU Type 2, the default is 265. The size of the MAXDATA value is based on the request/response header of 3 bytes, and a transmission header of 2 bytes for PU Type 1 and 6 bytes for PU Type 2. Thus the actual amount of user data transmitted is the maxdata value minus 5 bytes for PU Type 1 and minus 9 bytes for PU Type 2. The valid range of the MAXDATA value is from 5 to 65,535. The MAXOUT operand is

```
name    PU          ADDRESS=hex address
                     [,DISCNT=([YES|NO][,F|NF])
                     [,ISTATUS=ACTIVE|INACTIVE]
                     [,MAXDATA=size|261|265]
                     [,MAXOUT==n|1]
                     [,PASSLIM=n|MAXOUT]
                     [,PUTYPE=1|2|3|4|5]
                     [,RETRIES=n|7]
                     [,SPAN=(NCCF or NetView spanname]
                     [,SUBAREA=n]
                     [,TADDR=hex address|C1]
```

Figure 14.3 PU definition statement format for SDLC nonswitched channel-attached devices.

the maximum number of PIUs that VTAM will send to the PU before requesting a receive ready response from the PU. The range is 1 to 7. The PASSLIM operand defines to VTAM the number of consecutive PIUs to send to the PU before servicing other PUs on the same line. This allows you to provide more service to the more heavily used PUs on the same nonswitched line. The PUTYPE operand defines the type of PU to which this definition pertains. A value of 1 is for an SNA terminal-type controller, such as the IBM 3276; a 2 defines a cluster controller, such as the IBM 3274; a 4 defines the communications controller with an NCP, such as the IBM 3725; and a value of 5 defines the SSCP of VTAM. The RETRIES operand is the number of times VTAM will try to recover transmission errors to or from the PU before inactivating the PU. The default is seven times. The valid range is from 0 to 255. If NCCF or NetView is installed and you are using the span of control facility, you must supply the spanname from NCCF or NetView if you want this device to be under the specific control of the network operators. The SUBAREA operand is used for PU Type 4 and 5 only. The value coded here is the actual subarea of the VTAM or NCP you are connecting to over this line. The TADDR operand is the SDLC station address assigned to a PU Type 4 with an NCP. This is used by VTAM for communicating to the NCP because VTAM is always the secondary partner when communicating through a communications adapter.

14.1.4   The LU definition statement

Since we have previously discussed in detail the definition for a logical unit in Section 12.3.3, we will not discuss it here. All the operands from an MVS LU definition hold true for a VM or VSE operating environment.

14.2   SDLC Switched Channel-Attached Major Node

The VM and VSE SDLC switched channel-attached major node is quite similar to the MVS switched major node. The VM and VSE switched definitions contain GROUP, LINE, and PU statements. The clear differences between the two switched definitions are: (1) VM and VSE do not use the VBUILD statement and (2) LUs are not defined in a VM or VSE switched channel-attached major node definition.

Since many of the operands have the same functions as those in an MVS/VTAM environment, we will discuss only those operands that are specific to VM and VSE.

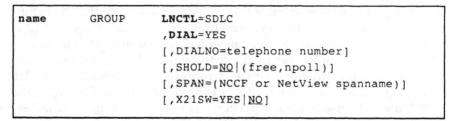

```
name        GROUP     LNCTL=SDLC
                      ,DIAL=YES
                      [,DIALNO=telephone number]
                      [,SHOLD=NO|(free,npoll)]
                      [,SPAN=(NCCF or NetView spanname)]
                      [,X21SW=YES|NO]
```

Figure 14.4   Format of the VM, VSE, SDLC switched GROUP statement.

14.2.1   GROUP statement for SDLC switched line

As you can see from Figure 14.4, you must have the line control operand (LNCTL) specified as SDLC along with the DIAL operand specifying NO in order for this major node to be considered by VTAM as a channel-attached SDLC switched major node in a VM or VSE environment.

The DIALNO operand is applicable to VSE only. The value coded is the actual telephone number used to connect the modems on the line. You must code the DIALNO operand if the SHOLD operand is coded.

The SHOLD operand is for X.21 Short Hold Mode/Multiple Port Sharing (SHM/MPS). The default NO means that this group is not defined for X.21 SHM/MPS. This feature is available to VTAM V3.1 and higher only. The operand X21SW must be set to YES if SHOLD is coded.

14.2.2   LINE statement for SDLC switched line

The important considerations for coding a switched SDLC line statement concern the procedural operands (Figure 14.5). The ANSWER operand determines if a PU can dial to VTAM. The PU may dial VTAM once the line is active if the default, YES, is taken. If NO is coded, the PU may not dial VTAM no matter what the status of the line.

The AUTO operand denotes the use of an autocall unit for this line. The address of the automatic calling unit is the same as the line address. If an X.21 switched interface is in use on a VSE system, code the AUTODL=YES operand to ensure X.21 connectivity.

The determination of whether VTAM and/or the remote station can initiate calls is coded on the CALL operand. The default IN allows only incoming calls; hence, VTAM cannot initiate the calls. If OUT is coded, only VTAM may initiate the calls. But if INOUT is coded,

```
name        LINE        [ACTIVTO=t]
                        [,ADDRESS=channel-unit address|030]
                        [,ANSWER=ON|OFF]
                        [,AUTO=address]
                        [,AUTODL=YES|NO]
                        [,CALL=IN|OUT|INOUT]
                        [,ISTATUS=ACTIVE|INACTIVE]
                        [,MAXBFRU=([norm|1] [,max|2]
                        [,PAUSE=t|0.1]
                        [,REPLYTO=t|1.0]
                        [,RETRIES=n|7]
                        [,RETRYTO=n|12]
                        [,SERVLIM=n|4]
                        [,SPAN=(NCCF or NetView spanname)
```

Figure 14.5   Format of the VM, VSE, SDLC switched LINE statement.

either VTAM or the station can initiate the calls. The RETRYTO operand is VSE specific, and only the ID, the X21SW=YES operand, had been coded.

14.2.3   PU statement for SDLC switched line

As we pointed out earlier, there are no LU statements for a channel-attached SDLC switched physical unit. The LUs are defined dynamically. The number of LUs, and consequently the number of LU control blocks, supported is defined by the MAXLU operand (Figure 14.6). All the other PU definition operands have been previously discussed.

14.3   BSC Channel-Attached Major Node

This major node defines to the VM/VTAM and VSE/VTAM all the characteristics, number of clusters, and terminals that are grouped into this major node.

```
name        PU ADDR = [,ISTATUS=ACTIVE|INACTIVE][,MAXLU=n|2]
                      [,SPAN=(NCCF or NetView spanname)]
```

Figure 14.6   Format of the VM, VSE, SDLC switched PU statement.

14.3.1 The BSC GROUP definition statement

For BSC channel-attached devices, this GROUP statement defines line characteristics for the lines defined following the GROUP statement and, if you desire, cluster and terminal-specific operands for the clusters and terminals associated with this group. Figure 14.7 contains the operands used for the GROUP definition statement.

Figure 14.7 identifies only one required operand for the GROUP definition statement. As with SDLC group definition statements, the LNCTL operand identifies to VTAM the protocol used for this grouping. In this case, BSC is coded to denote this group as binary synchronous.

14.3.2 The BSC LINE definition statement

The BSC LINE statement describes the channel to which the cluster controllers are attached. A LINE statement is coded for each BSC nonswitched line in your VM/VTAM or VSE/VTAM channel-attached network. Figure 14.8 outlines the operands for the LINE definition statement.

ADDRESS and SPAN have the same function here for BSC as they do for SDLC. However, the RETRIES and SERVLIM operands vary from their SDLC counterparts.

The RETRIES operand for BSC determines the number of recovery attempts VTAM tries during transmission to or from the cluster. Remember that SNA deals with physical units for this operand and the SERVLIM operand.

Under BSC protocol, each terminal has a polling address. This address is used by the communications adapter in the front end dur-

```
name    GROUP    LNCTL=BSC
                 [,SPAN=(NCCF or NetView spanname)]
```

Figure 14.7 GROUP definition statement format for BSC channel-attached devices.

```
name    LINE     [ADDRESS=cua|030]
                 [,ISTATUS=ACTIVE|INACTIVE]
                 [,RETRIES=n|7]
                 [,SERVLIM=n|4]
                 [,SPAN=(NCCF or NetView spanname]
```

Figure 14.8 LINE definition statement format for BSC nonswitched channel-attached devices.

ing polling sequences. For BSC, the SERVLIM specifies the maximum number of WRITE output operations to the 3270 screen before soliciting 3270 operator input. The value should be small enough to allow the end user to supply input before screen overwrites. The value is also application dependent, and value selection should be discussed with the application development team.

14.3.3 The CLUSTER definition statement

The CLUSTER statement is used to define the type of cluster controller, the cluster controller resource name, and the general polling address for the cluster controller (Figure 14.9).

The GPOLL operand is a hexadecimal representation of this cluster controller's polling address. For example, if the cluster controller's polling address were defined as the letter B, you would code the characters C2 for the GPOLL value. This GPOLL value is used by the VTAM during polling operations for requests to send or receive data to or from the cluster controller.

The CUTYPE operand defines to VTAM the type of station control unit. Usually, the default 3271 is taken. This value covers the IBM 3174, 3274, 3276, and 5937 BSC cluster control units. The 3275 value is coded for the IBM 5375 BSC controller.

14.3.4 The TERMINAL definition statement

A TERMINAL statement must be coded for each BSC terminal that is to be attached to the defined cluster control unit. The TERMINAL statement supplies VTAM with the resource name for the terminal, the type of terminal and its features, and a device address to be used by VTAM for polling.

We will discuss the three operands unique to this TERMINAL definition statement, ADDR, TERM, and FEATUR2.

The ADDR operand is determined just like the GPOLL operand of the CLUSTER statement. However, here the characters represent the device address instead of the station address. Figure 14.10 contains a table to assist you in determining the ADDR value for BSC cluster controllers using nonswitched lines.

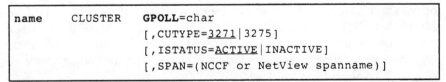

```
name      CLUSTER    GPOLL=char
                     [,CUTYPE=3271|3275]
                     [,ISTATUS=ACTIVE|INACTIVE]
                     [,SPAN=(NCCF or NetView spanname)]
```

Figure 14.9 The CLUSTER statement format for BSC nonswitched channel-attached devices.

```
name      TERMINAL   ADDR=character
                     ,TERM=3275│3277│3284│3286
                     [,ISTATUS=ACTIVE│INACTIVE]
                     [,DLOGMOD=default logmode entry]
                     [,FEATUR2=([MODEL1│MODEL2][,PRINTR│NOPRINTR])]
                     [,LOGAPPL=owning application program name]
                     [,LOGTAB=interpret table name]
                     [,MODETAB=logon mode table name]
                     [,SPAN=(NCCF or NetView spanname)]
```

Figure 14.10 TERMINAL statement format for BSC nonswitched channel-attached devices.

The TERM operand denotes the type of terminal attached to this device address for any IBM 3270-family-compatible display that is being defined with this TERM statement code 3277. For IBM 3287-family-compatible printers, code a value of 3286.

The FEATUR2 operand identifies the device's features. The MODEL1│MODEL2 operand tells VTAM the default screen or buffer size for this device. MODEL1 denotes a screen or buffer default size of 480 bytes. MODEL2 defines a screen or buffer default size of 1920 bytes. Usually, the default MODEL1 is not taken. Most interactive displays and printers have a buffer or screen size of 1920 bytes or more. The PRINTR│NOPRINTR option is valid for an IBM 3275-compatible terminal with an IBM 3284 MODEL3 compatible printer. This alleviates the need to code a separate TERM statement for the 3284 printer.

14.4 Summary

This concludes our discussion of VM and VSE VTAM definitions that need special attention. As you have seen, VM and VSE are unique in the definition of locally attached devices. Actually, the code necessary for defining these components is similar to the code used in an NCP. You will see the similarity later in Part 4 when we deal with coding an NCP. By now you should be aware of the repetitiveness of operands between definition statements. Operands such as DLOG-MOD, MODETAB, LOGAPPL, PACING, and VPACING are used again and again between the different resources. In the following chapters we will discuss these operands in depth and their relationship to VTAM and the network as a whole.

## Chapter 45

VTAM User Tables—USSTAB, MODETAB, COSTAB, and APPNCOS

VTAM has been designed to allow communications systems programmers to customize VTAM for a network's specific needs. The default constants, tables, modules, and exit routes supplied by IBM will indeed be suitable for some networks, but most will need some degree of customization. The management of your network, performance, and availability of network resources may prompt you to investigate the customization possibilities provided by VTAM. In this chapter we will discuss some of the more widely used customization practices. We will also point out why these areas are customized more than others.

15.1 Unformatted System Services Table (USSTAB)

A majority of SNA terminals need a character-coded command to be entered by the end user to request a LOGON or LOGOFF. Typically, the IBM 3270-family-compatible displays require unformatted system services to interpret the logon or logoff request. VTAM has two levels of unformatted system services, session and operation.

The session-level service provides the command and message handling for logical units. The command handling is used when the end user issues a logon or logoff request. Message handling is used when VTAM sends a message to the logical unit. IBM has supplied a default table named ISTINCDT. This is the table that is used if the operand USSTAB= has not been coded. When customizing the USS table, copy the default table to a new name and customize the new table. Also included in the default ISTINCDT is an upper- and lower-case translation table. This allows the end user to enter the unformatted request in upper- or lowercase. The name of the USS table you want the user to access first is coded on the USSTAB=operand of the LU or TERMINAL statements.

The operation level of USS handles VTAM operator requests and the messages generated from those requests. The IBM-supplied default table is named ISTINCNO. The specification of a USSTAB for a VTAM operator is coded in the ATCSTR00 member of VTAMLST by providing the start option USSTAB=. Figure 15.1 contains a table of some of the possible commands that ISTINCNO can modify.

The USSTAB ISTINCNO for operation level also contains the VTAM operator messages. If you decide to change the default message format from this table, make a copy of the table before altering the messages. There are two reasons for this procedure. The first is to provide recovery if the new messages fail to display properly. The second reason is NCCF and NetView usage of the messages. Message routing and the automatic start of some NCCF command lists key on the message ID and the format for the message. Be sure to investigate the network management operations functions before modifying the table.

IBM provides the source code for the default USSTABs that are supplied with VTAM. If your system is an MVS or MVS/XA operating system, the tables can be found in the SYS1.ASAMPLIB data set. For a VSE operating system, you may find the source code in the VSE B

VTAM OPERATOR COMMAND	SYSTEM		
	MVS	VSE	VM
All DISPLAY Commands	•	•	
All MODIFY Commands	•	•	•
VARY TERM	•	•	•
DISPLAY NCPSTOR			•
DISPLAY STATIONS			•
DISPLAY PATHTAB			•
DISPLAY ROUTE			•

Figure 15.1   USS table ISTINCNO default commands.

book. For a VM operating system, the tables are found in a file on the
VTM191 disk with a file type of ASSEMBLE.

Once you have coded your own USSTAB, the source code must be
assembled and then link-edited into the VTAM load library. The load
library contains all the VTAM-executable modules. For an MVS or
MVS/XA system, the USSTAB must be link-edited to the data set
named on the VTAMLIB DD statement from the VTAM start procedure. For a VSE system, the USSTAB must be link-edited to the VSE
core image library. In VM, a procedure tool named VMFLKED is provided to assemble and link-edit the USSTAB to the VTAMUSER
LOADLIB of VM.

Finally, before getting into the actual coding and examples of a
USSTAB, we should discuss the table search mechanism used to
resolve the unformatted system services request. The search for command verb, translation table, and message always begins with the
table named in the USSTAB= operand for the resource issuing the
request, that is, an end-user requesting a logon or a VTAM operator
requesting VTAM information. If VTAM searches the supplied table
name and the command verb and the translation table or the message is not found, it will search the IBM-supplied default tables. If
the request is not found in the default tables, an appropriate message
is sent to the requester describing the error. The parameters associated with a command verb, however, must be found in the table named
on the USSTAB= operand. No searching past the named table is
done for parameter.

We have reviewed the major components and described the uses of
the USS tables. Now, let's look at a coded example.

15.1.1 USSTAB macro

Each USS definition table must begin with the USSTAB macro
(Figure 15.2). This indicates to VTAM that this table can be used to
supply unformatted system services requests.

The name operand assigns a name to the assembled Control
Section (CSECT) for this USS table. As you can see, the CSECT name
is optional, and therefore, if it's not supplied, there will be no CSECT
name assigned to the table.

The other operand applicable to the USSTAB macro is the
TABLE=name operand. This operand is also optional. The value

```
[name]    USSTAB      [TABLE=name]
```

Figure 15.2 Format of USSTAB macro.

coded here is the name of the translation table that you are supplying to VTAM for command translation. Since it is optional, if we let the translation table look up the default, VTAM will use the IBM-supplied default translation table named STDTRANS from the IBM default USS table named ISTINCDT.

This describes the session-level USS table. If we code an operation-level USS table, the TABLE=name operand is ignored because command translation is not performed for the operation level.

15.1.2   USSCMD macro

The USSCMD macro is used to specify the actual command that can be entered by the end user or VTAM operator (Figure 15.3). The CMD operand supplies the user-defined command name. This name must be unique within the USS table being defined.

The FORMAT operand specifies the interpreted format of the command. It is usually allowed to default to PL1 in the majority of VTAM installations. For more information, see the IBM VTAM customization manual.

The REP operand specifies the valid USS command that this user-defined command is replacing. The five valid default VTAM terminal operator commands and how they are used are as follows:

1. LOGON. Used by the end user to request a session with a VTAM application.

2. LOGOFF. Used by the end user to request session termination with the current application session.

3. IBMTEST. Used by the end user to determine the physical connectivity between the terminal and VTAM. This command returns to the terminal display test data. The display of the test data verifies the connectivity to VTAM.

4. UNDIAL. Allows the VM end user to disconnect the terminal from the VTAM virtual machine and return control to the VM control program. This allows the end user to DIAL another virtual machine under the VM control program.

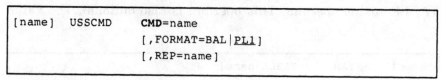

```
[name]    USSCMD     CMD=name
                     [,FORMAT=BAL|PL1]
                     [,REP=name]
```

Figure 15.3 Format of USSCMD macro.

```
USSSNA    USSTAB
LOGON     USSCMD    CMD=LOGON,FORMAT=PL1
CICSP01   USSCMD    CMD=CICSP01,REP=LOGON
```

Figure 15.4 Example of USSTAB and USSCMD macros.

5. VM. Allows a VTAM terminal to log onto the VTAM SNA Console Support (VSCS) facility of VM. This allows the end user to DIAL or logon to any virtual machine just as if the terminal were native to VM.

From Figure 15.4, we can see the immediate use of the USSCMD macro. In Chapter 12 we defined the applications that will be executing under VTAM. One of the applications is the CICS production region. The name that is assigned to this application is CICSP01. Remember that we discussed the importance of naming conventions and the continuity of those naming conventions throughout the network? Well, here is a prime example. By using the application program ID (APPLID) as the command to logon to the CICS application, we enhance the users' abilities to recognize what application they are logging onto, and this also assists in reporting problems with session establishment to the application.

15.1.3 USSPARM macro

The USSPARM macro further enhances the informality of session establishment (Figure 15.5). This instruction provides default values for VTAM LOGON and operator commands. The USSPARM macro must follow a USSCMD macro, and more than one USSPARM macro may follow.

The PARM operand supplies a user-defined keyword or positional parameter that can be entered by the end user for the previously coded USSCMD. The name parameter of the PARM operand specifies the keyword for this USSCMD. You may also code Pn for this USSPARM parameter. The n is a decimal integer starting with 1; the P

```
[name]    USSPARM    PARM=name|Pn
                     [,DEFAULT=value]
                     [,REP=name][,VALUE=value]
```

Figure 15.5 Format of USSPARM macro.

signifies that this parameter is positional, and the n value is the position of the parameter.

The DEFAULT operand value specifies the value for the PARM parameter. If either the keyword value or the positional parameter for the PARM is not entered by the end user, the value from the DEFAULT operand is used.

The REP operand value specifies a keyword that is found in the generated command. This allows you to replace the user-entered keyword value of the PARM operand with the value for the REP keyword value.

The VALUE operand provides a default value for the keyword specified on the PARM operand if the keyword is entered without a value. VALUE and DEFAULT are mutually exclusive. If you want both operands to be used for the same PARM keyword, use two USSPARM macros, the first macro using the VALUE operand and the second using the DEFAULT operand. This allows the end user to enter the keyword without a value, which processes the command using the VALUE keyword first; or, the end user may not enter the keyword, in which case, the DEFAULT operand value will be used.

In Figure 15.6, the USSPARM operand is used in four of the five possibilities. Under the LOGON USSCMD definition, the PARM value is specified as the first positional parameter entered by the end user. The value entered will replace the value for the keyword parameter DATA. The remaining two USSPARM macros define keywords that the end user may enter for this VTAM command. Notice that there are no default values assigned to these keywords. The user must enter the keyword and a value for this USSCMD for VTAM to generate the proper formatted command.

In the second USSCMD definition, the first USSPARM supplies both the keyword and the default value for the keyword if it is not

```
USSSNA    USSTAB
LOGON     USSCMD   CMD=LOGON,FORMAT=PL1
          USSPARM  PARM=P1,REP=DATA
          USSPARM  PARM=APPLID
          USSPARM  PARM=LOGMODE
CICSP01   USSCMD   CMD=CICSP01,REP=LOGON
          USSPARM  PARM=APPLID,DEFAULT=CICSP01
          USSPARM  PARM=LOGMODE
          USSPARM  PARM=DATA
```

Figure 15.6   Example of USSTAB and USSPARM macros.

entered. This allows the end user to enter CICSP01 on the VTAM terminal. This command will be processed by VTAM to initiate a session between the VTAM terminal and the CICS region.

Now that we can assist the end user in session initiation, we must provide a means of communicating possible application selections, session establishment errors, and VTAM terminal information.

15.1.4 USSMSG macro

VTAM provides informational messages to the terminal or network operator. These messages are called "USS" messages. The default messages can be modified to suit your network requirements. This is accomplished by using the USSMSG macro in the USSTAB definition table (Figure 15.7).

There are 14 possible USS messages. Each message is issued to the terminal when the corresponding error for the message has been encountered. We will discuss 3 of the 14 possible USS messages. MSG10, which is sent to an LU when the LU is activated and a SSCP-LU session is active, is often referred to as the VTAM logo. The second USS message we will discuss is MSG7, which is issued when the LU attempts to establish a session but an error has occurred which inhibits the session establishment. This session establishment is also referred to as "binding." The final message, MSG0, is issued to an LU when the USS command has been successfully executed.

As stated previously, USS MSG10 is VTAM's logo or greeting message. The IBM default table does not provide text for this message. This gives you the opportunity to create a message for the end user that can be useful to the terminal operator. You may elect to display logon selection options for the end user to choose or just a simple greeting. The information displayed is up to you. In our example, Figure 15.8, we have coded a simple greeting message. However, the display does describe to end users the name assigned to their LU.

In USS, a special substitutable parameter, @@LUNAME, can be used in the TEXT operand of the USSMSG macro. This parameter, if

```
[name]    USSMSG    MSG=n|(n1,n2,...)
                    [,BUFFER=buffer address]
                    [,TEXT='message text']
                    [,OPT=option]
                    [,SUPP=ALWAYS|NEVER]
```

Figure 15.7 Format of USSMSG macro.

```
USSSNA    USSTAB
LOGON     USSCMD  CMD=LOGON,FORMAT=PL1
          USSPARM PARM=P1,REP=DATA
          USSPARM PARM=APPLID
          USSPARM PARM=LOGMODE
CICSP01   USSCMD  CMD=CICSP01,REP=LOGON
          USSPARM PARM=APPLID,DEFAULT=CICSP01
          USSPARM PARM=LOGMODE
USSMSGS   USSMSG  MSG=10,TEXT='WELCOME TO THE NETWO         X
                  RK. YOUR SNA LU NAME IS @@LUNAME',         X
                  SUPP=NEVER
          USSMSG  MSG=0,BUFFER=MSG0
          USSMSG  MSG=7,TEXT='% SESSION NOT BOUND,           X
                  %(2), SENSE DATA %(3)',SUPP=NOBLKSUP
```

Figure 15.8   Example of USSTAB and USSMSG macros.

```
WELCOME TO THE NETWORK. YOUR SNA LU NAME IS LS320T02.
```

Figure 15.9   USSMSG10 TEXT message translated.

coded, notifies VTAM that the actual displayed message shall have the assigned LU name for this terminal (e.g., LS320T02) displayed on the screen. Figure 15.9 shows the processed message as it appears on the terminal screens.

The format for the TEXT operand is the standard national alphanumeric character set. The length of the displayed message may not exceed 255 characters. This includes all substituted strings. The percent (%) sign has special meaning when used in the TEXT operand. It represents the location of variable data that VTAM can issue with a USS message. The percent signs can be numbered if several variable data strings are issued by VTAM.

In Figure 15.8, message 7 (MSG7) supplies a good example for using the % variable character string insert capability of VTAM with USS messages. Message 7 is issued when the LU attempts to bind with an application but the request is denied. By using the %, VTAM can display some information about the reason for the bind failure. Figure 15.10 shows the inserted values that may appear during a bind failure by using the % substitution.

The OPT and SUPP operands are used only in conjunction with the TEXT operand. They are ignored if coded with the BUFFER operand.

```
The coded MSG 7 is:
%           SESSION NOT BOUND, %(2)              SENSE DATA %(3)
LS320T02 SESSION NOT BOUND, BIND FAILURE SENSE DATA 08010000
```

Figure 15.10   Using the % for VTAM variable string insertion.

The OPT operand specifies whether two or more blanks encountered will be suppressed into one blank. If the message coded contains spatial formatting on the display, code OPT=NOBLKSUP; otherwise code OPT=BLKSUP if you want two or more blanks compressed to one blank. The default is BLKSUP. The SUPP operand specifies whether the message is to be written to the terminal. If ALWAYS is specified, the message will not be written. The NEVER parameter indicates that the message shall be written. The default is the SUPP start option of the VTAM ATCSTR00 start list. It is best to code SUPP=NEVER to be on the safe side.

The last operand to be discussed, BUFFER, like the TEXT operand, can also be used to supply the USS message text to the terminal screen. Unlike the TEXT operand, the BUFFER operand cannot have variable data. VTAM will display the message text from the buffer just as it is coded. The parameter of the BUFFER operand is an assembler label for the storage area of the message text. The message text must be coded according to the rules for an assembler DC statement using C-type constants. The format of the buffer is as given in Figure 15.11.

Although the maximum length of the message is 65,535 bytes, it is not feasible to send a message that large to the terminal screen. Since most 3270 display terminals have a buffer of 1920 bytes, it is good practice not to exceed this value. The BUFFER storage area also allows you to store device-dependent write and control characters. In our example USSTAB for USS message 0 (MSG0), we have used the BUFFER operand. USS MSG0 is issued when a USS command has been successfully processed. The first two bytes of the message storage area, labeled MSG0 (Figure 15.12), contain the length of the mes-

```
   2 BYTE HEADER                  Message Text
MESSAGE TEXT LENGTH         Maximum length is:
  (Header + Message Text)    65,535 or X'FFFF'
```

Figure 15.11   Format of BUFFER storage area.

```
MSG0 DC   X'0063',X'15',C'The LOGON COMMAND entered has        X
          been successfully executed and sent to the app       X
          ropriate application'
```

Figure 15.12   The BUFFER operand storage area for MSG0.

sage. The next byte contains the write-type command for a carriage return to a new line X'15' followed by the actual message.

After coding all the USSCMD, USSPARM, and USSMSG macros and their operands for the USS table, the USSEND macro must be coded.

15.1.5   USSEND macro

The USSEND macro signifies the end of the code for the USSTAB being defined (Figure 15.13). It follows all the USS macros, including the buffer storage area for the BUFFER operand. Once you have finished coding the USSTAB, it must be assembled and link-edited into the proper VTAM executable library. Figure 15.14 documents the USSTAB that will be used throughout this book. This USSTAB definition contains the logon verbs for the applications that were defined in Chapter 12. The name of the USS table is USSSNA. This is the name that is used to link the USSTAB module to VTAMLIB and also the name that is coded for the USSTAB= operand for logical unit and non-SNA terminal definitions.

15.2   LOGON Mode Table

In the previous section we introduced the term "BIND" as being a request of an LU to establish a session with an application. The BIND contains session protocol information for the application from the LU. This session protocol information is also called the "BIND image." It consists of the session parameters that will be used between the LU and the application, that is, the protocol between the Primary Logical Unit (PLU) (e.g., application) and the Secondary Logical Unit (SLU) (e.g., LU). This protocol agreement will determine such factors as which LU can initiate a send or receive request, the size of the mes-

```
[name]  USSEND
```

Figure 15.13   Format of USSEND macro.

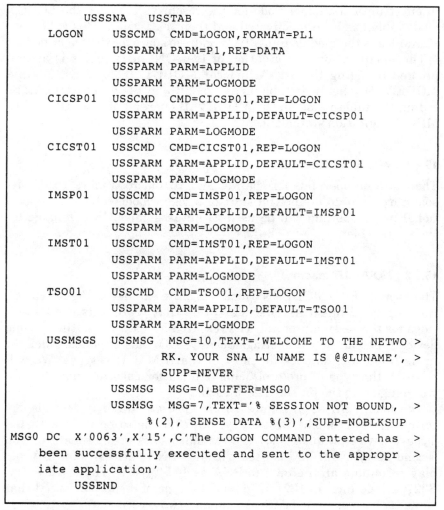

```
          USSSNA    USSTAB
   LOGON      USSCMD  CMD=LOGON,FORMAT=PL1
              USSPARM PARM=P1,REP=DATA
              USSPARM PARM=APPLID
              USSPARM PARM=LOGMODE
   CICSP01    USSCMD  CMD=CICSP01,REP=LOGON
              USSPARM PARM=APPLID,DEFAULT=CICSP01
              USSPARM PARM=LOGMODE
   CICST01    USSCMD  CMD=CICST01,REP=LOGON
              USSPARM PARM=APPLID,DEFAULT=CICST01
              USSPARM PARM=LOGMODE
   IMSP01     USSCMD  CMD=IMSP01,REP=LOGON
              USSPARM PARM=APPLID,DEFAULT=IMSP01
              USSPARM PARM=LOGMODE
   IMST01     USSCMD  CMD=IMST01,REP=LOGON
              USSPARM PARM=APPLID,DEFAULT=IMST01
              USSPARM PARM=LOGMODE
   TSO01      USSCMD  CMD=TSO01,REP=LOGON
              USSPARM PARM=APPLID,DEFAULT=TSO01
              USSPARM PARM=LOGMODE
   USSMSGS    USSMSG  MSG=10,TEXT='WELCOME TO THE NETWO >
                      RK. YOUR SNA LU NAME IS @@LUNAME', >
                      SUPP=NEVER
              USSMSG  MSG=0,BUFFER=MSG0
              USSMSG  MSG=7,TEXT='% SESSION NOT BOUND,  >
                      %(2), SENSE DATA %(3)',SUPP=NOBLKSUP
MSG0 DC  X'0063',X'15',C'The LOGON COMMAND entered has  >
     been successfully executed and sent to the appropr  >
     iate application'
          USSEND
```

Figure 15.14 USSTAB named USSSNA in its entirety.

sage, whether chaining or segmentation of message units is allowed, and more.

IBM has supplied a default logon mode table, ISTINCLM. The module can be found in the MVS or MVS/XA VTAMLIB library, SYS1.VTAMLIB, in the VSE core image library, and in the VM VTA-MUSER LOADLIB. If you create your own LOGON mode table, it must be assembled and link-edited into the appropriate VTAM library. If you decide to modify or create a LOGON mode table and relink it with the same module name or a new name, the mode table will not be in effect until the major node for the resources that use the LOGON mode table has been inactivated and then reactivated. This is because VTAM loads the module into VTAM address space for the major node.

The modification and creation of LOGON mode tables is accomplished by using the VTAM macros MOBETAB, MODEENT, and MODEEND. The mode table is assigned to an LU or TERMINAL by coding the table name on the MODETAB operand of the LU or TERMINAL definition statement.

15.2.1   MODETAB macro

There are no operands for the MODETAB macro (Figure 15.15). Its sole purpose is to define the start of a logon mode table. The operational name operand can be used as an assembler CSECT name for the mode table.

15.2.2   MODEENT macro

The format of the MODEENT macro is quite extensive (Figure 15.16). Each operand has a hexadecimal value that represents the bit settings for the session protocol. The research for each bit setting for one device type can be exhausting. Luckily, IBM has supplied a default mode table with entries for most of the IBM device types. We will describe the type of protocol being defined for each operand, but we will not go into the bit settings of each.

The mode entries in Figure 15.17 represent three typical display terminal devices. The first entry, S32702X, is to be used with LU Type 2 devices that support the extended data stream feature. The second entry, N32702X, is used with non-SNA devices. Non-SNA display terminals are defined to SNA as LU Type 0. The third entry, S32704X, defines an IBM 3278 model 4 type of display terminal that supports both extended data stream and two screen sizes.

The LOGMODE operand specifies the name of the logon mode entry. In this mode table we have three entries defined, S32702X, N32702X, and S32704X.

The FMPROF operand defines the data flow control (DFC) protocols used for the LU-LU session. For both of the SNA LOGMODE entries, the value coded is X'03'. This value signifies that function

```
[name]    MODETAB
```

Figure 15.15   Format of MODETAB macro.

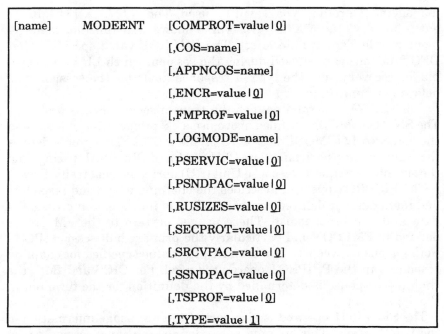

```
[name]        MODEENT        [COMPROT=value I 0]

                             [,COS=name]

                             [,APPNCOS=name]

                             [,ENCR=value I 0]

                             [,FMPROF=value I 0]

                             [,LOGMODE=name]

                             [,PSERVIC=value I 0]

                             [,PSNDPAC=value I 0]

                             [,RUSIZES=value I 0]

                             [,SECPROT=value I 0]

                             [,SRCVPAC=value I 0]

                             [,SSNDPAC=value I 0]

                             [,TSPROF=value I 0]

                             [,TYPE=value I 1]
```

Figure 15.16   Format of MODEENT macro.

```
MT01        MODETAB
S32702X     MODEENT   LOGMODE=S32702X,FMPROF=X'03',                    X
                      TSPROF=X'03',PRIPROT=X'B1',SECPROT=X'90',        X
                      COMPROT=X'3080',RUSIZES=X'87F8',                 X
                      PSERVIC=X'028000000000185000007E00'
N32702X     MODEENT   LOGMODE=N32702X,FMPROF=X'02',                    X
                      TSPROF=X'02',PRIPROT=X'71',SECPROT=X'40',        X
                      COMPROT=X'2000',RUSIZES=X'0000',                 X
                      PSERVIC=X'008000000000185000007E00'
S32704X     MODEENT   LOGMODE=S32702X,FMPROF=X03',                     X
                      TSPROF=X'03',PRIPROT=X'B1',SECPROT=X'90',        X
                      COMPROT=X'3080',RUSIZES=X'87F8',                 X
                      PSERVIC=X'02800000000018502B507F00'
```

Figure 15.17   MODEENT macro coding example.

management profile Type 3 will be used. The N32702X LOGMODE entry has a value of X'02'. This value shows that function management profile Type 2 is in effect. The FMPROF value specifies what DFC functions can be used during the session, which LU is responsible for recovery, and the type of response mode for the session, e.g., delayed or immediate.

The TSPROF operand defines the transmission services protocols. The SNA LOGMODEs defined above use a TS profile value of X'03', and the non-SNA LOGMODE uses a TS profile of X'02'. This profile defines the session rules for traffic flow PLU-SLU and SLU-PLU, pacing, and the size of the Request/Response Unit (RU) during normal traffic flow.

The PRIPROT operand describes the chaining usage and responses and immediate or delayed request mode and indicates compression of data and end bracket use. These usages pertain to the FM profile defined in FMPROF and particularly the primary half-session (PLU). For sessions between TSO and an LU, the value specified for chaining responses in the PRIPROT operand is used. For CICS and IMS, the chaining response is determined by the definition for the terminal in the application.

The SECPROT operand is also part of the FM usage information of the BIND image. This operand value defines the same session protocol parameters as the PROFPROT operand, except that they relate to the secondary half-session partner (SLU).

The final operand that is related to FM usage is the COMPROT. The value specified for this operand defines the common protocols that will be used between the primary and secondary half-session partners. This operand defines the use of Function Management Headers (FMH) and whether they can be exchanged; the use of brackets and the rules of their usage; the use of EBCDIC or ASCII code for the RU data; and the use of full-duplex (FDX), half-duplex flip-flop (HDX-FF), or contention protocol for normal send and receive flows for which LU is the contention loser and has recovery responsibility.

The SSNDPAC operand in our example is not coded. But a value is determined by the VPACING operand of the APPL definition statement which defines the primary logical unit. This operand defines inbound pacing from the SLU to the PLU. The VPACING value is used if the SSNDPAC value is not coded or a value of X'0000' is coded. Using this method provides greater flexibility for pacing modifications.

Outbound pacing is controlled by both the BIND image and another VTAM/NCP operand. The SRCVPAC operand defines the pacing count in the primary to secondary direction. If this value is not specified or is coded with a value of X'0000', the value used in the PACING operand of the SLU will be used. The indication of one- or two-stage pacing is also important here. Two-stage pacing will take place if the

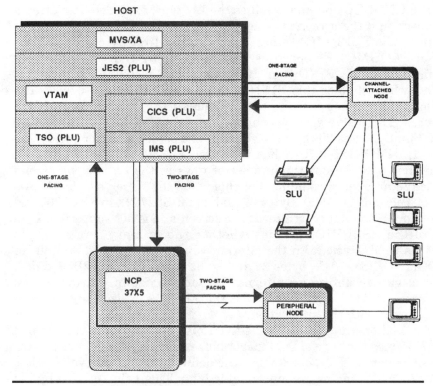

Figure 15.18 One-stage and two-stage pacing.

SLU is a resource of NCP. One-stage pacing is used when the SLU is a resource attached to the HOST. One- and two-stage pacing is diagrammed in Figure 15.18.

The RUSIZES operand defines the maximum size of the RU inbound and outbound to the PLU. The first byte defines the size of the RU inbound to the PLU from the SLU. For most LU Type 2 devices, the value is set at X'87', which equates to 1024 bytes. This equation is 8×2. The value coded here is used by the primary half-session if the PLU is TSO or IMS. If the PLU is CICS, the value coded in the CICS Terminal Control Table (TCT) operand RUSIZE for the entry that defines the LU secondary half-session partner is used. The second byte defines the maximum RU size outbound to the SLU. The peripheral node (e.g., cluster controller) will manage the buffers according to this value. In our example, the outbound RU size can be a maximum of X'F8' or 3840 bytes (15×2) in length. If IMS is the primary half-session partner, the value coded here is overridden with the IMS operand OUTBUF value. For CICS, it is application dependent. TSO will use the value specified in the RUSIZES operand.

The PSERVIC operand specifies the LU type, extended data stream support, and the screen sizes for the device. The LU type is the first byte of the PSERVIC. In our example, the first byte for the SNA device LOGMODEs specifies X'02', signifying that the LU being defined with this BIND is an LU Type 2. For N32702X, this value is X'00', which defines this BIND image for an LU Type 0 device. The following byte has a value of X'80' for all the defined LOGMODE entries. This value specifies support for extended data stream protocol. Finally, bytes 20 through 24 define the screen sizes for the device. The LOGMODE S32702X has the value X'1850' in bytes 20 and 21. The first byte defines the number of rows in the screen, the second byte the number of columns. For this value the screen size is 24 rows by 80 columns, or 1920. Bytes 22 and 23 for S32702X are X'0000', and therefore a secondary or alternative screen size is not supported. Byte 24 of the PSERVIC specifies whether the primary screen size is always used, static, or if the alternative size can be used as well. In LOGMODE S32704X, bytes 22 and 23 have the values X'2B50' coded. These values define an alternative screen size of 43 rows by 80 columns, or 3440. Byte 24 specifies that the screen size is dynamic, in accordance with the application.

The last operand to discuss is the TYPE operand. The value coded here defines the type of bind, negotiable or nonnegotiable. For LU-LU sessions with LU Type 2 devices, the default 1 is the only valid value. This default denotes a nonnegotiable bind. This value is used by the PLU when sending the bind to the SLU. In this case, the SLU must accept the session parameters presented by the PLU. A value of 0 requests a negotiable bind, but the application must also request the negotiable bind in the OPNDST macro by specifying PROC=NEG-BIND. For more detailed information on the session parameters discussed above, consult the following IBM manuals: *VTAM Programming* and *SNA—Sessions Between Logical Units*.

15.2.3 MODEEND macro

The MODEEND macro indicates the end of the logon mode table and follows the last modeent macro. There are no operands for the MODEEND macro (Figure 15.19).

After defining the logon mode table, you must assemble and link-edit the load module into the VTAM's executable library as defined in Section 15.2.

```
[name]    MODEEND
```

Figure 15.19 Format of MODEEND macro.

15.3   Class of Service Table

In Chapter 5 we discussed the routing capabilities of SNA in coordination with VTAM. The routes a PLU chooses for transmitting data to the SLU were described to be both physical (explicit routes) and logical (virtual routes). The ERs are mapped to the VRs, which define a specific route. In VTAM, priorities can be assigned to this mapping for use by the communications systems programmer to assist in meeting the service level agreement for the application. This prioritizing of routes is based on the VRs defined for the network. Named entries are specified in a table to define a class of service. The class can be identified as those PLUs that require a higher priority of transmission (e.g., CICS) versus PLUs that have a service level agreement requiring a lower level of transmission priority (e.g., RJE). These class entries are placed in a Class Of Service (COS) table. The entry name is associated with an SLU by using the COS operand in the mode table.

VTAM uses only one module name for the COS in a single network configuration, ISTSDCOS. This is the name you must use in the link-edit procedure when you want to add a COS table to VTAM executable library.

In ISTSDCOS, there are two required COS entry names. The first entry actually has no name. The name field consists of eight blanks. This entry is the default COS entry if the COS operand is not coded on the MODEENT macro that defines the LOGMODE for the SLU. The second entry must be named ISTSVTCOS. This entry defines the VRs and their priorities for the SSCP. This applies to all SSCP sessions in a single network. Special consideration must be given to the COS table for SNI.

15.3.1   COSTAB macro

The COSTAB identifies the start of the COS table (Figure 15.20). But, unlike the preceding tables, USSTAB and MODETAB, the name operand is required. In a single network, this name must be ISTSDCOS. Figure 15.21 defines the COS table that we will use as an example for our discussion.

15.3.2   COS macro

The name operand defines the COS name to be used on the COS operand of the MODEENT macro of the LOGON mode table entry.

```
name        COSTAB
```

Figure 15.20   COSTAB macro format.

```
ISTSDCOS COSTAB
ISTVTCOS COS VR=((0,2),(1,2),(2,2),(3,2),(4,2),(5,2),     X
              (6,2),(7,2),(0,1),(1,1),(2,1),(3,1),(4,1),  X
              (5,1),(6,1),(7,1),(0,0),(1,0),(2,0),(3,0),  X
              (4,0),(5,0),(6,0),(7,0))
INTERACT COS VR=((0,2),(1,2),(2,2),(3,2),(4,2),(5,2),     X
              (6,2),(7,2),(0,1),(1,1),(2,1),(3,1),(4,1),  X
              (5,1),(6,1),(7,1),(0,0),(1,0),(2,0),(3,0),  X
              (4,0),(5,0),(6,0),(7,0))
DEVELOP  COS VR=((0,1),(1,1),(2,1),(3,1),(4,1),(5,1),     X
              (6,1),(7,1),(0,2),(1,2),(2,2),(3,2),(4,2),  X
              (5,2),(6,2),(7,2),(0,0),(1,0),(2,0),(3,0),  X
              (4,0),(5,0),(6,0),(7,0))
BATCH    COS VR=((0,0),(1,0),(2,0),(3,0),(4,0),(5,0),     X
              (6,0),(7,0),(0,1),(1,1),(2,1),(3,1),(4,1),  X
              (5,1),(6,1),(7,1),(0,2),(1,2),(2,2),(3,2),  X
              (4,2),(5,2),(6,2),(7,2))
         COS VR=((0,0),(1,0),(2,0),(3,0),(4,0),(5,0),     X
              (6,0),(7,0),(0,1),(1,1),(2,1),(3,1),(4,1),  X
              (5,1),(6,1),(7,1),(0,2),(1,2),(2,2),(3,2),  X
              (4,2),(5,2),(6,2),(7,2))
         COSEND
```

Figure 15.21   Example of a COS table.

In Figure 15.21, five COS entries are defined. ISTVTCOS will be used by VTAM for SSCP sessions. The INTERACT COS name will be assigned to LOGMODE entries that are used for interactive applications such as CICS and IMS. The DEVELOP COS name will be used on the LOGMODE entries for application development PLUs such as TSO, CICS, and IMS development. The BATCH COS name will be used for JES/RJE and NJE sessions. The final COS entry name is blank and defines the default COS name for all LOG-MODE entries for which the COS operand is not coded. The COS name must be in each COS table of each VTAM in order for the name from the LOGMODE entry to be used. This is because the LUs owning SSCP will pass the logmode image along with the COS table name.

The VR operand of the COS macro (Figure 15.22) defines the ordered pair or ordered pairs of virtual routes (vr#) and their assigned transmission priority (tp#). As you know from Chapter 5 on SNA routing, there is a maximum of eight virtual routes, numbered 0 through 7. Each VR can have three transmission priorities assigned.

```
[name]    COS  VR=(vr#,tp#)|((vr#,tp#),...))
```

Figure 15.22 COS macro format.

```
COSEND
```

Figure 15.23 COSEND macro format.

Priority 0 assigns a low transmission priority, priority 1 assigns a medium transmission priority, and priority 2 assigns the highest transmission priority. This pairing allows for a maximum of 24 ordered pairs of VRs to TPs. The transmission priority is used by VTAM and NCP when transmitting data between subareas over a VR. Those SNA PIUs that display a higher TP number than others over the same VR will be serviced before PIUs on this same VR that have a lower TP. This means of prioritizing SNA data through the network has recently been extended to the peripheral SNA resources that are attached to NCP V4.3.1/V5.2.1. For more information on VRs and their role in network performance, consult *Advanced SNA Networking*.

15.3.3 COSEND macro

The COSEND macro (Figure 15.23) identifies the end of the COS table. It must follow the last COS macro defined in the table.

Now that we have discussed the coding options for the COS table, there are some points we would like to bring to your attention about the usage of the COS table.

As we stated, the COS table assigns transmission priorities for PLUs when sending data to an SLU. By using this scheme, we can effectively throttle the transmission of data through the FEP, telecommunications lines, and cluster controllers.

In Figure 15.21, the COS name INTERACT has the same ordered pairs as the SSCP COS name ISTVTCOS. This is because interactive applications, such as CICS and IMS, that provide large databases of information require the requested data to be sent to the end user as fast as possible. By assigning the highest priority of transmission to these applications, we can ensure quick network retrieval time.

We have defined a class of service that provides a medial transmission priority to the COS name DEVELOP. This is accomplished by assigning a transmission priority of 1 to the first set of ordered pairs in the list. The second set in the list assigns a TP of 2. We chose this scheme because development work on TSO, CICS, and IMS, although important, is not in demand by the end user community. Remember, overall, that it is the end user who must be supported.

The BATCH COS entry has been provided for applications that rely heavily on transmitting large amounts of print data over the network. Usually, this is aimed at JES/RJE and NJE facilities. But it can be of use for CICS or IMS background applications that also generate large amounts of data that must be transmitted over the network. Notice that the TPs for the BATCH COS entry are ordered from low priority in the first set of VRs to highest priority in the last set of VRs.

The last point concerns the actual assignment of COS names to the applications and SLUs. There is only one place where you can specify the COS name to use, on the MODEENT macro operand COS. The MODEENT macro defines a LOGMODE entry. There is only one operand that you can use to specify the default LOGMODE entry. That operand is the DLOGMOD, and it can be coded on the APPL definition; the LU, LOCAL, and the TERMINAL statements; and as a parameter default in the USSTAB definition by using the USSPARM macro. With all these possibilities, where do you code the LOGMODE entry name so you can take advantage of the COS table? This is a very good question. The answer can come only from experience. Examine your network's session mapping. Most terminals will usually have a regular session with the same application. This may be a good starting point. You can assign the COS name according to the most used application for a specific terminal, thereby using the DLOGMOD operand on the LU, LOCAL, or TERMINAL definition statements.

15.4 APPN COS Table

VTAM V4R1 maintains the topology database and the class of service database as a function on Topology Routing Services (TRS). Recall that TRS calculates optimal routes through an APPN network. APPN under VTAM V4R1 has its own COS table different from that found for traditional subarea COS. The APPN COS table, however, is not assembled and link-edited into modules, and it is not stored in VTAMLIB like the subarea COS table ISTSDCOS. Instead, the APPN COS table is defined in a member of VTAMLST. This member comes with VTAM V4R1 and is called COSAPPN. The APPN COS table defines the specifications for TGs and nodes.

APPC LU6.2 applications issue an ALLOCATE verb when the application wishes to establish an LU6.2 session with another LU6.2 application somewhere in the network. The ALLOCATE verb contains the mode name that is to be used for session characteristics and route selection. The mode defines the COS name to be used and is specified on the APPNCOS parameter in the mode table entry. The COS name is included in the APPN Locate request from the APPN EN to its NN server. The NN server determines the COS name and thereby uses the TG and node characteristics specified for the COS entry. The least-weight route will be the selected route.

The appropriate characteristics are determined by the TGP parameter value for the destination LU's PU definition. These characteristics are then compared with the characteristics found in the COS entry of the APPN COS table. The APPN COS table is defined using three definition statements, APPNCOS, LINEROW, and NODEROW.

15.4.1 APPNCOS definition statement

The APPNCOS definition statement is used to define the COS name and the priority of the session that uses this COS. The priority is specific to the session's placement on the TG queue for delivering data. The COS name can be up to eight characters in length. It is this name that is used in the LOGON mode table entry parameter APPNCOS. The PRIORITY parameter of the APPNCOS definition statement indicates one of the four possible priorities to be used for the session selecting this COS. The four possible values for this parameter are NETWORK, LOW, MEDIUM, and HIGH. The NETWORK value is the highest priority and is reserved for network management sessions such as the CP-CP session. The three remaining priorities are to be used for user-defined session requirements. In the sample definitions in Figure 15.24, two types of priorities are defined in the APPN COS table. The HIGHCOS entry may be used for interactive session traffic, and the LOWCOS entry may be assigned to LU applications that are not time dependent, such as batch updates.

15.4.2 LINEROW definition statement

Each COS entry in the APPN COS table has at least one row in the table defining TG link characteristics. This is done with the LINEROW definition statement. There can be up to eight sets of characteristic definitions defined by line row number and weight.

The NUMBER parameter of the LINEROW statement identifies the row in which the characteristics will reside. The value for NUMBER can range from 1 to 8, identifying eight possible rows. The WEIGHT parameter can range from 0 to 255, with 255 being the highest weight. The remaining characteristics of the LINEROW definition statement are similar to the definitions we previously discussed for the TG profile. The difference here is that CAPACITY, COSTTIME, COSTBYTE, UPARM1, UPARM2, and UPARM3 are

HIGHCOS	APPNCOS	PRIORITY=HIGH	Trans. priority
	LINEROW	NUMBER=5,	Line row number
		WEIGHT=150,	Line row weight
		CAPACITY=(1200,4M),	Effective capacity
		COSTTIME=(5,10),	Cost/connect time
		COSTBYTE=(10,20),	Cost/byte
		PDELAY=(MINIMUM,PACKET),	Prop. Delay
		SECURITY=(PUBLIC,ENCRYPT),	Security
		UPARM1=(0,100),	User defined parm
		UPARM2=(100,200),	User defined parm
		UPARM3=(0,10)	User defined parm
	NODEROW	NUMBER=5,	Node row number
		WEIGHT=100,	Node row weight
		CONGEST=(LOW,HIGH),	Congestion
		ROUTERES=(0,127)	Route resistance
LOWCOS	APPNCOS	PRIORITY=LOW	Trans. priority
	LINEROW	NUMBER=8	Line row number
	NODEROW	NUMBER=1	Node row number

Figure 15.24 APPN COS table format and sample definition.

coded with a minimum and maximum value. These minimum and maximum values are used by TRS to determine the current status and availability of a TG. The PDELAY and SECURITY parameters identify the two possible accepted values for this particular TG. In the example, the LOWCOS entry does not have any of these characteristics defined and hence will take defaults for all of them.

15.4.3 NODEROW definition statement

Recall that TGs are associated with links from one APPN node to an adjacent APPN node. The NODEROW definition statement comes into play when TRS is deciding which TG to use in conjunction with the adjacent APPN node characteristics. Like the LINEROW statement, the NODEROW definition statement can have up to eight rows defined, each with a specific weight.

The NUMBER parameter of the NODEROW statement identifies to TRS the row number for this node definition. The value for NUMBER can range from 0 to 8. The CONGEST parameter indicates to TRS the possible ways of viewing congestion for the adjacent node. A single value or two values may be coded for the CONGEST parameter. The possible values are LOW and HIGH. These two values are used with the values specified on the ROUTERES parameter. Recall from earlier that the ROUTERES parameter indicates the added route resistance to apply to the node. The ROUTERES parameter on the NODEROW statement can have one or two values defined. Each value can range from 0 to 255. Using ROUTERES in concert with the CONGEST parameter values, TRS can determine the router resistance for a node when congestion is low or high.

In conclusion, the mapping between the mode names and the COSs is accomplished for APPN just as in subarea SNA. The APPNCOS parameter in the mode table entry identifies the APPN COS entry to use. If routing through the existing subarea network is required, then the subarea COS entry specified by the COS keyword of the mode table entry will be used. When the routing is to pass through the

IBM-Supplied Mode Name	APPNCOS Keyword Default Value	COS Keyword Default Value
CPSVCMG	CPSVCMG	8 BLANKS
SNASVCMG	SNASVCMG	8 BLANKS
#BATCH BATCH BAT13790 BAT23790	#BATCH	8 BLANKS
#BATCHSC	#BATCHSC	8 BLANKS
#INTER INTERACT INTRACT INTRUSER	#INTER	8 BLANKS
#INTERSC	#INTERSC	8 BLANKS
DSILGMOD DSC4K D32901 EMUDPCX SCS etc. xxx3270x xxx3277x xxx3278x xxx3790x etc.	#CONNECT	8 BLANKS
ISTCOSDF	#CONNECT	8 BLANKS

Figure 15.25 IBM-supplied LOGON mode table names and the APPNCOS and COS keyword default values defining the COS entries for APPN and subarea class of service characteristics.

APPN network, then the APPN COS entry identified by the APPN-COS keyword of the mode table entry will be used.

IBM supplies new APPN mode names in the default LOGON mode table named ISTINCLM. A default APPN mode name called ISTCOS-DF is now included in ISTINCLM and is used when the requested mode name is not found in the mode table. Figure 15.25 gives a table identifying the IBM-supplied mode names and their equivalent APPNCOS and COS keyword values.

15.5 Summary

This concludes our discussion on using VTAM tables for session selection, session characteristics, and session class of service. The USS table provides a user-friendly interface to VTAM for the end user to initiate a session with an application. The LOGON mode table provides the session parameters that are to be agreed upon between the PLU and the SLU. The subarea COS table allows the systems programmer to assign priorities to virtual routes and map these into a class that can be specified on the LOGMODE entry for a device. The APPN COS table provides for added control on determining the optimal route for APPN sessions through the APPN network.

Now that we have gone into the specifics of VTAM and modifying user tables that assist in the access to the network, we can focus on the dynamic capabilities of VTAM to modify tables and resource definitions.

Dynamic Reconfiguration

The rapid growth of network dependency for corporate requirements has generated the need for more management capabilities within VTAM. The purpose of dynamic reconfiguration is to allow the deletion, addition, and modification of network resources and tables without requiring network downtime. The dynamic reconfiguration features are made possible by enhancements to VTAM and NCP.

16.1 Dynamic Reconfiguration Major Node

In VTAM and NCP you can dynamically reconfigure the network without an NCP generation by using dynamic reconfiguration statements. These statements let the systems programmer add, delete, move, or modify resources without affecting an entire NCP subarea node. Dynamic reconfiguration is available for both SNA and APPN resources. As with other VTAM major nodes, a VBUILD definition statement is coded that describes this definition as a dynamic reconfiguration major node. The VBUILD format used for this is

```
DRNODE VBUILD TYPE=DR
```

The statement above has the name DRNODE assigned to it, and the TYPE parameter indicates that this is a dynamic reconfiguration major node by the value DR.

16.1.1 ADD definition statement

The ADD definition statement defines to VTAM the name of the resource that is to have lower-level resources added to it. The format of the ADD definition statement is found in Figure 16.1. The name of the statement is optional. If it is coded, VTAM will use this name in

[name]	ADD	TO=resource name

Figure 16.1 Format of the dynamic reconfiguration ADD statement.

ADDPU01	ADD	TO=PU01
LUT11	LU	LOCADDR=11,USSTAB=USSSNA,PACING=0
LUP12	LU	LOCADDR=12,PACING=1,VPACING=2
ADDLN02	ADD	TO=LN02
PU02	PU	ADDR=C2,MAXDATA=265,USSTAB=USSSNA1
LUT03	LU	LOCADDR=03
LUT04	LU	LOCADDR=04

Figure 16.2 Sample ADD definition for dynamic reconfiguration.

error messages if the ADD fails. The resource name is the higher level of the resource(s) being added. For adding an LU, this is the name of the LU's associated PU. For the addition of a PU, this is the name of the PU's associated line.

In Figure 16.2, two LUs are to be added to physical unit PU01. The LUs will be assigned the LOCADDR of 11 and 12, respectively. LUT11 is a terminal and will not be paced; its USSTAB is named USSSNA. LUP12 is a printer, and it is assigned to LOCADDR 12. Since the LU is a printer, we will pace the data transmitted to the LU.

The second ADD statement defines LN02, line 2, as the higher-level resource. LN02 will have a PU (PU02) and two LUs (LUT03 and LUT04) added to it by using dynamic reconfiguration. The USSTAB operand on the PU statement will be sifted down to the LUs that are being added to LN02 along with this PU.

You can code VTAM-only operands on the PU and LU definition statements that are to be added dynamically. For a PU, however, you must code the MAXDATA parameter. For an LU, you must code the LOCADDR operand. All other VTAM-only and NCP-related operands will default if you do not include them in the dynamic reconfiguration major node definition.

16.1.2 DELETE definition statement

The DELETE definition statement defines to VTAM the name of the resource whose lower-level resources are to be disassociated from the

| [name] | DELETE | FROM=resource name |

Figure 16.3 Format of the dynamic reconfiguration DELETE statement.

named resource. The format of the DELETE statement is found in Figure 16.3.

The name parameter is optional, but if it is coded, VTAM will use it in error messages if the DELETE fails. The resource name is the resource whose associated lower-level resources are to be deleted. For deleting an LU, this is the name of the LU's associated PU. For a PU, this is the name of the PU's associated line.

In Figure 16.4, two LUs are to be deleted from physical unit PU02: LUs LUT11 and LUP12. PU PU01 is to be deleted from line LN01. Note that no PU or LU statements are required to delete a resource and that if a PU is deleted, all of its associated LUs are also deleted.

In summary, dynamic reconfiguration using the dynamic reconfiguration major node provides flexibility for reallocating network resources. The addition or deletion of resources can be accomplished without executing an NCP generation to change the network configuration. This can also allow the systems programmer to change VTAM-only operands or resources dynamically. However, if a session exists for the resource, the session must be broken before the dynamic updates take place. Dynamic reconfiguration allows you to make these VTAM-only changes without affecting the entire network. Remember, for dynamic reconfiguration, a DRDS DD statement is needed in the VTAM start procedure. The DRDS data set can be VTAMLST with a member defined for DR. The sift-down effect from the original definitions does not take effect on the dynamically added resources. Any operands that you do not want to default to the IBM standard defaults should be coded. However, sift-down does take effect within the ADD definition statements of the dynamic reconfiguration major node.

DELLUS	DELETE	FROM=PU02
LUT11	LU	
LUP12	LU	
DELPU01	DELETE	FROM=LN01
PU01	PU	

Figure 16.4 Sample DELETE definition for dynamic reconfiguration.

16.1.3   Enhanced dynamic reconfiguration

Dynamic reconfiguration prior to VTAM V3R2 uses preallocated control blocks in NCP V2 to V4R2. These control block allocations are created by the NCP PUDRPOOL, LUDRPOOL, and RESOEXT values. When resources are deleted, these control blocks and consequently the network addresses associated with them are not reusable. The dynamically reconfigured resources are assigned their network addresses from the predefined DRPOOLs. This limits the number of reconfigurable resources if several deletions and additions are performed, leading to unavailable network addresses. This limitation has been removed by VTAM V3R2 and NCP V4R3 and NCP V5.

Under VTAM V3R2 there are two types of dynamic reconfiguration for resources, explicit and implicit definitions. Using explicit definitions is quite similar to pre-VTAM V3R2. However, there have been modifications to the definition statements for dynamic reconfiguration major node. A new statement called MOVE has been added to eliminate the need to issue a DELETE definition statement and then an ADD definition statement in order to actually reallocate a resource from one location to another in the same NCP. The MOVE command is valid only for moving resources within the same NCP; that is, you cannot use this command to move a resource from one NCP to another. The old method of deleting and adding must still be used to perform moves from one NCP to another. The format of the MOVE statement is found in Figure 16.5.

Additionally, the PU definition statement in the dynamic reconfiguration major node has two new operands and the ability to modify the PU station address. The format of the new PU dynamic reconfiguration definition statement is seen in Figure 16.5.

The ACTIVE parameter tells VTAM to issue an activation request for the PU after the move has completed if it is coded with a value of YES. NO is the default value. The ADDR operand now allows you to change the station address used by the physical unit when responding to the SSCP. This is not available in pre-VTAM V3R2. The final

```
[name]       MOVE       TO=line name,FROM=line name

PU name      PU         ACTIVE=(NO I YES),

                        ADDR=station address,

                        DATMODE=(HALF I FULL)
```

Figure 16.5 Format of the MOVE and PU definition statements for dynamic reconfiguration.

parameter is the DATMODE operand. This operand tells VTAM and the NCP whether this PU can receive and transmit data concurrently. The default is HALF, meaning that the PU can only send or receive and does not possess the capabilities of concurrent send and receive data transmission.

This new functionality of dynamic reconfiguration leads to interesting possibilities. Because we can change the PU station address, we can now move a PU to a line that was previously defined as point-to-point. However, if this is going to be done, the SERVICE statement in the NCP for this line must have the MAXLST and MAXPU operands coded with sufficient values to handle extra PUs. Unlike pre-VTAM V3R2, the PU and its associated LUs carry with them all the definitions that are currently defined to them.

A new VTAM operator command to explicitly reconfigure the network resources is also available under VTAM V3R2. The MODIFY NET,DR operator command allows you to dynamically reconfigure the network without having a predefined DRDS member in VTAMLST. The format of this command is found in Figure 16.6.

Note that this command is performing only moves and deletions of network resources. Care should be taken by the VTAM operator when implementing this command.

In Figure 16.7, you can see that we are moving the PU that was added to line LU02. The move incorporates a delete and an add in one step. All the LUs associated with the PU are also moved to the new line. You should note that the resource names are the same after the move. This can lead to confusion if your resource names are based on line and station address.

16.2 Dynamic Table Updates

Since VTAM V3R2 and NCP V4R3 and V5R2, VTAM user tables have been included in dynamic reconfiguration for VTAM. In addition, VTAM V3R4 has enhanced the dynamic features of VTAM to also allow the systems programmer to modify the DLOGMOD keyword

```
MODIFY     procname,DR,

           TYPE=(MOVE I DELETE),ID=(puname I luname),

           FROM=(linename I puname)[,(TO=linename),

           ACTIVE=(NO I YES),ADDR=station address)]
```

Figure 16.6 VTAM dynamic reconfiguration MODIFY command.

[name]	VBUILD	TYPE=DR
[name]	MOVE	TO=LN01,
		FROM=LN02
PU02	PU	ACTIVE=YES,
		ADDR=C1,
		DATMODE=HALF

Figure 16.7 Sample format of the MOVE statement in the dynamic reconfiguration major node.

value for LUs without necessitating recycling the resource or its higher-level nodes.

VTAM now lets you dynamically change and modify the USS table, LOGON mode table, COS table, APPN COS table, and PATH tables. VTAM V3R4.1 has also provided the ability to change VTAM start options dynamically without the need to recycle VTAM itself. The advantage to this is that the entire SNA network does not need to come down to change a single one-line start option in VTAM's ATC-STR00 start list.

16.2.1 Dynamic USSTAB updates

The USS table used by a single resource or all resources can be dynamically changed through a VTAM MODIFY command. The format of the command is as follows:

```
MODIFY procname,TABLE,TYPE=USSTAB,OPTION=LOAD|ASSOCIATE,NEWTAB=name,
OLDTAB = name|*[,ID=resource name]
```

This command can be issued at any time, even while the resource is active. In the case of the MODIFY USSTAB command, the ID parameter should be included and may specify a specific LU or a PU. The * value for the OLDTAB parameter will indicate that the new table specified is to be used for resources subordinate to the resource specified in the command. As an example, if the ID parameter identified a PU, then all USS tables for resources under that PU would use the new USS table after their current LU-LU sessions end. The LOAD option is used to refresh a currently used USS table that resides in VTAM address space. The ASSOCIATE option will cause VTAM to load a new USS table and associate it with the resource defined by the ID keyword.

16.2.2 Dynamic MODETAB updates

The LOGON mode tables in VTAM can be dynamically updated and modified for resources. The format of the VTAM operator command for this is

```
MODIFY procname,TABLE,TYPE = MODETAB,OPTION = LOAD|ASSOCIATE, NEWTAB = name
[,OLDTAB=name|*,ID=resource name]
```

The OPTION=LOAD operand tells VTAM to load the link-edited module named by the NEWTAB keyword into VTAM's address space from VTAMLIB. The OLDTAB keyword is optional when the LOAD option is used. If OLDTAB is omitted, VTAM will assume that it is the same name as is found on the NEWTAB keyword. Optionally, you can use the OLDTAB parameter with the same name as is found on the NEWTAB parameter. Either way, VTAM will load a new copy of the table into memory. Resources associated with this new LOGON mode table will use the new parameters found in it during the next LU-LU session request. The ASSOCIATE option for LOGON mode tables is used in the same manner as discussed for the USS table MODIFY command.

16.2.3 Dynamic DLOGMOD updates

Prior to VTAM V3R4, an end user could override the default DLOGMOD value by using the long LOGON command format if that was available to them. At logon time the end user could issue the following command:

```
LOGON APPLID(TSO) LOGMODE(DYNAMIC)
```

The LOGMODE parameter of the LOGON command overrides the VTAM DLOGMOD parameter defined in VTAM for the LU used by the end user. VTAM V3R4 now provides an operator command to change the predefined DLOGMOD value for an LU, APPL, or independent LU. The command dynamically changes the DLOGMOD value, and the new DLOGMOD will be used during the next LU-LU session establishment process without the need to recycle the LU. The format for the new MODIFY command is

```
MODIFY NET,DEFAULTS,ID=name,DLOGMOD=name
```

The ID parameter identifies the VTAM resource that is to be modified, and the DLOGMOD value is the name of the LOGON mode entry that will be used as the default for the resource. VTAM does not validate the existence of the new DLOGMOD specified in the command. An invalid name could cause session establishment failure.

16.2.4   Dynamic COSTAB updates

The subarea COS table used by VTAM for prioritizing virtual routes through the SNA subarea network can also be dynamically modified. The dynamic reconfiguration of the COS table is provided by the following VTAM operator command:

```
MODIFY procname,TABLE,TYPE=COSTAB,OPTION=ASSOCIATE,NEWTAB=name,
NETID=name,ORIGIN=resource name
```

This command can be used to associate a new COS table with a gateway NCP. Recall that a gateway NCP is used in an SNI configuration. The gateway NCP is identified by placing its PU name in the ORIGIN keyword value. The NETID keyword identifies which network connected through a gateway NCP the COS table will be applied to. This command can also allow you to refresh the ISTSDCOS COS table in VTAMLIB by specifying VTAM's PU name in the ORIGIN keyword value. VTAM's PU name is defined on the HOSTPU start option. Prior to this, VTAM had to be recycled to reload the class of service table.

16.2.5   Dynamic APPN COS updates

Updates to the APPN COS table are not done with a MODIFY VTAM operator command like updates to the other tables. This is because the table is not assembled and link-edited into a loadable module. Instead, a VTAM VARY ACT command is used to refresh and modify the APPN COS table in use. Recall that VTAM V4R1 has a default table named COSAPPN in VTAMLST. Once activated, the APPN COS table cannot be inactivated. Instead, changes found in the VTAMLST member are used by VTAM to delete, add, or modify rows in the APPN COS. An example of refreshing an already active APPN COS table is

```
VARY NET,ACT,ID=COSAPPN
```

Upon successful execution of this command, the APPN COS table will be updated with the changes instituted by the code found in the member COSAPPN of VTAMLST.

16.2.6   Dynamic PATH table updates

This enhancement of VTAM V3R2 and NCP V4R3 and V5R2 allows the dynamic addition or deletion of SNA paths without the need for an NCP generation and subsequent loading of the NCP load module into the communications controller. There are two types of dynamic PATH table updates. You can delete the previously defined explicit routes that are inoperative from their TG, and you can add new explicit routes to TGs. The combination of the two allows you to move an explicit route from one TG to another. In addition, since the VTAM PATH table can be updated dynamically, you can now modify the virtual route pacing window size dynamically for inoperative routes. The only prerequisite for this capability is that the links used must be predefined. If the links have to be predefined, then why use dynamic path update? The answer to that question is shown in Figure 16.8.

Suppose our network was originally configured as in Figure 16.8a. The network has grown, and a new communications controller is coming in the door. Figure 16.8a shows us the links that currently exist between NCP11 and NCP21. With dynamic update we can utilize these existing links. To do this, we must first physically attach TG121 and TG122 to the new communications controller. Then we define the dynamic PATH statement in VTAMLST.

The PATH table member in VTAMLST is identified as a dynamic PATH table by having VPATH or NCPPATH definition statements on the first noncomment line of the member. VTAM will process the member from a VARY ACT command as a dynamic path update. As you can see from Figure 16.8, we have coded the dynamic path update members for the new configuration. The name of the VPATH definition statement must be the same as the value specified on the SSCP-NAME start list option. If the name does not match, VTAM will ignore the following PATH and DELETER statements until a VPATH name matches the SSCPNAME value. Similarly, the name of the NCPPATH definition statement must be the same as that specified on the NEWNAME operand of a BUILD definition statement in an NCP that this VTAM has an active session with. If the name does not match, VTAM will ignore any following PATH and DELETER statements until a matching NEWNAME value of an active NCP is found.

There are four ways to invoke the dynamic path update. The first is by issuing the following operator command:

```
VARY NET,ACT,ID=dynamic path update member name
```

The value for ID is the name of the dynamic path update member found in VTAMLST. The second approach is also an operator command:

```
VARY NET,ACT,ID=ncpname,NEWPATH=dpumember
```

where the NEWPATH parameter allows you to enter from one to three dynamic path update member names when activating an NCP. The third way to invoke dynamic path update is through the NEWPATH operand of the PCCU definition statement. This is applicable to VTAM

(a)

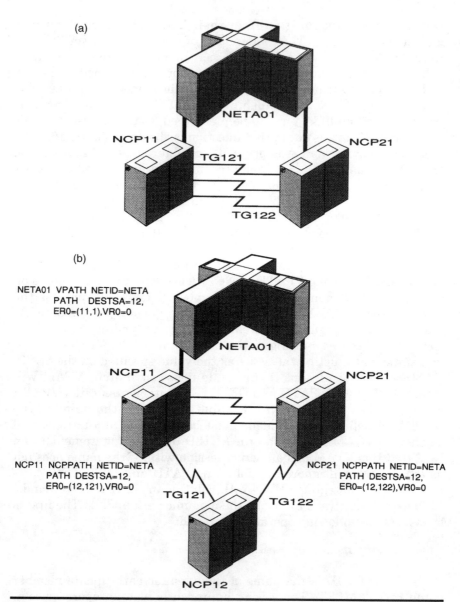

Figure 16.8 Diagram depicting the dynamic path definitions for the addition of a third NCP to the network.

V3R2 or higher. The updates are sent to the NCP before any of the peripheral links are activated. The member names found on the PCCU NEWPATH parameter are overridden by the operator if the NEW-PATH operand is specified on the VARY ACT command for this NCP. The fourth and final way is through the inclusion of the dynamic path update member names of VTAMLST in the member ATCCON00. In this case, you can specify as many dynamic path update members as necessary, but these are used only during VTAM initialization.

16.3   Dynamic PU and LU Configuration

VTAM V3R4.1 provides the functionality to dynamically define PU and LU resources when they power on and are connected to the mainframe through a channel attachment or through an NCP. This is in contrast to static definitions provided by VTAM and NCP major nodes. The feature is known as Dynamic Dependent LUs (DDLU). DDLU requires that the PUs support the Reply/Product-Set Identification (PSID) Network Management Vector Transport (NMVT) SNA request unit. The IBM 3174 with Configuration Support B4.1, C1.1, or later provides this support.

Figure 16.9 illustrates a configuration for using the dynamic definition of LUs. VTAM V3R4.1 includes a default user exit routine named Selection of Definitions for Dependent LUs (SDDLU). The exit is

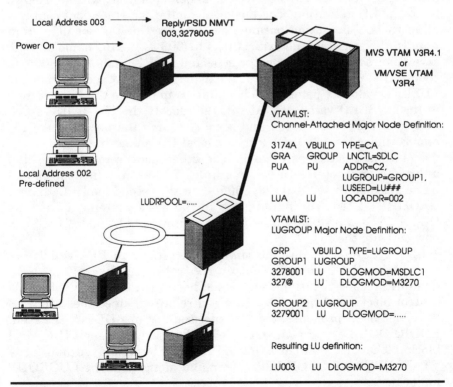

Figure 16.9 Illustration of a network configuration and the use of dynamic PU and LU definitions.

called when VTAM receives the Reply/PSID NMVT RU for undefined LUs at device power-on attached to PUs which have the LUGROUP keyword coded. The supplied exit is named ISTEXCSD, and it selects the LU name and LU model for each LU device. The exit selects the LU name by means of the LUSEED keyword found on the PU definition statement. The pattern provided by the LUSEED keyword is used as the root for the LU name.

As seen in Figure 16.9, the LU powers on and the PU sends to VTAM a Reply/PSID NMVT RU. This RU contains the local address for the LU and the machine type and model number of the powered-on device. For example, the local address may be 003, and the device type and model will be sent as 3278005. Upon receiving the Reply/PSID NMVT RU, VTAM calls the SDDLU exit routine if the device is not predefined and the PU is defined to VTAM with the LUGROUP keyword. In the example, the LUGROUP keyword identifies the LUGROUP definition found in a LUGROUP major node that is to be used for the LUs dynamically defined for this PU. The exit routine returns the name of the LU and chooses the name of the model LU found in the LUGROUP major node that will be used to build the LU definition. The model LU definition statement contains all of the LU definition parameters used for this type of LU definition. In the example, the exit returns LU003 as the LU name and the seven characters of the machine type and model number, 3278005, to choose the Model statement. The exit builds the LU name from the LUSEED value by using the local address to replace the # characters in the LUSEED value. VTAM will then locate the LUGROUP specified on the PU's definition and serially search the list of model LU names under the LUGROUP for a model LU name that matches the model name selected by the exit. The @ character represents a set of unknown characters. In the example, VTAM matches the model name selected by the exit (3278005) with the second model LU under GROUP1 (327@). The parameters in the 327@ statement, such as DLOGMOD=M3270, are used for the definition of LU003. VTAM can now activate the LU.

LUs dynamically defined that are attached to PUs which are attached to NCP communication controllers utilize VTAM/NCP dynamic reconfiguration facilities. These facilities obtain the LU control blocks and network addresses required, since the resources have not been pregenerated. To successfully use DDLU, the NCP LUDRPOOL keyword must be defined with sufficient LU control blocks. The LUs defined using DDLU can be displayed using the VTAM DISPLAY LUGROUPS command along with the LUGROUP major nodes.

[name]	VBUILD	TYPE=LUGROUP
name	LUGROUP	
name	LU	[ASLENT=associated LU table entry name]
		[,ASLTAB=associated LU table name]
		[,MDLENT=model name table entry name]
		[,MDLTAB=model name table name]

Figure 16.10 Keywords specific to LU definition for LUGROUP major node.

16.3.1 LUGROUP major node

The name parameter on the VBUILD definition statement (Figure 16.10) is optional, but when coded it is useful during error message displays. The TYPE keyword of the VBUILD definition statement identifies this major node to VTAM as an LUGROUP definition major node. The name parameter on the LUGROUP definition statement is required and is used as the identifier for the LUGROUP that is to be used for a PU's LUs that are to be dynamically defined. The value is specified on the PU keyword LUGROUP.

The name parameter on the LUGROUP LU definition statement is required and is used to identify the LU name template for dynamically defined LUs. The name can be one to eight characters and is used to select the model LU group that will be used to build the dynamic LU definition. The name value provided here is matched against the machine type and model number (model acronym). The model acronym is provided by the control unit of the device at power-on of the device. All alphanumeric characters are allowed in the name. The @ sign in the name represents a single unknown character. If the @ sign is at the end of the name, it represents any number of unknown characters. If the @ sign is the only character used in the name, it will then match any LU acronym. Each name in an LUGROUP definition must be unique. However, different model LU groups can contain model LU definitions with the same name.

The ASLENT keyword is used for DDLUs that do not initiate sessions on their own. That is to say that printers on cluster controllers need to be dynamically defined also. The ASLENT keyword specifies the name of the entry in the associated LU table that can be used to define the printer LU. The ASLTAB keyword is required if the ASLENT keyword is coded. However, if the ASLTAB keyword is coded and the ASLENT keyword is not defined, then VTAM will use the first entry in the specified table identified by the ASLTAB keyword.

The ASLTAB keyword specifies the name of the associated LU table that contains the list of associated LU entries that can be used by the PU to define LUs.

The MDLENT keyword is used to define the name of the model name table entry used to define the LU. The MDLTAB keyword is required if the MDLENT keyword is specified. If you omit the MDLENT keyword, then VTAM will use the first entry found in the MDLTAB specified table.

The MDLTAB keyword identifies the name of the model table that is to be used for defining the logical unit.

All other keywords normally found on the LU definition statement except for LOCADDR can be defined or defaulted on the LUGROUP LU definition statement.

16.3.2 Model name table

The MDLTAB keyword of the LUGROUP LU definition statement points to a model name table that resides as a member in SYS1.VTAMLST. The model name table has two required definition statements and one optional definition statement. Figure 16.11 details the format of the three statements.

The name parameter on the MDLTAB statement is optional and is used only in VTAM error messages. The statement itself is required to identify this member as a model name table.

The MDLENT definition indicates the beginning of a model name entry and optionally sets up the default model name. The end of a model entry is denoted by the presence of another model entry at the end of the file. The name on the MDLENT statement is required and can be one to eight characters. The MODEL keyword is the one- to eight-character model name expected by the subsystem for the terminal device. The MODEL keyword can be overridden during the LOGON process.

[name]	MDLTAB	
name	MDLENT	[MODEL=model name]
[name]	MDLPLU	PLU=plu name
		[,MODEL=model name]

Figure 16.11   Format of the model name table definition.

The MDLPLU statement defines the model name for a specific PLU within the MDLENT table entry. The PLU keyword is required if the MDLPLU statement is coded. The value specified here is the name of the PLU as it is known in the SLU's network. The MODEL keyword under the MDLPLU statement is optional and is the model name expected by the subsystem for the terminal device.

16.3.3 Associated LU table

The associated LU table is pointed to by the ASLTAB keyword on the LUGROUP LU definition statement. VTAM application programs use the associated LU names to create dynamic definitions for SLUs that cannot initiate a session. Typically these devices are printers attached to controllers. There are three statements that may be coded for the associated LU table, ASLTAB, ASLENT, and ASLPLU. The format of these statements is shown in Figure 16.12.

The ASLTAB statement identifies the beginning of the associated LU table member found in VTAMLST.

The ASLENT statement indicates the start of an associated LU table entry and optionally builds a set of defaults for the LU data. The name of the ASLENT statement is required and may be one to eight characters in length. The PRINTER1 and PRINTER2 keywords are optional. However, if they are coded, the PRINTER1 value identifies the primary printer associated with the terminal as it is known in the SLU's network. The PRINTER2 value is the alternative printer selection for the SLU. If PRINTER1 or PRINTER2 is coded, the printer must be in the same network as the terminal or must have a unique name across network boundaries without name translation taking place.

The ASLPLU statement identifies the PLU name controlling the printers specified. The PLU name must be the name of the application as it is known in the SLU's network.

[name]	ASLTAB	
name	ASLENT	[PRINTER1= lu name]
		[,PRINTER2= lu name]
[name]	ASLPLU	PLU=plu name
		[,PRINTER1= lu name]
		[,PRINTER2= lu name]

Figure 16.12 Format of the associated LU table definition.

16.4   Dynamic Additions via NCP Source Code

Since NCP V4R3 and V5R2, source code changes made for NCP can be used by VTAM to dynamically allocate control blocks in the NCP. This is valid only when adding resources to the NCP. Deletion of resources from NCP source code without executing an NCP generation process will result in an activation failure of the NCP. Addition of resources through source code changes only is known as implicit definition. Care should be taken when using this approach because the number of resources added can affect the amount of control blocks available for other resources, since a generation process did not occur.

16.5   Summary

In this chapter we examined the various dynamic enhancements found in VTAM and NCP. Dynamics are becoming increasingly important as corporations depend more and more on the availability of their networks. We discussed how to use dynamic reconfiguration to add, move, and delete predefined resources and reviewed dynamic definition of dependent LU resources. We also discussed how VTAM supports dynamic table replacement and modification for USS, LOGON mode, COS, and APPN COS tables. In addition, the four methods of dynamically changing VTAM subarea PATH statements were discussed.

NCP

NCP Macros—PCCU, BUILD, and SYSCNTRL

Now that the local domain has been defined to VTAM, we will describe the remote network. Its network is connected to the host processor via telephone circuits attached to a communications controller. In SNA, the communications controller is operated by a Network Control Program (NCP). This NCP is generated on the host processor. The NCP contains the definitions for the remote SNA resources of the VTAM domain. Figure 17.1 diagrams a sample single-domain network.

In this chapter we will concentrate on defining some of the more commonly used remote SNA resources found in today's SNA networks. Before we define resources in the NCP, you should have some understanding of the resources' physical and operational characteristics. The NCP must know its relationship to all SNA resources, which include the host VTAM, other communications controllers that are connected to this communications controller, the data links, and the resources that lie at the end of the data links. NCP definitions concerning LEN, enhanced subarea connectivity, dynamic table updates, and reconfiguration are discussed in detail in *Advanced SNA Networking*.

17.1 PCCU Definition

Although the PCCU definition statement is a VTAM-related statement, it is placed in the NCP generation code. The PCCU definition statement is the first statement found in the NCP generation code.

The PCCU definition statement describes the Programmed Communications Control Unit (PCCU) into which the NCP is loaded. This is a required definition statement. It defines the functions that

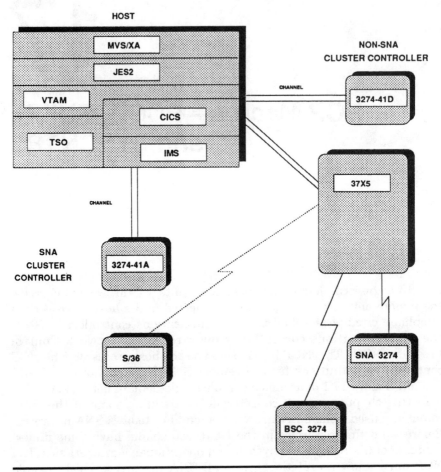

Figure 17.1   Sample single-domain network.

VTAM is supplying for this NCP. At least one PCCU definition statement is required for an NCP. Figure 17.2 shows the format of the operands that we will discuss for the PCCU definition statement.

In Chapter 11 we reviewed and coded the VTAM start options list which defines VTAM's environment in the host processor. In the definitions we coded the HOSTPU operand. This operand defines to the SSCP the name of VTAM's physical unit. For consistency we will use the same name for the name of the PCCU definition statement. This will assist you in locating the PCCU for a particular host VTAM when multiple hosts can activate the NCP.

The first operand listed in Figure 17.2 is SUBAREA. The value coded for this operand should match the HOSTSA operand value supplied in the VTAM start options list. For our example, 01 is used.

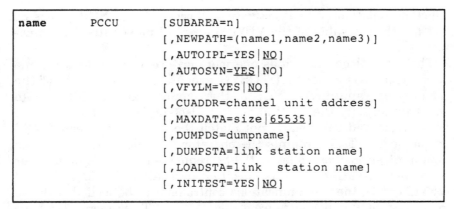

name   PCCU   [SUBAREA=n]
                        [,NEWPATH=(name1,name2,name3)]
                        [,AUTOIPL=YES|NO]
                        [,AUTOSYN=YES|NO]
                        [,VFYLM=YES|NO]
                        [,CUADDR=channel unit address]
                        [,MAXDATA=size|65535]
                        [,DUMPDS=dumpname]
                        [,DUMPSTA=link station name]
                        [,LOADSTA=link station name]
                        [,INITEST=YES|NO]

Figure 17.2   PCCU definition statement and operands.

This SUBAREA operand is used by VTAM to determine which PCCU definition statement to process when activating the NCP. If we were to omit the SUBAREA operand, this PCCU definition statement will be used by any VTAM that does not find a corresponding SUBAREA operand that matches its HOSTSA operand. We have chosen to code the SUBAREA operand because it also helps in documenting the NCP, although in this case it is not necessary.

The NEWPATH operand is specific to VTAM V3.2 and NCP V4.3/V5.2. These versions provide the means for dynamic path table updates to occur in both VTAM and NCP. This path table update is accomplished without executing the NCP generation procedure that was required in earlier versions of VTAM and NCP. The names specified on this parameter identify the names of the path table update members found in VTAMLST. You may define up to three names in this NEWPATH parameter. Dynamic path table updates are discussed in further detail in *Advanced SNA Networking*.

When a network operator issues the VARY NET,ACT,ID=NCP-name VTAM command, VTAM will use the AUTOSYN and VFYLM operands to determine if an NCP load module should be loaded in the communications controller.

During the activate process for an NCP major node, VTAM tries to match the NCP major node name specified on the VARY ACT command. If AUTOSYN=YES is specified, VTAM will immediately work with the NCP found in the communications controller if the NCP major node name matches the NCP load module name found in the communications controller. If NO has been specified, VTAM will ask the operator whether the NCP requested on the VARY ACT command should be loaded or if the NCP in the communications controller should be used. We will specify YES on the AUTOSYN operand to allow automation of NCP activation to occur. If the NCP names do not match, the value of the VFYLM operand makes the determination.

The VFYLM operand adds one more criterion to the load module name comparison that AUTOSYN executes. The presence of the VFYLM operand directs NCP in the communications controller to identify not only the NCP name but also the NCP subarea. If the NCP name and subarea do not match those of the requested NCP major node, VTAM prompts the operator for a decision to load the communications controller with the requested NCP major node or stop the activation. This is specified by coding VFYLM=YES. If VFYLM=NO, the communications controller will be loaded with the requested NCP major node from the VARY ACT command. We have specified VFYLM=YES, which will help guard us from activating and consequently loading the wrong NCP major node.

Prior to the actual load process, VTAM can request the communications controller to perform an initialization test and/or an Initial Program Load (IPL). The initialization test is performed before the IPL or before VTAM loads the communications controller.

The INITEST operand tells VTAM to run the initialization test if the value coded is YES. The test is not executed if you code NO for the INITEST operand. The initialization test performs diagnostics on the communications controller before it is loaded with an NCP. To run the INITEST program, a DD statement in MVS or a DLBL statement in VSE must be included in the VTAM start procedure. The name of the statement must be INITEST, and the data set defined by the statement must contain the INITEST load module. In most cases this module is found in the SYS1.SSPLIB data set.

The AUTOIPL operand determines whether the initial program load procedure is to execute after an unrecoverable failure in the NCP or the communications controller or after a dump of the communications controller has occurred. If YES is specified, the contents of the communications controller are refreshed, cleared, and awaiting the new copy of the NCP load module. This occurs only during an error situation and should be coded on only one PCCU in the NCP.

VTAM knows the path to the communications controller from the channel unit address. The channel unit address is obtained from the I/O GEN performed for the operating system. It can be specified with the U= parameter of the VARY NET,ACT command, or it can be found on the CUADDR operand of the PCCU definition statement that corresponds to the requestion VTAM. This channel unit address is used in conjunction with the LOADSTA and DUMPSTA operands.

The LOADSTA operand defines the link station VTAM will use to perform the load procedure of the communications controller. The

DUMPSTA operand defines the link station VTAM will use to perform the dump procedure for the communications controller.

In our example, the channel unit address is A01. We have coded CUADDR=A01. Note that the address is in hexadecimal format. The LOADSTA and DUMPSTA operands do not have to be coded. These values will be defaulted to the value of the CUADDR operand with a suffix of -S. For documentation purposes, however, we have coded LOADSTA=A01-S and DUMPSTA=A01-S.

If a situation occurred in the NCP that created a need to dump the contents of the communications controller, the value coded for the DUMPDS operand use will direct VTAM to the dump data set for that NCP. The value of the DUMPDS is the name of a DD or DLBL statement found in the VTAM start procedure. This statement defines a data set for receiving NCP dumps. This operand works with the AUTODMP operand. The AUTODMP operand specifies YES if the storage dump of the NCP is to be automated or NO if the operator is to have control of the dump procedure.

Lastly, the amount of data that can be transferred between the NCP and the VTAM represented by this PCCU definition statement cannot exceed the value of MAXDATA. The MAXDATA operand should be equal to the size of the largest PIU found in the network that may be transmitted between the NCP and VTAM. The default value is 65,535 and in most cases exceeds the largest PIU. It is best to determine the largest PIU by researching the applications and their output buffers used for transmitting data to a logical unit. As a guideline, the value coded here should not exceed the size of VTAM's IOBUF buffer pool in MVS and VM or the VFBUF buffer pool in VSE. Also, the product of the MAXBFRU and UNITSZ operand values of the NCP HOST definition statement should be less than or equal to the MAXDATA value on the PCCU. In addition, the BFRS and TRANSFR operand values of the NCP BUILD definition statement also supply a check against the MAXDATA operand. We will discuss these operands in greater detail in the next section. Figure 17.3 shows the completed PCCU definition statement used for our single-domain example.

17.2 BUILD Definition Statement

The BUILD definition statement is used to define to the NCP the physical makeup of the communications controller, the NCP's ability to communicate with VTAM, and information about the NCP itself. The BUILD definition statement consists of some 63 operands and their associated parameters. Many of these operands are used to create the Job Control Language (JCL) procedure used to generate the

```
VTAM01    PCCU    CUADDR=A01,                              X
                  AUTODMP=YES,                             X
                  AUTOIPL=YES,                             X
                  AUTOSYN=YES,                             X
                  DUMPDS=NCPDUMP,                          X
                  MAXDATA=4096,                            X
                  SUBAREA=01,                              X
                  VFYLM=YES,                               X
                  LOADSTA=A01-S,                           X
                  INITEST=YES,                             X
                  DUMPSTA=A01-S
```

Figure 17.3 Coded example of PCCU for a single-domain network.

NCP load module and will not be discussed here. Instead, we will focus our review on the BUILD definition statement operands that pertain to the NCP in the network control mode. The NCP in the communications controller is considered to be in network control mode when it performs dialing, answering, polling, and device addressing. This is in contrast to some of the other related software programs associated with a communications controller, such as Emulation Program (EP) or NCP Packet Switching Interface (NPSI). These programs and others are discussed in *Advanced SNA Networking*.

We will now define the communications controller to the NCP. In this example the communications controller is an IBM 3725. Figure 17.4 outlines the operands that define the physical make up of the 3725.

17.2.1 Defining physical characteristics to NCP

The format of the BUILD definition statement is as follows:

```
name BUILD operand, operand, ...operand
```

The name operand is symbolic and is not used by NCP. It is basically used for documentation only. When defining the physical characteris-

```
NCP11BLD    BUILD     MODEL=3725    comcontroller type      X
                      MEMSIZE=1024, installed memory         X
                      CA=(TYPE5,TYPE5-TPS)  CA types
```

Figure 17.4 Defining IBM 3725 physical characteristics to NCP.

tics of the 37x5, the MODEL operand specifies the type of communications controller. In Figure 17.4 the MODEL operand identifies the
communications controller type as a 3725. This indicates that the
NCP being generated is prior to NCP V5, because the 3725 communications controller supports only up to NCP V4R3.1. Any version of
NCP after this can execute only on 3745 communication controllers.
Valid MODEL values for NCP V5 and higher are listed in Figure
17.5. The MEMSIZE operand indicates the amount of memory
installed on the controller. In the example in Figure 17.4 we have
coded a value of 1024. The MEMSIZE value is interpreted in kilobytes when coded in this manner. A "K" may also follow the value to
specify the amount of memory in kilobytes. A second option for specifying the value is by coding 1M. This indicates that 1 megabyte (actually 1,048,576 bytes) of storage is installed. The maximum amount of
storage for the 3725 is 3 MB; for the 3745-31A and 3745-61A the maximum is 16 MB, and for all other 3745 models the maximum is 8 MB.
The NCP, however, will use either the full installed memory or the
amount specified on the MEMSIZE operand, whichever is smaller in
value. The CA operand defines to the NCP the number and position of
each channel adapter installed and the presence of a two-processor
switch for each. The channel adapter is the hardware interface

Value of MODEL operand	NCP Version
3745	V5 and higher
3745-130	V5 and higher
3745-150	V5 and higher
3745-160	V5 and higher
3745-210	V5 and higher
3745-21A	V6R2 and higher
3745-310	V5 and higher
3745 -31A	V6R2 and higher
3745-410	V5 and higher
3745-41A	V6R2 and higher
3745-610	V5 and higher
3745-61A	V6R2 and higher

Figure 17.5 Values for the BUILD definition operand MODEL.

between the mainframe and the communications controller. For a 3725 there is only one type of channel adapter. It is specified as TYPE. The CA operand is not used for NCP V5 and higher. From Figure 17.4 we can see that the 3725 has two channel adapters installed. They occupy positions 1 and 2 of the available size positions. The second channel adapter also has the two-processor switch feature installed.

17.2.2 Defining NCP-to-VTAM communications characteristics

Now that the hardware characteristics of the 3725 that will contain this NCP are defined, NCP must be made aware of its relationship with VTAM. We will begin by defining the channel characteristics NCP will impose on VTAM.

Figure 17.6 highlights the operands of the BUILD definition statement that define the channel between NCP and VTAM. The NCPCA operand specifies the status of the channel adapter's availability. For each channel adapter identified in the CA operand, the NCPCA operand will specify ACTIVE or INACTIVE. According to the configuration drawn in Figure 17.1, only one channel adapter is presently connected to the 3725. The NCPCA operand specifies that channel adapter position 1 is to be active. This is the channel adapter on the 3725 that will be used to load the NCP load module. Make sure that the physical channel connected to the 3725 is attached to the channel adapter you specified as being active in the NCPCA operand. Another check on the physical connection is that the IO address for the channel adapter at both ends of the physical channel, the 3725 end and the host processor end, are equal to the CUADDR operand of the PCCU definition statement. Failure of either of these situations will result in an error during NCP load attempts.

When the NCP has data (e.g., PIUs) to send to VTAM, it sends an attention interrupt signal to the host. VTAM will then read the data in the NCP buffers. This could cause serious performance problems if

```
NCP11BLD   BUILD      MODEL=3725,   comcontroller type         X
                      MEMSIZE=1024, installed memory            X
                      CA=(TYPE5,TYPE5-TPS), CA types            X
                      NCPCA=(ACTIVE,INACTIVE), active CAs       X
                      DELAY=(.2),   delay attn signal           X
                      TIMEOUT=420.0 default timeout/host
```

Figure 17.6 Operational operands for NCP-to-VTAM communications.

every read provided few PIUs in the NCP buffer retrieved. The DELAY operand is intended to assist in increasing the number of PIUs in the NCP buffer while reducing the number of reads for VTAM to perform when reading data from the NCP. The default value of 0 specifies that an attention interrupt will be issued to the host whenever there is a PIU queued in a buffer for a channel adapter. By specifying a larger value, the NCP can queue several PIUs destined for a host before issuing the attention interrupt. This saves the host and VTAM cycle time, which in turn helps performance. The DELAY value can be specified in tenths of a second between 0 and 6553.5 for each active channel adapter. We have chosen 0.2 s for the active channel adapter. The attention interrupt will be issued before reaching the DELAY value if the amount of data queued in the NCP will fill the receive buffers allocated in the host according to the MAXBFRU value of the NCP HOST definition statement.

After the NCP issues the attention interrupt, it will wait for a response from VTAM. The TIMEOUT operand of the BUILD definition statement specifies the amount of time the NCP will wait for that response before disconnecting its link with VTAM. The range for this value is 0.2 to 840.0 s. The value for each channel adapter installed will default according to the value defined on the NCPCA operand. In our example, channel adapter position 1 is active and will default to 420.0 s. Channel adapter position 2 is inactive and defaults to 0, indicating immediate channel discontact. The implication is quite clear. If for some reason VTAM cannot send its response to an attention interrupt from NCP, all sessions on this NCP major node will be terminated. The default of 420.0 s for an active channel adapter should be sufficient. If timeouts with the NCP do occur, they may be caused by channel activity rather than VTAM being too busy to respond.

In the previous section we discussed the maximum amount of data that VTAM will accept from this NCP. This limit is coded on the MAXDATA operand of the PCCU definition statement. We picked the value of 4096 bytes in a single transfer of data as a suitable number based on research of network traffic message unit sizes. In the BUILD definition statement two operands reflect on the MAXDATA value, BFRS and TRANSFR (Figure 17.7).

17.2.3 NCP buffer allocation operands

The BFRS operand tells NCP the size of the buffers in the buffer pool. These buffers will contain PIUs that pass through this NCP for resources in the network. The size should be a multiple of 4; other-

```
NCP11BLD   BUILD      MODEL=3725,  comcontroller type        X
                      MEMSIZE=1024, installed memory          X
                      CA=(TYPE5,TYPE5-TPS), CA types          X
                      NCPCA=(ACTIVE,INACTIVE), active CAs     X
                      DELAY=(.2),  delay attn signal          X
                      TIMEOUT=420.0, default timeout/host     X
                      BFRS=128,     NCP data buffer size      X
                      TRANSFR=32    # of BFRS xfrd to host
```

Figure 17.7   Buffer allocation operands for PIUs in the network.

wise, NCP will round the value up to a multiple of 4. The buffer size can range from 76 bytes to 240 bytes per buffer. The buffer pool in NCP is the remaining storage available after the NCP is loaded. Each buffer is formatted with a 12-byte management prefix. This buffer prefix is used by NCP to chain related buffers, and keep a data count and data offset of the PIU within the buffer. The default buffer size for the IBM 3725 and 3720 is 100 (88 + 12), and for the IBM 3745 it is 252 (240 + 12). For SDLC resources that support segmentation, a value of 128 is a good starting point.

The TRANSFR operand of the BUILD definition statement determines the maximum number of NCP buffers that can be transferred to the host at one time. Since the maximum amount of data in one PIU is 4096 bytes, as specified by the MAXDATA operand of the PCCU definition statement, the value for this TRANSFR operand times the BFRS value should not exceed the MAXDATA operand of the PCCU definition statement. In our example, a value of 32 buffers transferred in one read by VTAM will result in maximizing the data transfer to the host from NCP.

For the IBM 3720 and 3745 using NCP V5.1/V5.2, the value of TRANSFR times BFRS must be greater than 1296. The maximum value for TRANSFR is 255. The default for these NCPs is the number of NCP buffers required to hold a 4096-byte PIU. As you have probably surmised, the BFRS and TRANSFR operands play important management and performance roles in the network.

Prior to Extended Network Addressing (ENA), the MAXSUBA operand dictated the number of network addressable units available per subarea. The table in Figure 17.8 lists the possible values of MAXSUBA and the corresponding maximum number of resources per subarea prior to ENA. Our example supports ENA because VTAM and NCP are at version 3 and version 4 releases, respectively, which support ENA. Note that in the figure for NCP versions V4.3.1 and

Pre-VTAM V3 and NCP V4 (NON-ENA)

# of Subareas	Max. # of Elements/Subarea
3	16,384
7	8,192
15	4,096
31	2,048
63	1,024
127	512
255	256

VTAM V3 and NCP V4 ENA

# of Subareas	Max. # of Elements/Subarea
3	32,768
7	32,768
15	32,768
31	32,768
63	32,768
127	32,768
255	32,768

VTAM V3.2 and NCP V4.3.1/V5.2.1 ENA

# of Subareas	Max. # of Elements/Subarea
65,535	32,768

Figure 17.8 The relation of the number of subareas (MAXSU-BA) to the number of addressable elements in a subarea.

NCP V5.2.1 the number of subareas has been increased to 65,535, each with up to 32,768 elements. That allows for an astonishing total of 2,147,450,880 addressable resources in one SNA network.

17.2.4 NCP subarea considerations

The specification of MAXSUBA (Figure 17.9) in the BUILD definition statement for a network that supports ENA is optional. However, until all nodes, including NCP and HOST nodes, in the network are ENA capable, the MAXSUBA operand must be coded. It must match the value of MAXSUBA in the ATCSTR00 start options list of VTAM

```
NCP11BLD   BUILD      MODEL=3725,   comcontroller type      X
                      MEMSIZE=1024, installed memory         X
                      CA=(TYPE5,TYPE5-TPS), CA types         X
                      NCPCA=(ACTIVE,INACTIVE), active CAs    X
                      DELAY=(.2),   delay attn signal        X
                      TIMEOUT=420.0, default timeout/host    X
                      BFRS=128,     NCP data buffer size     X
                      TRANSFR=32,   # of BFRS xfrd to host   X
                      MAXSUBA=15,   maximum subareas         X
                      SUBAREA=11    Subarea of NCP
```

Figure 17.9   MAXSUBA and SUBAREA operand specifications.

and the MAXSUBA operand in the BUILD of each NCP for every subarea node in the network. For more information on ENA and its effect on network addressing schemes, refer to Chapter 9.

The SUBAREA (Figure 17.9) operand defines the subarea address assigned to this NCP. We have chosen subarea address 11. For an ENA networking scheme as in our example, the maximum subarea address can be 255 for this version of NCP. In a non-ENA networking scheme, the subarea address maximum value is limited to the value specified for the MAXSUBA of the BUILD definition statement. The SUBAREA operand is required for the BUILD definition statement.

17.2.5   Defining concurrent sessions support for NCP

When an NCP is activated by VTAM, an SSCP-PU session is established. An NCP can have several sessions with different hosts in a network. In a single-domain network the NCP has only controlling SSCP, and hence one SSCP-PU session. The MAXSSCP operand (Figure 17.10) defines to the NCP the number of SSCP-PU sessions that are conducted concurrently. The default and minimum value for MAXSSCP is the number of channel adapters specified as being active in the NCPCA operand. The maximum value of MAXSSCP is 8 (16 for NCP V5 in an IBM 3745 Communications Controller). The SSCP-PU session includes sessions established over channel or link attachments. If the MAXSSCP is left to default to the active channel adapters specified in the NCPCA operand, no SDLC link-attached host can be in session with this NCP.

The number of host subareas that can activate this NCP is specified by the NUMHSAS operand (Figure 17.10). Formally stated, the NUMHSAS is the number of host subareas that have virtual routes

```
NCP11BLD  BUILD     MODEL=3725,  comcontroller type      X
                    MEMSIZE=1024, installed memory        X
                    CA=(TYPE5,TYPE5-TPS), CA types        X
                    NCPCA=(ACTIVE,INACTIVE), active CAs   X
                    DELAY=(.2),  delay attn signal        X
                    TIMEOUT=420.0, default timeout/host   X
                    BFRS=128,    NCP data buffer size      X
                    TRANSFR=32,  # of BFRS xfrd to host   X
                    MAXSUBA=15,  maximum subareas          X
                    SUBAREA=11,  Subarea of NCP            X
                    MAXSSCP=1,   Concurrent SSCP-PU        X
                    NUMHSAS=1,   # of Host subareas        X
                    VRPOOL=1     # of VRs
```

Figure 17.10   Defining the number of concurrent HOST sessions.

ending in this NCP. This operand has a more important role in a multiple domain and in SNI. Since we have but one host and one virtual route, we can code the minimum value of 1.

The VRPOOL operand (Figure 17.10) specifies the number of virtual route and transmission priority number entries in the NCP virtual route pool. These entries are used during activation, restart, and VR clean-up after an ER failure. The value specified for VRPOOL should equal the number of virtual routes ending in this NCP. To ensure this, make the VRPOOL value equal to the NUMHSAS value. For more information on this, see the chapter on SNI for VRPOOL and NUMHSAS values concerning a gateway NCP in *Advanced SNA Networking*. The default for the VRPOOL operand is 6 times the NUMHSAS value, but cannot exceed 5000. There is a second parameter for the VRPOOL operand that is used for reserving table entries for VRs that are dynamically added by NCP V4.3/V5. Dynamic path table updates are discussed in length in *Advanced SNA Networking*.

17.2.6   NCP load module characteristics

The final operands of the BUILD definition statement specific to NCPV4.2 concern the characteristics of the NCP load module that are produced at the end of the generation process. In defining these characteristics, we tell the NCP the host operating system environment that will load this NCP load module. The TYPSYS operand (Figure 17.11) provides this information. There is a value for each host operating system environment. The MVS value coded in Figure 17.11 specifies that the host operating system is MVS or MVS/XA. The

```
NCP11BLD  BUILD     MODEL=3725,  comcontroller type          X
                    MEMSIZE=1024, installed memory            X
                    CA=(TYPE5,TYPE5-TPS), CA types            X
                    NCPCA=(ACTIVE,INACTIVE), active CAs       X
                    DELAY=(.2),  delay attn signal            X
                    TIMEOUT=420.0, default timeout/host       X
                    BFRS=128,    NCP data buffer size         X
                    TRANSFR=32,  # of BFRS xfrd to host       X
                    MAXSUBA=15,  maximum subareas             X
                    SUBAREA=11,  Subarea of NCP               X
                    MAXSSCP=1,   Concurrent SSCP-PU           X
                    NUMHSAS=1,   # of Host subareas           X
                    VRPOOL=1,    # of VRs                     X
                    TYPSYS=MVS,  host operating system        X
                    TYPGEN=NCP,  operation configuration      X
                    VERSION=V4R2, version of NCP              X
                    NEWNAME=NCP11 NCP load module name
```

Figure 17.11 Defining NCP load module characteristics.

default for the TYPSYS operand is OS. This also is interpreted by the NCP to mean an MVS or MVS/XA host environment.

The TYPGEN operand (Figure 17.11) defines to NCP the type of operational configuration; that is, it defines whether the communications controller this NCP will execute in is channel-attached or link-attached, if the emulation program is included, and if NCP V4 Subset is being used. In our example from Figure 17.1, we can see that the communications controller is channel-attached only. In this case the TYPGEN operand will be set to NCP.

The NCP is made aware of the version of NCP you are defining by the VERSION operand (Figure 17.11). This operand assists the generation process in resolving default values for many of the NCP definition statements and their associated operands. The default for this operand is V3 for version 3. We have declared that the NCP in our scenario is version 4, release 2. Therefore, we have coded V4R2 for the VERSION operand. For NCP V4.3, code V4R3; for NCP V5.1, V5R1; and for NCP V5.2, code V5R2.

There are two other operands for NCP V4.3 and V5.2.

The NETID operand of the BUILD definition statement identifies the owning network for this NCP. This name must match the NETID operand value specified on the VTAM start list statement NETID. The other operand for NCP V4.3 and V5.2 is the PUNAME operand. This operand allows you to assign a PU name to the NCP. In previous releases the NEWNAME operand is used as the NCP PU name. Since the NEWNAME value is also the name of the VTAM NCP major node, the name is usually changed any time a new NCP generation is performed, thus causing operation and configuration management confusion. The addition of the PUNAME operand allows for ease of operations by having the PU name value remain constant while the VTAM NCP major node name and the NEWNAME values can be changed with each new NCP generated. This is particularly useful for automating network operations.

Last, but not least, is the NEWNAME operand (Figure 17.11) of the BUILD definition statement. The value coded in this operand is the name of the NCP load module created at the end of the generation process. The name you specify cannot be greater than seven characters in length and must conform to the naming convention standards for partitioned data set members. This is because the resulting NCP load module will reside in a partitioned data set that is referenced by the NCPLIB DD statement in the VTAM start procedure. The name is limited to seven characters because the generation procedure will append the letter R to the name to create the NCP Resource Resolution Table (RRT). The default name chosen by the generation process is NCP001 for TYPGEN=NCP. We have coded NCP11 as the name of the NCP load module. This NEWNAME parameter is also the NCP PU name for NCP V4.2 and under.

The operands coded for the BUILD definition statement in our scenario provide NCP with the basis for communicating with VTAM and characteristic information about the NCP load module. BUILD definition statements concerning dynamic path updates are discussed in *Advanced SNA Networking*.

17.2.7 APPN Node T2.1 considerations

Several parameters have been added to the BUILD definition statement to support APPN and Type 2.1 nodes. Figure 17.12 lists the new BUILD parameters. The ADDSESS parameter defines a pool containing Session Control Blocks (SCBs) for LU-LU boundary sessions

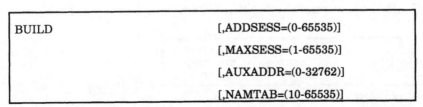

BUILD	[,ADDSESS=(0-65535)]
	[,MAXSESS=(1-65535)]
	[,AUXADDR=(0-32762)]
	[,NAMTAB=(10-65535)]

Figure 17.12 BUILD parameters to support APPN T2.1 nodes.

between independent LUs. These blocks are used in conjunction with the SCBs reserved by the RESSCB parameter of the LU definition statement. The range for ADDSESS is 0 to 5000. If RESSCB is coded for an LU, the SCB pool is accessed only when the LU control point must increase the number of LU-LU sessions between the two T2.1 nodes. The CP issues a Change Number Of Sessions (CNOS) command. CNOS can be used to increase or decrease the number of sessions between any T2.1 nodes.

The MAXSESS parameter specifies the number of LU-LU sessions any ILU can have through this NCP. This restricts the number of sessions to less than the architected value of 64K PLU and 64K SLU. This is done to ensure NCP resource availability for other ILUs that may exist in the network.

Each ILU is assigned a network address. However, for the NCP and VTAM to manage the multiple sessions an ILU can have, a dynamic address is created. The number of additional addresses assigned to an ILU is specified by the AUXADDR parameter.

The final parameter is NAMTAB. This parameter reserves entries in a network name table. This table contains the names of all networks, SSCPs, and CPs that establish a session using this NCP. There is one entry for each APPN network, SNI network, and SSCP and CP.

17.2.8 Token-ring considerations

The NCP defines the physical as well as the logical token-ring network connections associated with a IBM 37X5 Communication Controller. The OPTIONS and BUILD definition statements of an NCP specify generation parameters specific to an NCP supporting token-ring interfaces. This text will discuss only those parameters necessary for token-ring connectivity.

The OPTIONS statement, shown in Figure 17.13, indicates to the NCP generation process certain options specific to the NCP. The NEWDEFN keyword of the OPTIONS definition statement specifies whether NTRI definitions should be generated even if they have not been coded. The BUILD definition statement in an NCP has two keywords that are specific to token-ring. These are the MXRLINE and MXVLINE keywords. The MXRLINE keyword identifies the number

```
NCPOPTS              OPTIONS    NEWDEFN=YES
NCPBUILD             BUILD      MXRLINE=...,
                                MXVLINE=...,
```

Figure 17.13 NCP definition statements to support token-ring resources.

of physical lines attached to the TICs. There is one line for each phys-ically attached TIC. The MXVLINE keyword indicates the number of logical lines associated with the physical lines. These keywords and values are automatically generated when using NCP V5R3 and the System Support Program (SSP) V3R5 and higher. Prior to these releases, each physical and logical line definition would have to be counted manually and the totals then entered on these keywords.

17.2.9 Ethernet considerations

Support for Ethernet connectivity is available with NCP V6R1. There are four BUILD parameters that are specific to Ethernet support. These parameters assist in controlling congestion caused by connec-tionless traffic over the SNA backbone network. The values of these four parameters are used to limit the IP traffic queued for SNA ses-sions or for Ethernet adapters. The four parameters are CNLSQMAX, CNLSQTIM, IPRATE, and IPPOOL. Figure 17.14 lists the BUILD parameters and possible values for Ethernet support.

The CNLSQMAX parameter is used to specify the maximum size in bytes allowed on an outbound session queue for an IP interface when NCP reaches slowdown. The valid range for the CNLSQMAX value is 1000 to 65535. The CNLSQTIM parameter indicates the maximum number of seconds a datagram can remain on the outbound session queue used by an IP interface. A PIU is discarded after sitting on the outbound queue for the time specified by this parameter. The range is 1 to 255. The default is 10 seconds. The IPRATE parameter is specific to traffic that passes over an Ethernet adapter that is destined for another Ethernet device. The rate operand is the number of IP data-grams that will be accepted by the NCP in a tenth of a second. The rate operand value can range from 1 to 65535. The burst operand is used to provide for a maximum number of Internet frames that the NCP will accept in a tenth of a second. It will default to the rate operand value if it is not coded. Again, the burst value can range from 1 to 65535. These values for the IPRATE parameter are specific to the NCP as a whole and not to a particular Ethernet interface. If the IPRATE parameter is not specified, then no control is applied, and the NCP may be overrun

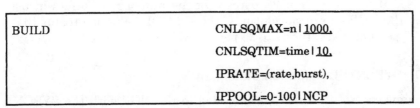

```
BUILD                        CNLSQMAX=n I 1000,

                             CNLSQTIM=time I 10,

                             IPRATE=(rate,burst),

                             IPPOOL=0-100 I NCP
```

Figure 17.14 BUILD parameters specific to Ethernet support with NCP V6R1.

| | Maximum Frame Rate (Frames per 0.1 second) | |
Controller Model	68-Byte Packets	1500-Bytes Packets
3745-130	90	70
3745-150	90	70
3745-160	90	70
3745-170	90	70
33745-210	110	90
3745-21A	110	90
3745-410	110	90
3745-41A	110	90
3745-310	180	140
3745-31A	180	140
3745-610	180	140
3745-61A	180	140

Figure 17.15 Suggested initial values of maximum Internet frame rates to use for the IPRATE parameter.

with datagrams from the Ethernet interfaces. The values coded here are of importance because the NCP gives Ethernet-type Internet frames priority over SNA traffic. The table found in Figure 17.15 lists suggested values for IPRATE for the various communications controllers. The IPPOOL parameter specifies the percentage of NCP buffer space that can be used for Ethernet IP traffic. The IPPOOL value can be defined with a value of 0 to 100. This defines the percentage of buffers in use when NCP enters slowdown. If 0 is entered, the NCP will guarantee that the IPPOOL will be big enough to hold 6000 bytes of IP datagram data. If the IPPOOL value is specified as NCP, the NCP will use a formula to determine the IPPOOL size. The resulting value will be used if it is less than 25 percent of the number of buffers in use when the NCP enters slowdown. If the IPPOOL value is greater than the 25 percent mark, then the 25 percent value will be used.

17.3 SYSCNTRL Definition Statement

Dynamic control facilities are included in the NCP with the SYSCN-TRL definition statement. These facilities allow VTAM to issue requests to the NCP to modify specific BSC and S/S parameters defined in the NCP. The SYSCNTRL definition statement is required and must follow the BUILD definition statement even if there are no BSC or SS devices in your NCP. The format of the definition statement is

```
[name] SYSCNTRL OPTIONS=(entry,…)
```

The name operand is a symbolic name for the definition statement and is optional. The OPTIONS operand is the only operand for the SYSCNTRL definition statement and is required. The entry parameters for VTAM are shown in Figure 17.16.

The sample single-domain network in Figure 17.1 shows that a BSC 3274 remote cluster controller is attached to the communications controller that will contain this NCP. The first two entries in the OPTIONS operand, MODE and RIMM, are required if BSC 3270 devices are included in the NCP.

The MODE parameter allows VTAM to set the destination mode of the BSC device. The RIMM parameter allows VTAM to reset the BSC device immediately after an error. The remaining parameters are used by VTAM for operator control functions. The NEGPOLL parameter allows VTAM to issue the MODIFY NEGPOLL command. This operator command modifies the NEGPOLL operand of a BSC LINE definition statement. The SESSION parameter supplies the code to VTAM for the MODIFY SESSION operator command. This command is used to modify the number of concurrent sessions on a BSC LINE. The SSPAUSE parameter provides VTAM with the ability to modify the service-seeking pause operand PAUSE of a BSC or SS LINE. The MODIFY POLL operator command is used for this dynamic change. The STORDSP parameter is required for VTAM to display the NCP storage using the operator command DISPLAY NCPSTOR. This allows viewing of the NCP storage without interrupting the NCP.

These entries are the basic SYSCNTRL options needed for VTAM for supporting BSC and SS devices. There are many more options that may be coded. During the NCP generation process, the options

```
SYSCNTRL   OPTIONS=(MODE,RIMM,NAKLIM,SESSION,   X
                    SSPAUSE,STORDSP,BACKUP,DLRID,X
                    BHSASS,DVSINIT,ENDCALL,       X
                    LNSTAT,RCNTRL,RCOND,RDEVQ,    X
                    RECMD,SESINIT,XMTLMT)
```

Figure 17.16 Dynamic control facility options for VTAM.

coded are compared with the NCP resource definitions for this NCP. If resources have not been defined that correlate with the SYSCNTRL option specified, the code for that dynamic control facility is not included in the NCP load module. Therefore, it is better to be on the safe side and code all the possible options for the SYSCNTRL definition statement.

17.4 Summary

In this chapter, we discussed coding PCCU, BUILD, and SYSCNTRL macros of NCP. The PCCU macro defines the functions that VTAM is supplying for this NCP. Although it is a VTAM-related statement, it is placed in the NCP code. The BUILD macro defines the physical makeup of the communications controller and the information about the NCP itself. It also establishes NCP's ability to communicate with VTAM. The SYSCNTRL macro allows VTAM to issue requests to the NCP to modify specific BSC and SS parameters defined in the NCP. It is a required statement and must follow the BUILD macro.

NCP Macros—HOST, LUDRPOOL, PATH, and GROUP

18.1 HOST Definition Statement

The NCP must allocate buffers used in sending and receiving data between the host and the channel-attached communications controller. The HOST definition statement provides the number of buffers and the size of the buffers used for this communication. There must be one HOST definition statement for each VTAM that communicates to this NCP over a channel. The statement must also appear before any GROUP definition statements are encountered. The format of the HOST definition statement is

```
[name] HOST operand[,operand,...operand]
```

The name operand is any symbolic name you choose to code. The remaining operands will be discussed in conjunction with Figure 18.1. The operands related to the buffers in VTAM and NCP play a major role in tuning your network.

```
HOST01          HOST        INBFRS=10,                      X
                            MAXBFRU=16,                     X
                            UNITSZ=256,                     X
                            BFRPAD=0,                       X
                            SUBAREA=1
```

Figure 18.1 HOST definition statement for a single-domain in NCP.

When defining the buffer characteristics for the HOST definition statement, there are several factors to consider. One is the amount of storage required in the NCP buffer pool when allocating buffers for communications with the HOST. The INBFRS operand defines to the NCP the number of buffers that are initially required to handle data received from VTAM and is a required operand. If the data being received from VTAM is greater than this initial allocation, NCP will allocate more buffers in the multiple specified with the INBFRS operand. The BUFSIZE operand of the BUILD definition statement is used with the INBFRS value to allocate storage. The BUFSIZE we specified is 128 bytes. The amount of NCP storage used is BUFSIZE times INBFRS. We have allocated 10 buffers of the NCP buffer pool for the transfer of data from the host. Therefore the amount of storage initially allocated is 2560 bytes. This is an acceptable value until further analysis of message unit size can be determined.

In the PCCU definition statement we defined the size of the largest PIU that can be transmitted between this NCP and VTAM as 4096 bytes in length. Here in the HOST definition statement we define the amount of data VTAM can receive from this NCP. This value is determined by the MAXBFRU and UNITSZ operands. Their product is the largest amount of data this NCP will transfer to the host defined by this HOST definition statement. This does not mean the largest PIU, but rather the largest amount of data in a single transfer of multiple PIUs. The product should not exceed the MAXDATA value coded on the PCCU definition statement that corresponds to this HOST definition statement.

The MAXBFRU operand value specifies the number of buffers allocated in VTAM to receive data from this NCP. This value cannot exceed the base number of IO buffers allocated by the IOBUF start option in ATCSTR00 of VTAMLST. Our IOBUF specified 32 buffers of 256 bytes each to be the base allocation of the IO buffer pool for this VTAM. We have coded a value of 16 based on the value specified for the operand UNITSZ.

The UNITSZ operand specifies the buffer size in the host. This value must match the IOBUF BUFSIZE operand in the ATCSTR00 start options list of VTAMLST. The UNITSZ operand is required and is compared by the NCP with the IOBUF BUFSIZE value passed to it in the XID format during activation. The NCP will not activate if these values do not match.

The BFRPAD operand is optional. This buffer pad is used by the access method to insert the message header and message text prefixes. VTAM under VSE uses a 15-byte buffer pad. TCAM uses a minimum of 17 bytes, and VTAM under MVS and VM uses 0 bytes. We must specify the BFRPAD operand and assign a value of 0 because the default value is 28 bytes.

Each subarea in an SNA network must have a unique subarea address. The subarea of the host described by this HOST definition statement is supplied to the NCP by the SUBAREA operand. The subarea number assigned to the host according to the HOSTSA option of the start options list should be coded here. For our sample network the host subarea is 1.

18.2 LUDRPOOL Definition Statement

In our sample single-domain network we have configured a switched SDLC line. The switched line will connect an S/36 emulating an SNA/3770 RJE device. The LUDRPOOL operand is used to specify the number of LUs that are expected to have a session on this switched line. This allows NCP to allocate the appropriate number of LU control blocks in the LU pool for dynamic resources. The LUDR-POOL is also used in dynamic reconfiguration.

The LUDRPOOL definition statement is positional and must precede the first GROUP definition statement. Figure 18.2 details the usage of the LUDRPOOL definition statement.

The LUDRPOOL definition statement consists of two operands. The first operand, NUMTYP1, defines the number of expected LUs that are associated with a PU Type 1 SNA device. The switched line for our example is a PU Type 2 SNA device. In this case the NUM-TYP1 operand could have been omitted, since the default for the LUDRPOOL operands is 0. The NUMTYP2 operand specifies the number of expected LUs to be dynamically allocated in this NCP that are associated with a PU Type 2 SNA device. The value of 2 is coded here in accordance with the switched major node definition SWS36RJE that was previously defined in Chapter 13.

18.3 PATH Definition Statement

Like VTAM, the NCP has a routing capability for transmitting data through the network. In NCP, the PATH definition statement is used in the same manner as the PATH definition statement in VTAM. The destination subareas NCP can route data to and the control of the data flow can be defined in the PATH definition statement. By using the adjacent subarea and transmission group number pair associated with an explicit and virtual route, NCP can direct data to a destina-

```
LUDR01          LUDRPOOL   NUMTYP1=0,                          X
                           NUMTYP2=16
```

Figure 18.2 Allocating LU pool entries for switched SDLC links.

```
PATH1201        PATH        DESTSA=1,        VTAM SUBAREA 01        X
                            ER0=(1,1),       ER0, ADJSA1, TG1       X
                            VR0=0            VR 0 MAPPED TO ER 0
```

Figure 18.3   Defining the route between subarea 01 and subarea 12.

tion subarea. The PATH definition statement is positional in the NCP. It must follow the GWNAU definition statements in a gateway NCP that defines the network addressable units of cross-network resources. In addition, the PATH definition statement must precede the first GROUP definition statement. Figure 18.3 outlines the coding of a PATH definition statement.

The label PATH1201 is a symbol defined for documentation purposes only. In this instance, it denotes the route table between subareas 11 and 01. The DESTSA operand can specify more than one destination subarea. The format for the DESTSA operand is

```
DESTSA=(sa1[,sa2,sa3,…,sa255])
```

As you can see, DESTSA can range from 1 to 255 subarea addresses pre-VTAM V3.2 and NCP V4.3.1/V5.2.1. For these versions the maximum DESTSA value is 65535. It is the only required operand of the PATH definition statement. At least one subarea address must be coded. Any subarea specified in this DESTSA operand cannot appear in any other DESTSA operand of this NCP. Therefore, it is best to code one PATH definition statement for each destination subarea to avoid routing discrepancies. The value coded for the DESTSA operand must be numeric.

In our sample single-domain configuration, we are concerned with two subareas. The NCP must define a route to the only destination subarea possible with this configuration. That destination subarea is the controlling VTAM subarea 01. The Explicit Route (ER) between these subareas is defined in VTAM as ER0. The ER operand format for NCP is

```
[ER0=(adjsa,[tgn],[lothresh],[medthresh],[hithresh],[totthresh])]
         1     5000        5000          5000         20000
```

The ER operand is optional but is usually coded. The default ER is ER0. You can code ER0 through ER7 for explicit routes to the destination subarea specified in the DESTSA operand for pre-VTAM V3.2 and NCP V4.3.1/V5.2.1. These versions will support ER8 through ER15 as well. The ER operand defines the adjacent subarea (*adjsa*) and transmission group number (*tgn*) pair used to map this explicit route. The *adjsa* is a subarea directly connected to this NCP that can be used to route data bound for the destination subarea(s) defined in DESTSA. The *adjsa* range permitted is 1 to 255. VTAM V3.2 and NCP V4.3.1/V5.2.1 support an *adjsa* range of 1 to 65,535. For our scenario the only adjacent subarea is the destination subarea. So we code a 1 for the destination subarea.

The *tgn* is optional within the ER operand. It defines the transmission group that can be used to route the data to the adjacent subarea. The *tgn* defaults to 1 if it is not specified. Since the only route is a channel-attached host connection, the *tgn* must be defined as 1. If more ERs were to be specified in the PATH definition statement, this *tgn* could be used again. This *tgn* should be used in the adjacent subarea when defining the adjacent subarea's ER to this NCP.

The Transmission Group (TG) is the highway between subareas. During rush hour the highway may become congested. The ER operand allows us to control the data flow to avoid the congestion. These flow-control thresholds are optional within the ER operand. There are four threshold specifications possible. Each has a corresponding default value. The *lothreshold* value is the threshold used by NCP for low transmission priority traffic. The *medthreshold* value specifies the threshold for transmission priority 1 of the VR. The *hithreshold* value is the threshold for transmission priority 2 of the VR. Each threshold defaults to a value of 5000. The value represents the number of bytes queued for that transmission priority over the TG. The *totthreshold* is the total number of bytes that can be queued for that TG before NCP stops queuing PIUs for this TG. The queuing is ceased until the number of bytes falls below the threshold value.

The virtual route mapping is used here because NCP V4.2 supports activation of VRs to adjacent subareas. The VR format is

```
[VR0=ern] [,VR1=ern] ... [,VR7=ern]
```

The VR operand is optional for the PATH definition statement. The VR operand maps the virtual route to the explicit route. This map is used by NCP when activating VRs. The Explicit Route Number (*ern*) must be an ER previously defined on an ER operand of this PATH definition statement. The VR mapping used in this NCP should agree with the VR mapping used in the destination subarea PATH table for the return explicit route.

Each VR defined has a corresponding pacing window. This pacing window is used to determine the number of PIUs that can be transmitted over the VR. The format of the VRPWS pacing window size operand is

```
[VRPWS00=([min][,max]), ...,VRPWS72=([min][,max])]
```

The VRPWS operand has two digits suffixed. The first represents the VR number, and the second represents the transmission priority. There are three transmission priorities (TPs) for each VR. You can specify up to 24 VR pacing window size operands. The *min* and *max* values represent the minimum and maximum number of PIUs for this VR and the associated TP.

For the most part, the threshold values of the ER operand and the VRPWS operands are not coded. However, default VRPWS *max* values are often too low when using line speeds of 56 Kbps or higher. Usually in-depth performance studies are needed to determine appropriate values. Performance implications of the thresholds and VR pacing window size are discussed in *Advanced SNA Networking*.

18.4   GROUP Definition Statement

Many of the data links and devices attached to the NCP have a common set of characteristics. These NCP resources can be grouped together by the GROUP definition statement. The operands used on the GROUP definition statement allow you to specify certain characteristic functions that are common to the group. Most of these common functions are determined by the communications protocol. BSC communications protocol over lines will have different characteristics and operational options than a line that uses SDLC protocol. Hence you will often hear that the GROUP definition statement describes the line group. The format of the GROUP definition statement is

```
name GROUP [operand,operand...operand]
```

There is only one required operand for the GROUP definition statement, the LNCTL operand. LNCTL identifies the type of line control for all resources within this group. The name label is required and provides a symbolic name for the line group. A benefit of the GROUP definition statement is the ability to code several operands here that directly influence the capabilities of many of the resources defined under this group. This saves the communications systems programmer from repetitive coding for each resource. This is possible because the generation procedure uses the sift-down effect described in Chapter 11. For our sample single-domain configuration there will be three GROUP definition statements: a BSC group, an SDLC non-switched group, and a switched SDLC group, each having a different characteristic flavor.

18.4.1   BSC GROUP definition statement

Line group definitions of BSC resources must be defined to the NCP before any SDLC line groups are defined. Operands coded in this

GROUP definition statement reflect values for all lines under this group unless they are explicitly coded on successive LINE definition statements. We will discuss the group characteristics in three categories. The first will focus on the line characteristics of the group, the second will concern error recovery, and the third will center on servicing the data links defined for this group. Figure 18.4 details the BSC GROUP definition statement for our example.

When defining line groups, all lines under the group must use the same type of line control. BSC line groups must follow all SS line groups and precede all SDLC line group definitions. In the line group definition it is customary to code the line characteristics for the group. The LNCTL operand specifies the type of line control for this group. Figure 18.4 shows that the BSC line control is prevailing for all lines defined under this group. We code BSC because the devices at the end point of the physical link use BSC protocol for communicating with the host.

Because our configuration is a BSC 3270 family device, it is supported under SNA without the use of an emulation program. BSC 3270 family devices are the only non-SNA devices supported under SNA in their native mode. This permits the NCP to act in network control mode with this link. We specify this to NCP by coding the TYPE operand. The value NCP denotes the use of network control mode for the lines in this group.

To differentiate between switched and nonswitched data links, the DIAL operand is coded. This operand specifies whether switched line

```
BSC3270   GROUP     LNCTL=BSC,        BSC LINE CONTROL            X
                    TYPE=NCP,         LINE OPERATION MODE         X
                    DIAL=NO,          NON-SWITCHED LINK           X
                    CLOCKNG=EXT,      EXTERNAL CLOCKING/MODEM     X
                    DUPLEX=FULL,      4-WIRE CIRCUIT              X
                    SPEED=9600,       SPEED OF MODEM              X
                    CODE=EBCDIC,      EBCDIC CHARACTERS USED      X
                    RETRIES=(5,10,3), 3 RETRIES 5/10 SECS        X
                    REPLYTO=3,        3 S BEFORE POLL T.O.        X
                    TEXTTO=23.5,      23.5 S B4 MSG T.O.          X
                    PAUSE=1,          SERVICE PAUSE               X
                    POLLED=YES,       POLL ALL DEVICES            X
                    SERVPRI=OLD,      SESSION SERVICE PRIORITYX
                    SERVLIM=8,        SERVICE SEEK LIMIT          X
                    CUTOFF=1,         # OF TRANSFR SEQUENCE       X
                    TRANSFR=3         3 OF NCP BUFFERS XMIT
```

Figure 18.4 BSC GROUP highlighting line characteristics.

control procedures are required for the lines in this group. By specifying NO, nonswitched line control procedures are enforced.

Another characteristic of the links in this group is the responsibility for clocking the line speed. The CLOCKNG operand relays this information to NCP. The specification of CLOCKNG=EXT tells NCP that the modem attached to the link will supply the clocking. If CLOCKNG=INT is specified, the communication scanner for that line will provide the clocking within the communications controller if the proper business machine clock is installed. See Chapter 2 for more information on the IBM 3725 communications controller.

The physical medium of the link is also described to NCP. The DUPLEX operand specifies whether the facility is physically capable of half-duplex or full-duplex transmission. These are also referred to as "two-wire" or "four-wire" facilities. The NCP uses the value specified to control the Request-To-Send (RTS) signal to the modem. The specification of FULL keeps the RTS signal active whether the NCP is sending or receiving data. Specifying HALF activates the RTS signal only when the NCP is sending data. This operand should not be confused with half-duplex and full-duplex protocol, the transfer of data between session partners.

The speed of the link is specified on the SPEED operand. The SPEED operand is required when CLOCKNG=INT is coded. This is necessary because the communications scanner must know how to clock the data bits for sending and receiving data. If CLOCKNG=EXT is coded, the generation procedure ignores the SPEED operand since the modem is responsible for the clocking. However, the SPEED operand is used by some network performance tools. It is also useful for documentation purposes.

The transmission code is used by the NCP for communicating to the devices defined under this line group. Using the CODE operand, NCP can determine how the data destined for the device is to be translated. NCP's internal processing code is EBCDIC, and it translates the outgoing data to the code specified on the CODE operand. For BSC devices in network control mode, the default is EBCDIC.

These line characteristic operands are the more widely used operands for BSC lines in conjunction with BSC 3270 devices. Let's now turn our attention to the operands concerned with error detection and recovery (Figure 18.5).

During the course of a session between two partners over the BSC link, NCP keeps a timer for response time to a poll of a device and a timer for measuring the time between message characters received from a device. For the latter, if the interval between two successive characters in a message received from a device exceeds the time spec-

```
BSC3270    GROUP      LNCTL=BSC,         BSC LINE CONTROL              X
                      TYPE=NCP,          LINE OPERATION MODE           X
                      DIAL=NO,           NON-SWITCHED LINK             X
                      CLOCKNG=EXT,       EXTERNAL CLOCKING/MODEM       X
                      DUPLEX=FULL,       4-WIRE CIRCUIT                X
                      SPEED=9600,        SPEED OF MODEM                X
                      CODE=EBCDIC,       EBCDIC CHARACTERS USED        X
                      RETRIES=(5,10,3),  3 RETRIES 5/10 SECS           X
                      REPLYTO=3,         3 S BEFORE POLL T.O.          X
                      TEXTTO=23.5,       23.5 S B4 MSG T.O.            X
                      PAUSE=1,           SERVICE PAUSE                 X
                      POLLED=YES,        POLL ALL DEVICES              X
                      SERVPRI=OLD,       SESSION SERVICE PRIORITYX
                      SERVLIM=8,         SERVICE SEEK LIMIT            X
                      CUTOFF=1,          # OF TRANSFR SEQUENCE         X
                      TRANSFR=3          3 OF NCP BUFFERS XMIT
```

Figure 18.5 Error detection and recovery for BSC links.

ified in the TEXTTO operand, NCP will end the read or invite operation for the message with a text time-out error indication. Errors of
this kind can lead to suspecting the cluster controller, modem, or line.
The value coded is in seconds and is a nominal value. The actual
timed interval may be between the nominal value specified and twice
the nominal value. The default 23.5 seconds is recommended for most
BSC devices.

During a polling sequence the NCP will wait for a response to the
poll, selection, or message text for the interval specified in the
REPLYTO operand. If the response is not received in the time specified by the REPLYTO operand, the NCP will indicate to the host that
a time-out error has occurred on this link. For BSC lines, a default
value of 3 seconds is recommended. The value may be coded to the
nearest tenth of a second. If your BSC session is using conversational
replies, ensure that the REPLYTO operand specifies a value large
enough to accommodate the conversational text to be received.

Data link errors occurring during the transmission of data can be
handled by the RETRIES operand. This operand defines a retry
sequence to be used by NCP to recover from the transmission error.
The first value is the maximum number of retries in the sequence.
Our definition calls for five retries per sequence. This value can range
from 0 to 255. The second parameter in the RETRIES operand is the
number of seconds the NCP will pause between sequences. This value
can range from 0 to 255 seconds. We have specified 10. The third

parameter specifies the maximum number of times the retry sequence is executed. The range for this parameter is 1 to 255. If the maximum retry value is reached, the device is marked inoperative.

In the retry sequence during a send operation, NCP will retransmit the block of data in which the error occurred until it is successfully transmitted or until the maximum retry value is reached. If the error occurred during an NCP receive operation, the NCP will issue a negative response to the sending device. This causes the device to send the block in error until the NCP successfully receives it, the device sends an End-Of-Text (EOT) character, or the NCP has sent the device the number of negative responses that match the maximum number of retries. The values coded in Figure 18.5 are the recommended values.

When NCP services the lines defined in this group, it uses several operands. These operands determine the rate and type of service, which devices get service, and the amount of data the NCP will receive after servicing the line.

For BSC devices, the PAUSE operand (Figure 18.6) specifies the number of seconds for delaying successive service cycles when no sessions exist. The value can range from 0 to 255. The value must be an integer with no decimal extension. This operand saves the NCP cycles when sessions on a BSC line do not exist. The NCP will wait 1 second between each polling service cycle for each device on a line when no sessions exist. The PAUSE operand can be coded for BSC devices only

```
BSC3270    GROUP    LNCTL=BSC,          BSC LINE CONTROL            X
                    TYPE=NCP,           LINE OPERATION MODE         X
                    DIAL=NO,            NON-SWITCHED LINK           X
                    CLOCKNG=EXT,        EXTERNAL CLOCKING/MODEM     X
                    DUPLEX=FULL,        4-WIRE CIRCUIT              X
                    SPEED=9600,         SPEED OF MODEM              X
                    CODE=EBCDIC,        EBCDIC CHARACTERS USED      X
                    RETRIES=(5,10,3),   3 RETRIES 5/10 SECS         X
                    REPLYTO=3,          3 S BEFORE POLL T.O.        X
                    TEXTTO=23.5,        23.5 S B4 MSG T.O.          X
                    PAUSE=1,            SERVICE PAUSE               X
                    POLLED=YES,         POLL ALL DEVICES            X
                    NEGPOLP=.2,         NEGATIVE POLL PAUSE         X
                    SERVPRI=OLD,        SESSION SERVICE PRIORITYX
                    SERVLIM=8,          SERVICE SEEK LIMIT          X
                    CUTOFF=1,           # OF TRANSFR SEQUENCE       X
                    TRANSFR=3           3 OF NCP BUFFERS XMIT
```

Figure 18.6 Line service operands for a BSC line group.

when the line is in network control mode and the POLLED operand equals YES.

If POLLED=YES is specified, the NCP must poll and address each BSC device on the line individually. The default is NO. All lines in the line group must use the same POLLED operand. Both options cannot be included for different lines within the line group.

Nonproductive polling performed by an NCP for devices of a line can create unnecessary overhead and can waste NCP cycles. The NEGPOLP operand for BSC lines provides a means for reducing the overhead. Nonproductive or negative polls are responses received from a device that has been polled by the NCP but does not have data to send to the NCP. The value specified on the NEGPOLP operand causes the NCP to wait before polling the device again for data. The range is 0.1 to 23.5. A large NEGPOLP value can increase the end-user response time. The default is 0, but we have coded the recommended value of 0.2 seconds. This operand can also be changed by issuing the VTAM operator command MODIFY NEGPOLP. Section 18.3 further discusses modifying this operand.

When servicing a line, it is important to provide the best response available to existing sessions. This can be defined by the SERVPRI operand. The specification of OLD as seen in Figure 18.6 tells NCP to give priority service to existing sessions before establishing new sessions (SERVPRI=NEW).

The SERVLIM operand specifies the number of devices the NCP will check during service seeking of the service order table for a cluster on the line. An optimal value for this operand is half the entries in the largest service order table. NCP can then have more frequent opportunities to service existing sessions, and better link response time; however, a higher value optimizes service seeking. Once a device has been serviced, the NCP can limit the amount of data it will receive from devices on the line with the CUTOFF and TRANSFR operands.

The CUTOFF operand defines the number of subblocks the NCP will accept from a device on this line. A subblock is the message text sent by the device that fills the number of NCP buffers specified in the TRANSFR operand for this line. If this number is reached before the NCP receives an end-of-block character from the device, the NCP breaks off the transmission. It is highly recommended that CUTOFF=1 and TRANSFR=3 be coded for BSC 3270 devices. Using these values avoids a "HOT I/O" situation that can bring down VTAM or NCP.

This ends our review of line characteristic definitions for a BSC line attached to this NCP. Now let's look at the definitions for an SDLC line group.

18.4.2   SDLC GROUP definition statement

Line group definitions of SDLC resources must be defined to the NCP after all BSC and SS line groups have been defined. Operands coded in this GROUP definition statement reflect values for all lines under this group unless they are explicitly coded on successive LINE definition statements. We will discuss the group characteristics in three categories. The first will focus on the line characteristics of the group, the second will concern error recovery, and the third will center on servicing the data links defined for this group. Figure 18.7 details the SDLC GROUP definition statement for our example.

The line group definitions for SDLC lines are quite similar to those for BSC lines. Many of the operands reviewed for BSC lines are used in the same fashion for SDLC lines. The LNCTL operand specifies the type of line control for this group. Figure 18.7 shows that the SDLC line control is prevailing for all lines defined under this group. SDLC is coded because devices at the end point of the physical link use SDLC protocol for communicating with the host.

Since the remote devices under this line group are using SDLC protocol, they are SNA resources; therefore, the NCP acts in network control mode with this link. We specify this to NCP by coding the TYPE operand. The value NCP denotes the use of network control mode for the lines in this group.

Just as in the BSC line group, the DIAL operand differentiates between switched and nonswitched data links. This operand specifies whether switched line control procedures are required for the lines in this group. By specifying NO, nonswitched line control procedures are enforced. For more information about switched line control procedures, consult *Advanced SNA Networking*.

```
SDLC3270 GROUP    LNCTL=SDLC,        SDLC LINE CONTROL             X
                  TYPE=NCP,          LINE OPERATION MODE           X
                  DIAL=NO,           NON-SWITCHED LINK             X
                  CLOCKNG=EXT,       EXTERNAL CLOCKING/MODEM       X
                  DUPLEX=FULL,       4-WIRE CIRCUIT                X
                  SPEED=9600,        SPEED OF MODEM                X
                  NRZI=NO,           NON-RETURN to ZERO MODE       X
                  RETRIES=(5,10,3),  3 RETRIES 5/10 S              X
                  REPLYTO=1,         1 S    BEFORE POLL T.O.X
                  TEXTTO=1,          1 S    B4 MSG T.O.            X
                  PAUSE=.2,          SERVICE PAUSE                 X
                  SERVLIM=254        SERVICE SEEK LIMIT
```

Figure 18.7   SDLC group highlighting line characteristics.

As with the BSC line group, the responsibility for clocking the data falls under the jurisdiction of the modems. The CLOCKNG operand relays this information to NCP. The specification of CLOCKNG=EXT tells NCP that the modem attached to the link will supply the clocking. If CLOCKNG=INT is specified, the communications scanner for that line will provide the clocking within the communications controller if the proper business machine clock is installed.

The DUPLEX operand for SDLC links is dependent upon the ADDRESS operand of the LINE definition statement. If the second parameter of the ADDRESS operand specifies FULL, the DUPLEX operand has no effect on the SDLC link. On the other hand, if HALF is specified on the ADDRESS operand, the DUPLEX operand affects the RTS signal the same way as for a BSC line. The specification of FULL keeps the RTS signal always active. Specifying HALF activates the RTS signal only when the NCP is sending data. Again, this operand should not be confused with half-duplex and full-duplex protocol, the transfer of data between session partners.

The speed of the link is specified on the SPEED operand, which is required when CLOCKNG=INT is coded. This is necessary because the communications scanner must know how to clock the data bits for sending and receiving data. If CLOCKNG=EXT is coded, the generation procedure ignores the SPEED operand, since the modem is responsible for the clocking. However, the speed operand is used by some network performance tools. It is also useful for documentation purposes.

The NRZI operand is pertinent to SDLC lines only. This operand specifies the use of non-return-to-zero change-on-ones (NRZI) mode used by all SNA cluster controllers attached to this SDLC link. For the most part, the specification of NRZI=YES is dependent on the modems used for the SDLC line. This is because most SNA cluster controllers can support NRZI. The modems are in question because many modems are sensitive to the bit pattern used in NRZI and may lose synchronism. The guideline used for this operand is, if the modem is a non-IBM modem, code NRZI=NO; if it is an IBM modem, code NRZI=YES. Your best bet is to consult the maker of the modem to get a definitive answer.

These line characteristic operands are the more widely used operands for SDLC lines. Let's now turn our attention to the operands concerned with error detection and recovery (Figure 18.8).

Time-outs for polling responses and message text are also used for SDLC lines. If the interval between two successive characters in a message received from a device exceeds the time specified in the TEXTTO operand, NCP will end the read or invite operation for the message with a text time-out error indication. Errors of this kind can

```
SDLC3270 GROUP        LNCTL=SDLC,      SDLC LINE CONTROL            X
                      TYPE=NCP,        LINE OPERATION MODE          X
                      DIAL=NO,         NON-SWITCHED LINK            X
                      CLOCKNG=EXT,     EXTERNAL CLOCKING/MODEM      X
                      DUPLEX=FULL,     4-WIRE CIRCUIT               X
                      SPEED=9600,      SPEED OF MODEM               X
                      NRZI=NO,         NON-RETURN TO ZERO MODE      X
                      RETRIES=(5,10,3), 3 RETRIES 5/10 SECS        X
                      REPLYTO=1,       1 S BEFORE POLL T.O.         X
                      TEXTTO=1,        1 S B4 MSG T.O.              X
                      PAUSE=.2,        SERVICE PAUSE                X
                      SERVLIM=254      SERVICE SEEK LIMIT
```

Figure 18.8 Error detection and recovery for SDLC links.

lead to suspecting the cluster controller, modem, or line. The value coded is in seconds and is a nominal value. The actual timed interval may be between the nominal value specified and twice the nominal value. In SDLC, the time-out occurs only when the link station is the primary link station. The recommended value for SDLC lines is based on the line speed. In our case, the recommended time-out value is 1, since the line speed is greater than 2000 bps. For line speeds between 0 and 1199 bps, the value is about 4. For line speeds between 1200 and 2000 bps, the recommended value is approximately 3 seconds. You will have to do some experimentation on your network to find the appropriate values for your SDLC lines.

The interval specified on the REPLYTO operand is the time NCP will wait for a response to a poll, selection, or message text response from a device on an SDLC line. The default for an SDLC line is 1 and is recommended. But again, realization of this value on your network must be determined by examining the configuration.

The RETRIES operand is used in the same manner as was previously discussed for the BSC GROUP definition. The difference is the range of values that can be coded for the parameters of the operand. The first value is the maximum number of retries in the sequence. Our definition calls for five retries per sequence. This value can range from 0 to 128. The second parameter in the RETRIES operand is the number of seconds the NCP will pause between sequences. This value can range from 1 to 255; we have specified 10. The third parameter specifies the maximum number of times the retry sequence is executed. The range for this parameter is 1 to 127. If the maximum retry value is reached, the device is marked inoperative.

In the retry sequence during a send operation, NCP will retransmit the block of data in which the error occurred until it is successfully transmitted or until the maximum retry value is reached. If the error occurred during an NCP receive operation, the NCP will issue a negative response to the sending device. This causes the device to send the block in error until the NCP successfully receives it, the device sends an End-Of-Text (EOT) character, or the NCP has sent the device the number of negative responses that match the maximum number of retries. The values coded in Figure 18.8 are the recommended values.

When NCP services the lines defined in this group, it uses several operands. These operands determine the rate and type of service, which devices get service, and the amount of data the NCP will receive after servicing the line (Figure 18.9).

The PAUSE operand for an SDLC line is the polling interval of resources in the service order table. The value specified here is the amount of time NCP will wait before polling all resources in the service order table. That time includes the period from the moment the NCP polls the first entry to the time the NCP services that first entry again. The cycle includes time for polling, reading, and writing to resources on the line. If the time spent in servicing all the active resources in the service order table equals or exceeds the values of PAUSE, the service cycle begins again. An advantage to this algorithm is that if the SDLC link is in a poll-wait state (i.e., waiting for the PAUSE value to elapse), any data ready for transmission to resources on the link is sent during the pause. The default value is 0.2 and is recommended.

The SERVLIM operand for SDLC line resources provides the NCP with the number of times it will scan the service order table for normal service before attempting to scan for special services. The maximum value for an SDLC link is 255. The default is 4 and is recom-

```
SDLC3270 GROUP     LNCTL=SDLC,        SDLC LINE CONTROL          X
                   TYPE=NCP,          LINE OPERATION MODE        X
                   DIAL=NO,           NON-SWITCHED LINK          X
                   CLOCKNG=EXT,       EXTERNAL CLOCKING/MODEM    X
                   DUPLEX=FULL,       4-WIRE CIRCUIT             X
                   SPEED=9600,        SPEED OF MODEM             X
                   NRZI=NO,           NON-RETURN TO ZERO MODE    X
                   RETRIES=(5,10,3),  3 RETRIES 5/10 SECS        X
                   REPLYTO=1,         1 S BEFORE POLL T.O.       X
                   TEXTTO=1,          1 S B4 MSG T.O.            X
                   PAUSE=.2,          SERVICE PAUSE              X
                   PASSLIM=254,       max # of BLUs              X
                   SERVLIM=4          # of SOT scans
```

Figure 18.9 Line service operands for a SDLC line group.

mended. This means that the NCP will scan the service order table for normal service four times before scanning the table for special service. Normal service is that service required by existing active sessions. Special service includes those LUs and PUs seeking to establish a session or the issuance of activate and inactivate commands for PUs and LUs on this SDLC line.

To optimize the service received from the NCP, the PASSLIM operand is used to define the number of SDLC frames that the NCP can send to a PU in one service cycle. The default for this operand is one SDLC frame for a PU Type 2 device. The normal practice is to match the MAXOUT operand of the PU definition statement. For most PU Type 2 devices, this is set to 7. Setting PASSLIM = 7 for a PU Type 2 device is fine for a mixed message set and particularly for a multipoint line configuration. For performance considerations, on a point-to-point SDLC line with a PU Type 2 device, the value of 254 should be coded. As with most of the NCP operands and their parameters, analysis of your network's performance is the only real measure for defining these operands.

18.4.3   Switched SDLC GROUP definition statement

Applications executing under VTAM are unaware of the LU's link state. VTAM and NCP provide the capability for a remote SNA resource to use a nondedicated line to gain access to applications on the host. This switched connection is more often than not used for Remote Job Entry (RJE). Associated with switched lines is the VTAM switched major node definition. The switched major node defines the PUs and LUs that may access the network using a switched line.

RJE is used with most applications for batch processing. In our example, the IBM S/36 is use during the day for remote data entry processing. Interaction with the host is not necessary. However, at some point during the day, the information entered on the S/36 is needed for processing on the host. At this time, either the S/36 can dial in to gain access to the network or VTAM can initiate the connection by a dial-out procedure. The information from the S/36 is then transmitted to the host for processing by the appropriate application. Figure 18.10 details the switched SDLC GROUP definition statement for our example.

When defining switched line group definitions, the distinction between switched and dedicated is determined by the DIAL operand. By specifying DIAL = YES, the NCP generation process sets the lines defined under this GROUP definition statement for switched link procedures. Many of the operands coded on the SDLC GROUP definition statement are also used for a switched group. LNCTL, NRZI,

```
SWSDLC    GROUP      LNCTL=SDLC,        SDLC LINE CONTROL           X
                     TYPE=NCP,          LINE OPERATION MODE         X
                     DIAL=YES,          SWITCHED LINK               X
                     CLOCKNG=EXT,       EXTERNAL CLOCKING/MODEM     X
                     DUPLEX=HALF,       2-WIRE CIRCUIT              X
                     SPEED=4800,        SPEED OF MODEM              X
                     NRZI=NO,           NON-RETURN TO ZERO MODE     X
                     RETRIES=(7,4,5),   5 RETRIES 7/4   S           X
                     REPLYTO=3,         1 S     BEFORE POLL T.O.    X
                     TEXTTO=3,          1 S     B4 MSG T.O.         X
                     PAUSE=1,           SERVICE PAUSE               X
                     XMITDLY=2.2        delay xmiting XID
```

Figure 18.10 Switched SDLC GROUP definition.

CLOCKNG, and SPEED are also used and have the same implications on the line. Some of the operands are also used with the same context but with more consideration.

The DUPLEX operand for switched SDLC links is dependent upon the modems and the facility circuit. Recently switched lines have become able to use full-duplex transmission. New modems have allowed dial-up access to occur with full-duplex capabilities. But we have coded DUPLEX=HALF to show you the norm for a switched line.

In conjunction with full-duplex transmission, the SPEED operand has also recently been affected. Now with the newest modems, a dial-up line can run not only full duplex but at speeds up to 9600 bps. However, in keeping with the norm, we have defined the speed in the NCP SPEED operand to be 4800 bps.

During a dial-up connection, the facility being used is in the public networks. There is no special line conditioning performed over the circuits such as is found in most private and dedicated circuits. The lines used for transmission are the same telephone lines we use when talking on the phone. For this reason we should give special consideration to the time-out and polling operands.

Only real measurements can give you appropriate values. What we have done with the time-out operands is to give ample time for the slower transmission speed and lower circuit grade than that used for the nonswitched SDLC lines (Figure 18.11). We have also reduced the polling rate by specifying the PAUSE operand value at 1. This may hinder response-time performance, but if response time is a consideration, using a switched line for this resource is an error. Remember, switched lines are primarily used for short access times to the host or for backup procedures.

```
SWSDLC    GROUP    LNCTL=SDLC,      SDLC LINE CONTROL          X
                   TYPE=NCP,        LINE OPERATION MODE        X
                   DIAL=YES,        SWITCHED LINK              X
                   CLOCKNG=EXT,     EXTERNAL CLOCKING/MODEM    X
                   DUPLEX=HALF,     2-WIRE CIRCUIT             X
                   SPEED=4800,      SPEED OF MODEM             X
                   NRZI=NO,         NON-RETURN TO ZERO MODE    X
                   RETRIES=(7,4,5), 5 RETRIES 7/4  SECS        X
                   REPLYTO=3,       1 S BEFORE POLL T.O.       X
                   TEXTTO=3,        1 S B4 MSG T.O.            X
                   PAUSE=1,         SERVICE PAUSE              X
                   XMITDLY=2.2      delay xmiting XID
```

Figure 18.11   Switched group operands of special consideration.

The XMITDLY operand is new in NCP V4.3/V5. During initial contact the two resources will exchange identification. In SNA the NCP link station is always the primary link station when the switched device is a peripheral node. The values can be NONE|n|2.2. Defining NONE will negate the XMITDLY function, and the NCP will transmit its XID as soon as contact is made. This may cause XID collisions over the link, resulting in retransmissions of the XIDs. The n value is a number you specify in seconds that is used by the NCP to wait before sending an XID. The default value 2.2 is more than adequate for most dial-up circuits. By specifying a value other than NONE, the modems have a chance to equalize with each other before the NCP sends an XID. Again, this reduces the possibility of transmission errors. The XMITDLY value is also used as the reply-time-out value if (1) it is larger than the REPLYTO specification, (2) the line is a switched line, or (3) XIDs are being exchanged. The XMITDLY operand can also be defined on leased SDLC links. In particular it is used with T2.1 nodes using Low-Entry Networking (LEN). LEN and T2.1 nodes are discussed in detail in *Advanced SNA Networking*.

18.4.4   Channel-attached GROUP definition statement

In NCP V5.1 and NCP V5.2, an added feature of the NCP is the definition of a channel-attached HOST as a special line group. The IBM 3720 Communications Controller with this version of NCP can define the channel adapter in the traditional way in the BUILD statement or here in the CA group statement. If the CA line group is used, the HOST definition statement becomes a VTAM-only definition statement and is not used by the NCP. The IBM 3745 communications controller, however, defines all channel-attached nodes with the CA line group. As such, in an IBM 3745, the HOST definition statement is always a VTAM-only definition statement.

There are two flavors to channel-adapter links. A channel subarea link available with NCP V5.1 defines the channel between the NCP and a host access method, namely VTAM. The channel peripheral link attaches a T2.1 peripheral host node to the NCP through the channel. This node is seen by the SSCP as a peripheral T2.1 node attached to the NCP via an SDLC peripheral link in the NCP's subarea. Programs residing on System/370, 30XX, and 4300 IBM processors can function as the T2.1 peripheral node. One such program is the Transaction Processing Facility (TPF). This program can enter the NCP's subarea through the channel peripheral link as a T2.1 peripheral HOST node, but instead of utilizing LU6.2, TPF uses the standard LU2 session protocols.

18.4.5   Token-ring GROUP definition statement

In an NCP, GROUP definition statements are used to define characteristics and parameters commonly found for many different network resources. Token-ring resources connected to an IBM 37x5 Communications Controller use the LAA number assigned to the TIC as a destination station address. The physical GROUP definition statement has three keywords that define the physical attributes of the TIC. These three keywords are ECLTYPE, ADAPTER, and TRSPEED.

The ECLTYPE keyword identifies the type of connection associated with the TIC being defined in the group. As shown in Figure 18.12, the ECLTYPE keyword has two variables. The first variable indicates that the group is defining the physical interface attributes to the NCP.

The second variable specifies the type of traffic that will be traveling over the physical connection. The specification of PERIPHERAL in the second variable of the ECLTYPE keyword indicates that the TIC will be supporting peripheral node traffic. Peripheral or Boundary Network

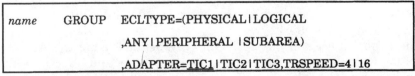

```
name      GROUP    ECLTYPE=(PHYSICAL I LOGICAL

                    ,ANY I PERIPHERAL  I SUBAREA)

                    ,ADAPTER=TIC1 I TIC2 I TIC3,TRSPEED=4 I 16
```

Figure 18.12 The GROUP definition statement and format of the ECLTYPE keyword for token-ring specification.

Nodes (BNN) support SNA Node Type 2.0 and Type 2.1 connectivity. In a token-ring network, these devices are typically IBM 3174 Establishment Controllers with token-ring support or workstations with software that emulates SNA Node Type 2.0 devices. The SUBAREA value specified for the second variable of the ECLTYPE keyword indicates that the TIC will be supporting communication between SNA subareas. This subarea-to-subarea connection is also referred to as Intermediate Network Node (INN) traffic when the connection involves two or more NCP subareas. Each NCP subarea is therefore referred to as an intermediate network node. The last possible value for the second variable of the ECLTYPE keyword is ANY. Coding ANY indicates that either BNN or INN traffic will be used over TIC Type 2 or 3. ANY is the default for an ADAPTER value of TIC2 or TIC3.

The ADAPTER keyword specifies the type of token-ring adapter associated with the GROUP definition. The default is TIC1. TIC1 indicates that TRA Type 1 is installed. TRA Type 1 can run only at a data rate of 4 Mbps. A value of TIC2 indicates that the IBM 37X5 has a Type 2 TRA installed. Coding TIC3 indicates that the 3745-900 extension is installed and that Type 3 TRAs are being defined for use. The Type 3 adapter can have a mixture of both BNN and INN traffic like the Type 2 TIC, but operates only at 16 Mbps. If ANY was coded as the second variable of the ECLTYPE keyword, then the ADAPTER value would default to TIC2.

The TRSPEED keyword defines to the NCP the data rate for the TIC. The possible values are 4 and 16. The default is 4 Mbps. Token-ring adapter Type 1 can support only 4 Mbps, while Type 2 can support both 4 Mbps and 16 Mbps. The Type 3 TRA on the 3745-900 can only have 16 specified for the TRSPEED value. The value coded here must equal the data rate of the other token-ring adapters on the same LAN segment. The downstream physical unit definitions are defined in the NCP through a logical GROUP definition.

Again an ECLTYPE keyword is used to define the characteristics of the group. Downstream physical units are defined as boundary network nodes, and therefore the ECLTYPE keyword will specify LOGICAL as the type of connection being defined and PERIPHERAL as the type of traffic over the TIC being defined. This is shown in Figure 18.13.

The PHYPORT keyword is the link back to the physical LINE definition associated with the TIC. The association is made by coding the PORTADD value specified on the physical LINE definition as the value for the PHYPORT keyword.

The logical lines and physical units are defined using the AUTOGEN keyword. This keyword generates logical line and physical unit pairs based on the number defined on the AUTOGEN keyword. The maximum value is 3000. The NCP builds control blocks to associate VTAM definitions with the logical line and physical unit definitions.

```
TRBNNGRP        GROUP       ECLTYPE=(LOGICAL,PERIPHERAL),
                            PHYPORT=1,CALL=INOUT,AUTOGEN=140
```

Figure 18.13 Token-ring GROUP definition example for downstream physical units.

18.4.6 Ethernet GROUP definition statement

As of NCP V6R1, Ethernet traffic can be passed through an SNA backbone network between two Ethernet-attached workstations. Figure 18.14 illustrates a use for Ethernet transport over the SNA backbone network. Recall that Ethernet LANs typically use the TCP/IP protocol to communicate. Using NCP V6R1 and higher, Ethernet LANs can be attached to an Ethernet adapter on the 3745 communications controller. IP datagrams can then be transported over the SNA backbone network from one Ethernet LAN to another. One distinctive advan-

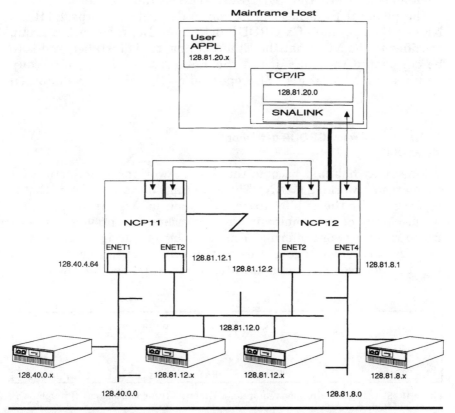

Figure 18.14 An Ethernet transport configuration over an SNA backbone.

tage to this is that the IP datagram is encapsulated into an SNA frame and then transported via an SNA LU-LU session utilizing SNA's inherent ability to guarantee ordered and reliable delivery of data and class of service. Two types of GROUP statements are required to define Ethernet connectivity transport through the NCP V6R1 and higher. The first defines the NCP connectionless SNA transport (NCST) session, and the second defines the Ethernet connectivity to the Ethernet Subsystem (ESS) of the NCP by defining the physical line between the NCP and an Ethernet V2 or IEEE 802.3 network.

The two GROUP definition statements for Ethernet connectivity are very straightforward. Figure 18.15 lists the two GROUP definition statement formats for the Ethernet connectivity definition. The first GROUP definition statement identifies the group for the NCP connectionless SNA transport (NCST) support. The only value on the GROUP definition statement for this is NCST=IP. Defining this identifies this group to the NCP as providing the LU definitions for IP datagram transport. Only one line group can be defined in an NCP for NCST=IP support. When defining NCST support, the OPTION definition statement must also have the NEWDEFN=YES operand specified.

The physical Ethernet connection is defined using the ETHERNET=PHY operand of a GROUP definition. Defining such a group specifies to the NCP that the Ethernet adapter is installed and is to be connected to a physical Ethernet LAN. Coding this group requires the NEWDEFN=YES operand on the OPTIONS definition statement.

18.4.7   Frame relay GROUP definition statement

Frame relay has fast become the new networking architecture for wide area networking (WAN). The advantage of frame relay is that it is provided by the public carriers and can be used in private networks. Many of the dominant public carriers now provide a robust frame relay network offering. You as a network designer and architect can take advantage of this in two ways. First, there is a good potential for reducing communications line costs. Second, the carrier

```
NCSTGRP      GROUP      NCST=IP         IP xport support


ENETGRP      GROUP      ETHERNET=PHY    Physical Enet support
```

Figure 18.15 GROUP definitions required for Ethernet transport support through NCP V6R1 and higher.

provides for redundancy in the network, and therefore recovery of a private T1 backbone network is not mandated.

Support for frame relay requires two GROUP definition statements. The first GROUP definition statement defines the physical line group for the frame relay connection, and the second GROUP definition statement defines the logical line group associated with the physical group. Frame relay support in communications controllers provides for dynamic and redundant connections between communication controllers and allows for logical point-to-point INN connections even though the physical connection is multipoint.

Figure 18.16 lists the two GROUP definition statement formats to support frame relay connectivity through communications controllers using NCP V6R2. NCP V6R1 supports frame relay also, but does not provide for frame relay switching support. We will discuss the features of NCP V6R2, since both types of support are included.

The FRELAY operand on the GROUP definition statement specifies frame relay support and identifies this group for defining the physical or logical frame relay line definitions. A value of FRELAY=PHYS identifies this group for the physical line definitions. A value of FRELAY=LOG identifies this group for defining logical lines. Logical lines for frame relay pertain to virtual circuits defined in the frame relay connection to the communications controller.

18.5 Summary

In this chapter we learned about HOST, LUDRPOOL, PATH, and GROUP macros of NCP. The HOST macro provides for defining buffers and buffer sizes. These buffers are used for sending and receiving data between the host and the channel-attached communications controller. The LUDRPOOL macro specifies the number of LUs that are expected to have a session on a line. The PATH macro uses the ER and VR statements to direct data to a destination subarea. It uses the adjacent subarea and the TG number pair associated with an ER and VR. The GROUP macro allows you to specify certain characteristic functions that are common to a group of links and devices attached to an NCP.

```
FRPGRP        GROUP        FRELAY=PHYS        Physical frame relay


FRLGRP        GROUP        FRELAY=LOG         Logical frame relay
```

Figure 18.16 GROUP definitions required for frame relay support through NCP V6R1 and higher.

NCP Macros—LINE, SERVICE, CLUSTER, AND TERMINAL

19.1 LINE Definition Statement

A LINE definition statement provides the NCP with information about the functional characteristics of the link being defined. These include the speed of the line, the type of facility (full or half-duplex), and the recovery procedure for the link from transmission errors. The characteristics just outlined have been discussed under the GROUP definitions for BSC, nonswitched SDLC, and switched SDLC.

The functional characteristics that we will discuss here are specific to the type of line being defined, the assignment of an address in the NCP, the type of attachment, and the modems used for the link. We will also begin to introduce to you the VTAM-only operands that are used in an NCP for SNA resources. These operands are coded in the NCP source generation deck and are ignored by the NCP generation process.

19.1.1 BSC LINE definition statement

The definition of the GROUP for the BSC line that we will describe here contains many of the LINE operands that are common to all lines regardless of their line control (e.g., BSC and SDLC. In this section we will concentrate on operands specific to BSC line control (Figure 19.1).

The one required operand of the LINE definition statement is the ADDRESS operand. This operand defines to NCP which Line Interface Coupler (LIC) port these definitions will affect. The description of the ADDRESS operand that follows is valid for all line definitions. The format of the ADDRESS operand is

```
N11L000   LINE      ADDRESS=(000,HALF), PORT 0, HDX PROTOCOL    X
                    ATTACH=MODEM,    MODEM ATTACHMENT            X
                    POLIMIT=(1,QUEUE), MAX # -RSP/ACTION         X
                    AVGPB=256,        AVG # BYTES/POLL           X
                    PU=YES,           BSC 3270 = PU              X
                    ISTATUS=ACTIVE    INITIAL STATUS=ACTIVE
```

Figure 19.1 BSC LINE definition.

```
ADDRESS=[(lnbr, [HALF|FULL][sa1,..san])]
[channel adapter position]
```

The port parameter identifies the even or even/odd pair of ports that will be used by the NCP. The value is the position of the physical port of the LIC. The addresses assigned are logically defined within the NCP in accordance with the second parameter. See the Appendix for specifics about defining the *lnbr* parameter for the IBM 3720 and IBM 3745 communications controller using NCP V5 and higher.

By specifying HALF, NCP will assign a logical even-numbered address to the port. This value also defines the data transfer mode as half-duplex. This means that the data will travel over the assigned address for sending and receiving data. However, only one direction can operate at a time. If FULL is coded in this position, the data transfer mode uses full duplex; that is, the NCP can send and receive data over this port simultaneously. This occurs because the NCP logically associates an even/odd pair of addresses with the physical port. The even address is used for sending or transmitting data from the NCP, and the odd address is used for receiving the data.

The *sa1* through *san* parameter pertains to the use of this port in emulation mode. The *sa1* through *san* corresponds to the Emulation Subchannel (ESC) address for the port. The ESC address is the hexadecimal digits specified in the Unit Control Block (UCB) of MVS or the Physical Control Block (PCB) of VSE. Each ESC address is followed by the position of the channel adapter. The ESC address is the last two hexadecimal digits of the UCB. This type of address definition is used most often with the emulation program.

The usage of the channel-adapter position parameter comes into play when user-written channel-adapter code is loaded into the 3725. For an IBM 3725 this value ranges from 0 to 5; in an IBM 3720 the range is 0 to 1. Using the ADDRESS operand in this fashion is exemplified by the Non-SNA Interconnect (NSI) program from IBM.

In Figure 19.1 we see that line N11L000 is connected to port 0 of the IBM 3725 and transmits in half-duplex mode. The ATTACH operand tells NCP that the line attachment to the remote resource is through a modem. There are three other possibilities for this operand: DIRECT, LODIRECT, and DIR3275. Use the DIRECT parameter when the line is attached to the remote resources of this NCP without the use of modems. You must also specify DIAL=NO and CLOCK=EXT and omit the use of the SPEED operand. In this case, the clocking is performed by the remote resource. The LODIRECT parameter is used when the SPEED operand is coded for a remote resource attached to this NCP without modems. In addition, the DIAL=NO and CLOCK=INT operands must be specified. This indicates that the communications scanner for the port to which this direct attachment is connected will provide the clocking. The final ATTACH parameter option is the DIR3725. This is specified on a LINE definition statement when the link is between two IBM 3725 communications controllers without modems.

When NCP polls devices, it expects each device to return a positive response to the poll whether or not the device has data to send to the host. POLIMIT (Figure 19.2) is used to tell the NCP the action it should take when it receives a negative response to a poll. Under VTAM, the only NCP action that is valid is to (1) break the logical connection to the device, (2) notify VTAM, and (3) queue the most current read request for the device at the top of the queue. The specification of the QUEUE parameter determines these events.

After receiving at least one message block from the device, the actions outlined above will go into effect after the NCP receives a specified number of consecutive negative responses to a poll. This number is assigned in the first parameter of the POLIMIT operand. Its value can range from 0 to 255. The default is 1 and is also the recommended value.

The receipt of data in NCP from a device in response to a poll can be governed by the AVGPB operand. For BSC or SS devices, the LINE definition statement is the lowest-level definition statement that the AVGPB operand can be coded on. For an SDLC line, the PU

```
N11L000   LINE      ADDRESS=(000,HALF), PORT 0, HDX PROTOCOL      X
                    ATTACH=MODEM,     MODEM ATTACHMENT             X
                    POLIMIT=(1,QUEUE), MAX # -RSP/ACTION           X
                    AVGPB=256,        AVG # BYTES/POLL             X
                    PU=YES,           BSC 3270 = PU                X
                    ISTATUS=ACTIVE    INITIAL STATUS=ACTIVE
```

Figure 19.2 BSC LINE polling operands.

definition statement can specify the AVGPB. The value specified here assists the NCP in determining if there is enough buffer space to receive the expected data transmitted from a device after a poll. This protects the NCP from an overrun error in the buffer pool. If the buffer space is not available, the NCP will delay polling the device until it can ensure that enough buffer space exists for receiving the data. The range for the value is 1 to 65,535 bytes. The default is the value specified on the BUILD BFRS operand of this NCP. The selection of the AVGPB value is most practical if the value reflects the link buffer of the device and is a multiple of the BFRS operand value. We have chosen the value 256 for two reasons: (1) the link buffer of an IBM 3274 is 256 bytes and (2) 256 is a multiple of 128 (size of NCP buffers).

In SNA the only non-SNA device that is supported without the use of an emulation program or line control program is the BSC 3270 device. The PU operand (Figure 19.3) on the LINE definition statement for a BSC line control protocol specifies to VTAM (not NCP) that this line contains BSC 3270 cluster controllers and BSC 3270 terminals that are to be treated as PUs and LUs, respectively. PU=YES is the default for a BSC 3270 device. For all other BSC devices, the default is PU=NO. In this case, the resources on this line must be supported by a line control program similar to IBM's Network Terminal Operator (NTO) program. The PU operand cannot be coded on the CLUSTER definition statement. It may be coded on the GROUP or LINE statement of the CLUSTER devices.

The ISTATUS operand performs the same function here as in the definitions for all SNA resources. In this statement we specify to VTAM that the line defined by this statement is to be activated whenever the NCP major node is activated. The activation pertains only to the LINE definition statement if the succeeding CLUSTER and TERMINAL definition statements have specified their own ISTATUS operand. However, the value specified here can sift down to the CLUSTER definitions for this line and in turn to the TERMINAL definitions associated with each cluster.

```
N11L000   LINE        ADDRESS=(000,HALF),  PORT 0, HDX PROTOCOL        X
                      ATTACH=MODEM,      MODEM ATTACHMENT              X
                      POLIMIT=(1,QUEUE), MAX # -RSP/ACTION             X
                      AVGPB=256,          AVG # BYTES/POLL             X
                      PU=YES,             BSC 3270 = PU                X
                      ISTATUS=ACTIVE     INITIAL STATUS=ACTIVE
```

Figure 19.3 LINE VTAM operands.

19.1.2 SDLC LINE definition statement

The definition of the GROUP for the SDLC line described here focuses on a nonswitched point-to-point SDLC line. Many of the operands coded on the GROUP definition statement are actually LINE operands and may also be coded on the LINE definition statement. Here in this section we will concentrate on prioritizing polling and transmission functions.

As in the BSC LINE definition, we must supply the physical port number of the IBM 37x5 communications controller used to connect the physical line defined by this LINE definition. The ADDRESS operand in Figure 19.4 identifies port number 001 of the IBM 3725 as the attachment point for the physical line. It also defines the data transfer protocol as half duplex.

Here in the SDLC LINE definition for N11L001, modems are used on the facility to attach the remote resources to the IBM 37x5 communications controller. This is specified on the ATTACH operand by the presence of the MODEM parameter.

When the data arrives in the NCP with a destination for an LU off of the NCP, the NCP can expedite the delivery of the data to the LU. This is determined by the value specified on the HDXSP operand (Figure 19.5). The operand is valid only if (1) LNCTL=SDLC, (2) PAUSE=0, and (3) the ADDRESS operand has HALF specified. If these requirements are met, this operand will go into effect when more than one device is active on the link. The specification of NO (the default) tells the NCP to send the data to the device during the

```
N11L001   LINE      ADDRESS=(001,HALF), PORT 1, HDX PROTOCOL     X
                    ATTACH=MODEM,     MODEM ATTACHMENT           X
                    HDXSP=NO,         SEND PRIORITY OVER POLL    X
                    AVGPB=256,        AVG # BYTES/POLL           X
                    ISTATUS=ACTIVE    INITIAL STATUS=ACTIVE
```

Figure 19.4 SDLC LINE definition statement for nonswitched resources.

```
N11L001   LINE      ADDRESS=(001,HALF), PORT 1, HDX PROTOCOL     X
                    ATTACH=MODEM,     MODEM ATTACHMENT           X
                    HDXSP=NO,         SEND PRIORITY OVER POLL    X
                    AVGPB=256,        AVG # BYTES/POLL           X
                    ISTATUS=ACTIVE    INITIAL STATUS=ACTIVE
```

Figure 19.5 Polling sensitive operands for the SDLC LINE definition.

polling service for that device. The NCP sends the data just prior to the actual polling sequence unless the link is in a poll-wait state. We have defaulted and coded HDXSP=NO because our PAUSE operand specifies PAUSE=.2 and therefore negates the use of HDXSP. The specification of YES on the HDXSP operand tells NCP that data destined for LUs on this link may be sent before the normal polling service of the destination device. However, if you have NCP cycles to burn, code PAUSE=0 and HDXSP=YES. This will decrease end-user response time for the receipt of data.

The AVGPB operand for SDLC links has differing considerations if the link being defined is for Boundary Network Node (BNN) SDLC devices or for Intermediate Networking Node (INN) links. This discussion concerns BNN SDLC devices. The AVGPB operand for SDLC lines is really specific to PUs defined for this line. We have placed it in the LINE definition statement so that it may sift down to the PUs defined for this line. At the PU level, we may override this value if measurements indicate that value for a specific PU on the line warrants modification. Again, the appropriate value for PU Type 2 devices represented by an IBM 3274 cluster controller is optimized by the value 256. The default for this value is the BFRS value times 7. The minimum value is the BFRS value; the maximum value is 65,535. Usage for INN links will be discussed in *Advanced SNA Networking*.

19.1.3 Switched SDLC LINE definition statement

In defining the switched SDLC LINE definition, many of the operands that are specific to the LINE definition statement have been coded on the GROUP definition statement in Section 18.4.3. Here we will concentrate on the dial procedure. Figure 19.6 contains the outline of the switched SDLC LINE definition statement used for our sample single-domain network.

The physical port number of the IBM 37x5 Communications Controller used that will accept the switched facility link is designat-

```
N11L002   LINE      ADDRESS=(002,HALF),  PORT 2, HDX PROTOCOL       X
                    ATTACH=MODEM,      MODEM ATTACHMENT             X
                    ANSTONE=NO,        NCP SUPPLIES ANSWER TONE     X
                    ANSWER=ON,         ALLOW INCOMING CALLS         X
                    CALL=IN,           END-USER INITIATES CALL      X
                    ISTATUS=ACTIVE      INITIAL STATUS=ACTIVE
```

Figure 19.6 Switched SDLC LINE definition statement.

ed by the ADDRESS operand. The ADDRESS operand in Figure 19.6 identifies port number 002 of the IBM 37x5 as the switched facility entry point into the SNA network. It also defines the data transfer protocol as half duplex.

Since this is a switched LINE definition, N11L002 must use modems to gain access to the network. These modems must be capable of handling a switched facility. Most modems have an auto-answer feature or generate the answer tone to the calling modem. If, however, the modem does not provide this feature, the NCP can supply the answer tone. The operand ANSTONE allows for this specification by coding the YES parameter. We have auto-answer-capable modems, and therefore we have coded the default value of ANSTONE=NO.

Figure 19.7 highlights two VTAM-only operands that affect the switched SDLC line dial capabilities. The CALL operand is valid only if the DIAL=YES operand is specified on the GROUP definition statement. The function of this operand is to notify VTAM whether the end user, VTAM, or both can create the switched link connection for the line being defined. The specification of CALL=IN identifies the end user as the initiator of the connection. Specifying CALL=OUT notifies VTAM that it can initiate the connection in accordance with the switched major node definition for this switched SDLC line. If CALL=INOUT is coded, either the end user or VTAM can establish the connection.

The ANSWER operand is valid for dial-in capability of switched SDLC lines only. The end user can dial into the NCP over this switched line definition only if the line is active. To allow this, we have coded ANSWER=ON and ISTATUS=ACTIVE. You can also specify ANSWER=OFF, which prohibits the dial-in capability for a connection regardless of the active or inactive status of the line. You may want to use this for security purposes. The ANSWER=OFF operand can be modified by the VTAM operator command VARY ANS. This command gives the VTAM operator or network manager the ability to enable or disable the dial-in capability.

```
N11L002   LINE       ADDRESS=(002,HALF),  PORT 2,  HDX PROTOCOL   X
                     ATTACH=MODEM,     MODEM ATTACHMENT            X
                     ANSTONE=NO,       NCP SUPPLIES ANSWER TONE    X
                     CALL=IN,          END-USER INITIATES CALL     X
                     ANSWER=ON,        NCP ALLOWS DIAL IN CALL      X
                     ISTATUS=ACTIVE      INITIAL STATUS=ACTIVE
```

Figure 19.7 VTAM-only operands that affect switched SDLC LINE definitions.

```
HOST01CA LINE        ADDRESS=0,        CA position 0                    X
                     CA=TYPE6,         3745 CA TYPE 6                   X
                     DELAY=0.2,        WAIT .2 S B4 ATTN                X
                     TIMEOUT=440,      TIME CA OUT AFTER 440 S          X
                     NCPCA=ACTIVE,     CA COMES UP ACTIVE               X
                     CASDL=10,         CA SLOW DOWN LIMIT               x
                     TRANSFR=32,       # OF-ncp BUFFERS 2 SEND          x
                     INBFRS=16         # OF ncp BUFFERS 2 RECEIV
```

Figure 19.8   Sample CA LINE definition for a channel-adapter link on an IBM 3745.

19.1.4   Channel-adapter LINE definition

As you can see from Figure 19.8, the parameters used for the CA LINE definition are the same as those used in the BUILD definition statement. In fact, they are all used and defined in the same manner by the NCP. The CA line group is always used for the IBM 3745 and is optionally used for the IBM 3720 when executing NCP V5 and higher. If you define your channel link on the IBM 3720 in a CA line group, be sure to remove the channel link definitions from the BUILD definition statement. Channel links defined in the line group must be defined in ascending order according to the channel-adapter logical address specified on the ADDRESS operand.

The ADDRESS operand defines the logical address of the channel-adapter by specifying the CA position on the controller greater than the unit address of the channel. The value for the channel adapter for the IBM 3720 is either 0 or 1; for the IBM 3745, the value range is 0 through 15. For the 3745 Model 900 and the assigning of the physical channel-adapter position to the value on the ADDRESS operand, consult Appendix D.

The CASDL is used by the NCP to block inbound traffic to the NCP before indicating that the station has gone inoperative. The value is specified in seconds. The lowest possible value is 10.0 s, and the highest is 840.0 s. The default is the TIMEOUT value specified for this line. If TIMEOUT is NONE, the CASDL operand is invalid. A low value will reduce the amount of lost data.

The INBFRS operand is usually defined on the HOST definition statement. Since the HOST definition statement is now a VTAM-only statement, post-NCP V4R2, the INBFRS value is now specified on the CA LINE definition statement. Recall that the INBFRS value indicates to the NCP the number of NCP buffers that must be available to receive data from VTAM. The CA operand on the CA LINE definition statement can have the value TYPE6, TYPE6-TPS, TYPE7, or TYPE7-TPS. The TPS channel adapter indicates that the channel provides the two-processor switch option. If you are using a TPS channel adapter, specify an even number on the ADDRESS operand. TYPE6 is the default value for CA. The Type 7 channel adapter can operate in emulation mode for a Type 6 adapter. The CA operand is not valid for ESCON channel definitions.

19.1.5 ESCON channel definition

The IBM 3746 Model 900 extension for the IBM 3745 communications controller provides Enterprise Systems Connection (ESCON) links for logical connections to subarea and peripheral hosts. The ESCON links can be defined for direct connection to the host processor or through an ESCON channel director.

Figure 19.9 illustrates the use of an ESCON channel link from a mainframe to a communications controller. The ESCON channel provides for a physical link with logical link connections over the physical link. These logical links define the connection of PU Type 5 or PU Type 2.1 devices.

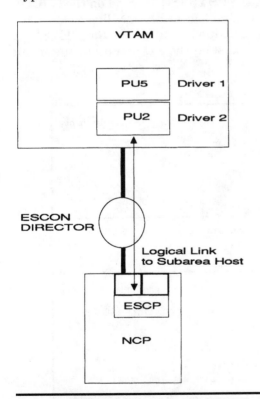

Figure 19.9 Diagram depicting an ESCON channel connection from the mainframe to an IBM 37x5 using an ESCON Director.

Two line groups are needed to define an ESCON channel connection. A physical and a logical line group are defined and associated with each other. The physical ESCON line definition provides the information needed to identify the ESCON processor or an ESCON Director (ESCD) connected to the communications controller. The ADDRESS operand specifies the physical address of the ESCON channel adapter on the communications controller. The SPEED operand is used only by NPM to determine data transmission rates over this link. This is used to compute line utilization statistics. Figure 19.10 provides a sample definition of the ESCON physical line group definition.

Following the physical ESCON line definition, ESCON logical line definitions are specified. The logical ESCON line definition establishes the logical connection from the communications controller to the ESCON processor.

The logical line definition identifies the logical line number associated with the physical line. The ESCON logical lines are associated with the physical line by using the PHYSRSC operand on the GROUP definition statement that defines the logical ESCON line group. This is shown in Figure 19.11. The value specified on the PHYSRSC operand must match the label coded on the PU definition statement

```
ESCONPGR    GROUP    LNCTL=CA,              Define channel adapter group

                     SRT=(32768,32768),

                     XMONLINK=YES

ESCONPLN    LINE     ADDRESS=2240,

                     SPEED=144000000
```

Figure 19.10   ESCON physical line definition sample.

```
ESCONLGR    GROUP    LNCTL=CA,              Define channel adapter group

                     PHYSRSC=ESCONPPU,      Name of physical ESCON PU

                     MONLINK=YES

ESCONLN1    LINE     ADDRESS=NONE,

                     HOSTLINK=1,            Logical line number

                     SPEED=144000000
```

Figure 19.11   ESCON logical line definition sample.

of the ESCON physical line definition. The PU definition statement is discussed in Chapter 20.

The HOSTLINK operand assigns a logical line number to an ESCON logical line. The logical lines must be assigned in a sequential order from 1 to 16. A value from 1 to 32 can be specified, but ESCON V1R0 will not activate host links greater than 16. The ADDRESS operand on the logical line definition normally identifies the relative line number for the physical line. For ESCON channel links, the ADDRESS operand can be omitted and a value of NONE specified. This is because ESCON logical lines are associated with a link address through the MOSS-Extended function. The relationship between the IOCDS, NCP, and MOSS-E definitions for the ESCON LINE definitions are shown in Figure 19.12.

19.1.6 Token-ring LINE definition statement

Every GROUP definition statement has associated LINE definitions. The LINE definition statement for token-ring resources is outlined in Figure 19.13. For a token-ring connection, the physical LINE definition specifies the physical location of the TIC on the IBM 37x5. The physical location is defined on the ADDRESS keyword. In an IBM 3725, this value is related to the TIC's position in the Line Attachment Base (LAB) Type C. In an IBM 3745, the value can range

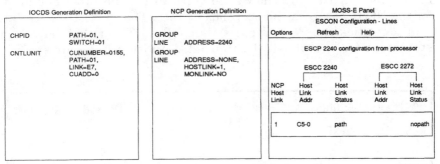

Figure 19.12 The IOCDS, NCP, and MOSS-E definitions needed to support ESCON connectivity to an IBM 37x5 communications controller.

> *name*   LINE   ADDRESS=(n,FULL),PORTADD=n
>
> ,LOCADD=4000abbbbbbb
>
> [,MAXTSL=692 l n][,RCVBUFC=n]

Figure 19.13 The NCP LINE definition statement for defining token-ring connectivity.

from 1088 to 1095. For a complete list of the ADDRESS values for the actual TIC physical positions, refer to Appendix D. The FULL parameter of the ADDRESS keyword indicates that the TIC can send and receive data simultaneously. This value must be coded for token-ring connections.

An associative mechanism is used to relate the physical token-ring connection to a logical token-ring connection. This is accomplished through the PORTADD keyword. This keyword associates a logical number to the physical line definition for the token-ring connection. The value can range from 0 to 99.

The LOCADD keyword is the token-ring MAC address assigned to the TIC on the IBM 37x5 Communications Controller. The format for the value is 4000abbbbbbb. The a value can be in the range of 0 to 7. The bbbbbbb value can be a combination of numbers with values ranging from 0 to 9. The value coded here is the destination MAC address for token-ring stations to gain access to VTAM and its applications on the SNA mainframe. The LOCADD value of the BNN TIC should be coded on question 107 of an IBM 3174 Establishment Controller configuration.

19.1.7   Ethernet LINE definition statement

A LINE definition statement must follow the NCST GROUP definition and the Ethernet physical definition statements. The LINE definitions for Ethernet are shown in Figure 19.14. The LINE definition for the NCST group has no operands defined. The NCST LINE definition identifies the PU-LU pairs that follow. The LINE definition statement for the physical Ethernet definition has four specific

```
NCSTGRP    GROUP   NCST=IP                      IP xport support
NCSTLINE   LINE


ENETGRP    GROUP   ETHERNET=PHY                 Enet support
ENETLINE   LINE    ADDRESS=(1056,FULL),         line number
                   FRAMECNT=(100000,5000),      xmit/receive cnt
                   INTFACE=(ETH1,1500),         Interface name
                   LOCADD=70001056FFFF          Enet MAC address
```

Figure 19.14 LINE definitions required for Ethernet transport support through NCP V6R1 and higher.

operands that pertain to Ethernet, ADDRESS, FRAMECNT, INT-FACE, and LOCADD.

The ADDRESS operand of the Ethernet LINE definition statement is specified as previously discussed for all other lines. The first value specified on the ADDRESS operand identifies the relative line number of the Ethernet adapter on the communications controller. This relative line number is the sequential even-odd pair on IBM 3745 x10 and x1A communications controllers with a valid range from 1056 to 1071. For IBM 3745 models 130, 150, 160, and 170, this range may be 1060 to 1063. In our example, the Ethernet adapter is installed at address 1056 of the communications controller. The second value of the ADDRESS operand indicates that full duplex transmission is used out of the adapter.

The FRAMECNT operand for the Ethernet LINE definition statement indicates the total number of frames transmitted and received and the total number of frames that can be discarded because of transmission and receive errors. The default for the first value is 100000. The minimum value allowed is 1, and the maximum is 9999999. This value is the threshold for transmit and receive frame counts. The second value coded is the threshold for transmit and receive errors resulting in discarded frames. The minimum value for this is 1, and the maximum is 99999. The default value is 5000. Care should be taken when defining these values. Low FRAMECNT values may reduce NCP performance and generate an excessive number of alerts that are sent to NetView. It is recommended that values larger than 300 be coded.

The INTFACE operand of the Ethernet LINE definition statement assigns to the Ethernet adapter. The name defined can be no longer than eight characters. This allows the communications controller Ethernet adapter to be referenced by other Ethernet devices using the name defined on the INTFACE operand. The name defined here must be unique within the NCP and may be referenced by the IPLOCAL and IPGATE definition statements. The second value shown for the INTFACE operand in Figure 19.14 is optional. This value specifies the Maximum Transmission Unit (MTU) size that will be supported by this Ethernet adapter. Figure 19.15 lists the default values for the MTU suboperand of the INTFACE operand. The defaults for MTU are determined by the value defined on the LAN-TYPE operand of the PU definition statement for the physical Ethernet definition.

As with the token-ring definitions, the LOCADD operand of the Ethernet LINE definition statement allows you to assign a MAC address to the Ethernet adapter. This address must be unique on the LAN to which the adapter is attached. If the LOCADD operand is not

LANTYPE value	Default	Minimum	Maximum
ENETV2	1500	68	1500
802.3	1492	68	1492
DYNAMIC	1492	68	1492

Figure 19.15   Values for the INTFACE suboperand MTU.

defined, the NCP will use the hardware address of the Ethernet adapter. The format of the LOCADD operand is xyyyyyyyyyyy. The x value can be a number from 4 to 7. The values allowed for y can range from hexadecimal x'0' to x'F'. However, a LOCADD value of 7FFFFFFFFFFF is invalid.

19.1.8   Frame relay LINE definition statement

Frame relay uses permanent virtual circuits over the same physical communication line. Connections to NCP V6R2 can be between Terminal Equipment Subports (TESPs) off of frame relay devices and Frame Handler Subports (FHSPs) found on the 3745 communications controller. Figure 19.16 gives an example of a frame relay LINE definition. The MAXFRAME operand is the only operand on the LINE definition statement that is specific to frame relay. The MAXFRAME operand defaults to 2106 and allows a range from 285 to 8250. The MAXFRAME operand value defines the maximum size for frames that can be transmitted over the physical line. The frame size includes the 6-byte frame header of the frame relay frame as well as the user data. When coding the PU MAXDATA value for the PU asso-

```
FRPGRP      GROUP    FRELAY=PHYS              Phys frame relay
FRPLINE     LINE     ADDRESS=(1036,FULL),
                     CLOCKNG=EXT,
                     MAXFRAME=2106,
                     NRZI=NO,
                     SPEED=1544000
FRLGRP      GROUP    FRELAY=LOG               Log frame relay
FRLLINE     LINE     ISTATUS=ACTIVE
```

Figure 19.16   LINE definition for frame relay physical and logical groups.

ciated with this LINE, make sure that the value does not exceed the MAXFRAME value. In addition, the MAXFRAME value coded for all physical line definitions along a permanent virtual circuit must be the same. All other LINE operands have been previously discussed and also pertain to a frame relay line.

19.2 SERVICE Definition Statement

Nonswitched data link devices attached to your NCP must be serviced by the NCP to send and receive data. The SERVICE definition statement causes the generation procedure to create a service order table for the resources associated with this line. The SERVICE definition statement must immediately follow the LINE definition statement and is valid for nonswitched links only. Each link must have a corresponding SERVICE definition statement. If the statement is omitted, a default service order table is created for a line operating in network control mode. The service order table is serviced by the NCP in accordance with the SERVLIM operand defined in the previous GROUP definition statements.

19.2.1 BSC Service Order Table

For each BSC device associated with the LINE definition, this SERVICE definition statement belongs to an entry and must be included in the service order table. The service order table is defined on the SERVICE definition statement by coding the ORDER operand. Figure 19.17 contains the service order table for the devices associated with the BSC line N11L000.

In the service order table you can specify up to 255 entries. The name of the SERVICE definition statement is required for BSC and SS definitions. In Figure 19.17 the name is defined as L000SOT. Each entry can be unique, or a resource name can be specified several times within the service order table. The ORDER operand gives us

```
L000SOT   SERVICE   ORDER=(N11BC01,B1101T00,B1101T01,B1101T02,    X
                     B1101T03,B1101T04,B1101T05,B1101T06,B1101T07,    X
                     B1101T08,B1101T09,B1101T10,B1101T11,B1101T12,    X
                     B1101T13,B1101T14,B1101T15,B1101T16,B1101T17,    X
                     B1101T18,B1101T19,B1101T20,B1101T21,B1101T22,    X
                     B1101T23,B1101T24,B1101T25,B1101T26.B1101T27,    X
                     B1101T28,B1101T29,B1101T30,B1101T31)
```

Figure 19.17 Service order table for BSC line N11L000.

this option. The specification of a resource several times in the service order table provides more service for that resource. If you choose to take advantage of this capability, review performance statistics relating to this cluster and its devices. You may wish to code the terminals that have a high access to high-priority interactive applications on the host rather than those terminals that primarily use batch and print applications.

Each device defined under a LINE definition statement must be included in the service order table at least once. The generation procedure checks each resource name against the service order table to make sure that (1) the association with the LINE this SERVICE definition statement is coded for and (2) each resource name defined under this LINE definition statement is included in the service order table.

From Figure 19.17 we can see that there are nine entries in the service order table for this SERVICE definition statement. The table is scanned according to the value specified on the SERVLIM operand. See the BSC GROUP definition statement in Section 18.4.1 for a description of the SERVLIM operand as it relates to a BSC line.

19.2.2   SDLC Service Order Table

The service order table definition for SDLC lines needs only the PU names of the clusters that are associated with the LINE definition for which this service order table is created (Figure 19.18). Each PU name associated with the LINE definition statement must be coded in the service order table. The importance of order in the table is greater because the PUs are obtaining the service from the NCP and not the individual LUs.

The service order table can specify up to 255 entries. Each entry can be unique, or a resource name can be specified several times within the service order table. In a point-to-point configuration such as ours, there is only one possible entry. It does not buy you or the resource additional service to replicate the resource name in the service order table. Including a PU name several times in the service order table is feasible for a multipoint configuration. Refer to Section 18.4.2 for the SERVLIM impact on an SDLC service order table definition.

```
L001SOT   SERVICE    ORDER=(N11SCC1)
```

Figure 19.18   Service order table for a dedicated SDLC line.

```
N11BC01    CLUSTER    GPOLL=40407F7F,       GENERAL POLLING        X
                      CUTYPE=3271,          CONTROL UNIT TYPE      X
                      FEATUR2=(MODEL2),     BUFFER SIZE OF 1920    X
                      DLOGMOD=R3270,        LOGMODE ENTRY          X
                      MODETAB=MT01,         MODE TABLE NAME        X
                      USSTAB=USSNSNA,       NON-SNA USSTAB         X
                      ISTATUS=ACTIVE
```

Figure 19.19   BSC 3274 cluster controller definition.

19.3   CLUSTER Definition Statement

For each BSC cluster controller representing an IBM 3270 series type of controller, you must code a CLUSTER definition statement (Figure 19.19). The definition is valid only if the DIAL=NO and POLLED=YES operands are coded on the GROUP definition statement of the LINE definition statement this cluster controller is attached to. The CLUSTER definition must be coded directly after the SERVICE definition statement, if it is coded, and must precede any TERMINAL or COMP definition statements that pertain to this cluster controller.

Remember that earlier we stated that VTAM supports BSC 3270-type devices such as PUs and LUs by coding the PU=YES operand on the GROUP or LINE definition statement for a BSC line discipline. This treatment for a PU is made possible by the use of the GPOLL operand. The BSC 3270 cluster controller examines the status of all attached resources upon receiving a general poll.

The GPOLL operand provides the cluster control units with a unique station address on the line. This address is directly related to the cluster's position on the line. The first cluster attached to (or dropped off) the line is considered to be position 0. Position 0 is represented in the GPOLL operand as the hexadecimal value 40. A general poll is composed of two fields, a station address and a general poll character of hexadecimal 7F. Together they are coded on the GPOLL operand of the CLUSTER definition statement as 40407F7F.

The CUTYPE operand identifies the control unit as an IBM 3271 or 3275 cluster controller. The IBM 3x74 cluster controller series is classified as CUTYPE=3271. This is also the default value for the CUTYPE operand.

The remaining operands for the cluster N11BC01 are VTAM-only operands. These are used in the same manner as described in Section 12.2.3 and will not be reviewed at this time.

19.4   TERMINAL Definition Statement

A TERMINAL definition statement must be coded for each terminal or printer attached to a BSC cluster controller. This statement allows you to specify the name of the terminal, the type of terminal, and the polling and address characters used by the NCP to contact the device (Figure 19.20). For the TERMINAL definition statements coded in this discussion, all the VTAM-only operands coded on the CLUSTER definition are sifted down to the TERMINAL definitions.

The name of each TERMINAL definition statement is required. This name can be any valid assembler-language name. However, the name must be unique within this SNA network. The name assigned to the device in Figure 19.20 is B1101T00. The convention used for this name is as follows: The B identifies the device as a BSC device. The next four characters specify the subarea of the NCP for this cluster (11) and the cluster control unit station address (01) that owns this terminal. The last three characters identify the class of device (T for display terminal and P for printer) and the physical port number (00) used to attach the device to the cluster controller.

The POLL operand is used to define the polling characters that are assigned to this device. These characters are made up of the GPOLL control unit position character and the device position on the cluster controller. Using the GPOLL and device addressing table for BSC devices in Figure 19.21, we can equate the control unit position on the line and the device position on the cluster to the proper EBCDIC polling characters. The polling characters must be repeated for BSC protocol.

In Figure 19.20 we have assigned the value 40404040 to the POLL operand. The first set of numbers, 40, represents the cluster control unit's position on the line. Looking at the GPOLL and device addressing table, we see that position 0 (actually the first cluster position) is equated to a 40. The second set of 40 in the POLL operand identifies the specific address for the device on the cluster controller. Here again the device is the first terminal on the cluster. The position of the terminal is directly related to the physical port used for attachment. In BSC 3274 cluster controllers, the port numbers range from 0

```
B1101T00  TERMINAL  POLL=40404040,      POLLING ADDRESS      X
                    ADDR=60604040,      SELECTION ADDRESS    X
                    TERM=3277           DEVICE TYPE
```

Figure 19.20   Sample TERMINAL definition statement for the BSC 3270 cluster controller.

Control Unit or Device Position	Control Unit GPOLL or Device POLL Character	Control Unit Selection Character
0	4040	6060
1	C1C1	6161
2	C2C2	E2E2
3	C3C3	E3E3
4	C4C4	E4E4
5	C5C5	E5E5
6	C6C6	E6E6
7	C7C7	E7E7
8	C8C8	E8E8
9	C9C9	E9E9
10	4A4A	6A6A
11	4B4B	6B6B
12	4C4C	6C6C
13	4D4D	6D6D
14	4E4E	6E6E
15	4F4F	6F6F
16	5050	F0F0
17	D1D1	F1F1
18	D2D2	F2F2
19	D3D3	F3F3
20	D4D4	F4F4
21	D5D5	F5F5
22	D6D6	F6F6
23	D7D7	F7F7
24	D8D8	F8F8
25	D9D9	F9F9
26	5A5A	7A7A
27	5B5B	7B7B
28	5C5C	7C7C
29	5D5D	7D7D
30	5E5E	7E7E
31	5F5F	7F7F

Figure 19.21   GPOLL and device address table for BSC 3270 cluster controller.

to 31. Looking at the GPOLL and device addressing table, we see that port 0 is also represented by the character 40. The values together detail the NCP-specific poll address for the device.

The ADDR operand specifies the selection address for the device. This address is formulated by replacing the GPOLL characters of the POLL operand with the selection character of the control unit. This character is also based on the position of the control unit on the line. In our sample configuration, the line is point-to-point, and therefore there is only one drop off of the line, the first drop. The selection character for the first drop off of a line is 60, as determined in the GPOLL and device address table. By replacing the first set of 4040 of the POLL operand value with 6060, we can specify the selection address of the ADDR operand as 60604040 for the terminal.

The TERM operand specifies to the NCP the type of device this TERMINAL definition represents. This is a required operand for the

```
B1101T00 TERMINAL POLL=40404040,ADDR=60604040,TERM=3277,VPACING=0
B1101T01 TERMINAL POLL=4040C1C1,ADDR=6060C1C1,TERM=3277,VPACING=0
B1101T02 TERMINAL POLL=4040C2C2,ADDR=6060C2C2,TERM=3277,VPACING=0
B1101T03 TERMINAL POLL=4040C3C3,ADDR=6060C3C3,TERM=3277,VPACING=0
B1101T04 TERMINAL POLL=4040C4C4,ADDR=6060C4C4,TERM=3277,VPACING=0
B1101T05 TERMINAL POLL=4040C5C5,ADDR=6060C5C5,TERM=3277,VPACING=0
B1101T06 TERMINAL POLL=4040C6C6,ADDR=6060C6C6,TERM=3277,VPACING=0
B1101T07 TERMINAL POLL=4040C7C7,ADDR=6060C7C7,TERM=3277,VPACING=0
B1101T08 TERMINAL POLL=4040C8C8,ADDR=6060C8C8,TERM=3277,VPACING=0
B1101T09 TERMINAL POLL=4040C9C9,ADDR=6060C9C9,TERM=3277,VPACING=0
B1101T10 TERMINAL POLL=40404A4A,ADDR=60604A4A,TERM=3277,VPACING=0
B1101T11 TERMINAL POLL=40404B4B,ADDR=60604B4B,TERM=3277,VPACING=0
B1101T12 TERMINAL POLL=40404C4C,ADDR=60604C4C,TERM=3277,VPACING=0
B1101T13 TERMINAL POLL=40404D4D,ADDR=60604D4D,TERM=3277,VPACING=0
B1101T14 TERMINAL POLL=40404E4E,ADDR=60604E4E,TERM=3277,VPACING=0
B1101T15 TERMINAL POLL=40404F4F,ADDR=60604F4F,TERM=3277,VPACING=0
B1101T16 TERMINAL POLL=40405050,ADDR=60605050,TERM=3277,VPACING=0
B1101T17 TERMINAL POLL=4040D1D1,ADDR=6060D1D1,TERM=3277,VPACING=0
B1101T18 TERMINAL POLL=4040D2D2,ADDR=6060D2D2,TERM=3277,VPACING=0
B1101T19 TERMINAL POLL=4040D3D3,ADDR=6060D3D3,TERM=3277,VPACING=0
B1101T20 TERMINAL POLL=4040D4D4,ADDR=6060D4D4,TERM=3277,VPACING=0
B1101T21 TERMINAL POLL=4040D5D5,ADDR=6060D5D5,TERM=3277,VPACING=0
B1101T22 TERMINAL POLL=4040D6D6,ADDR=6060D6D6,TERM=3277,VPACING=0
B1101T23 TERMINAL POLL=4040D7D7,ADDR=6060D7D7,TERM=3277,VPACING=0
B1101T24 TERMINAL POLL=4040D8D8,ADDR=6060D8D8,TERM=3277,VPACING=0
B1101T25 TERMINAL POLL=4040D9D9,ADDR=6060D9D9,TERM=3277,VPACING=0
B1101T26 TERMINAL POLL=40405A5A,ADDR=60605A5A,TERM=3277,VPACING=0
B1101T27 TERMINAL POLL=40405B5B,ADDR=60605B5B,TERM=3277,VPACING=0
B1101T28 TERMINAL POLL=40405C5C,ADDR=60605C5C,TERM=3277,VPACING=0
B1101T29 TERMINAL POLL=40405D5D,ADDR=60605D5D,TERM=3277,VPACING=0
B1101T30 TERMINAL POLL=40405E5E,ADDR=60605E5E,TERM=3277,VPACING=0
B1101T31 TERMINAL POLL=40405F5F,ADDR=60605F5F,TERM=3277,VPACING=0
```

Figure 19.22   TERMINAL definitions for BSC 3274 cluster controller.

TERMINAL definition statement. If the line this terminal attaches to operates in emulation and network control modes, the TERM operand must also be specified on the LINE or GROUP definition statement. For the IBM 3X78 and 3X79 display stations, code TERM=3277. Figure 19.22 contains the entire code for the full 32-port BSC cluster controller.

19.5   IP-Related Definition Statements

NCP V6R2 support for TCP/IP allows the 3745 to act as a TCP/IP router using Routing Information Protocol (RIP). This is done through several IP-related definition statements. These are the IPLOCAL, IPSUB, and

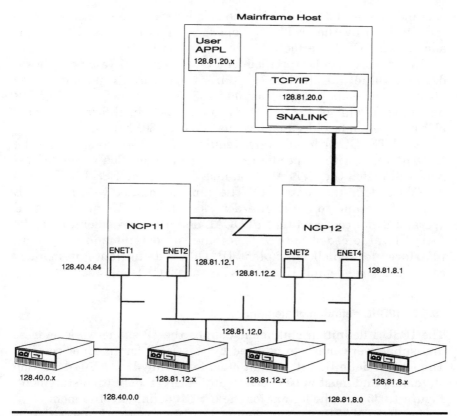

Figure 19.23 Ethernet configuration to an IBM 37x5 with routing support provided by the mainframe using TCP/IP for MVS.

IPGATE definition statements. Figure 19.23 diagrams a sample Internet configuration with two NCPs and three Ethernet LANs.

19.5.1 IPLOCAL definition statement

The IPLOCAL definition statement is required for Internet configuration support. This definition statement is used to define the IP address of the local Ethernet interface. The definition statement has two operands. Its format is shown in Figure 19.24. The LADDR

name	IPLOCAL	LADDR=n0.n1.n2.n3 l hhhhhhhh
		[,INTFACE=name]

Figure 19.24 Format for the IPLOCAL definition statement.

operand is required and defines the IP address for the Ethernet interface defined by the INTFACE operand or is used as the default address for NCST interfaces.

The LADDR may be specified as a dotted-decimal value or a hexadecimal value. Recall that IP address are 4 bytes in length. The n0–n3 numbers may range from 0 to 255. As an example, the LADDR may have a value of 128.255.16.64 in dotted-decimal form. This may also be coded in hexadecimal form as LADDR=80FF1040.

The INTFACE operand is not required. It is used to associate the LADDR value with a specific Ethernet definition. The name defined here will match the INTFACE value defined on the ESS LINE or the NCST LU definition statement. The name associated with INTFACE can be from one to eight characters in length. If the INTFACE operand is not coded on the IPLOCAL definition statement, then the LADDR value coded here will become the default address for all interfaces for which no explicit IPLOCAL definition statement is coded. Only one default IP address for each NCP is allowed.

19.5.2 IPSUB definition statement

The IPSUB definition statement defines the IP subnetwork address and subnetwork mask to be used by this NCP for Internet routing. The IPSUB definition statement must follow any IPLOCAL definition statement and must appear before any IPGATE definition statement. Figure 19.25 lists the format for the IPSUB definition statement.

The SNETMASK operand defines the IP subnetwork mask that is used to identify the subnetwork boundary used on the IP network. The mask specified here is associated with the network address defined on the NETADDR operand. The address defined on the SNETMASK operand can be in either dotted-decimal or hexadecimal form. The 4-byte IP address is subnetted according to the value defined here. For example, since a Class A network address uses the first full byte, the hexadecimal value would be FFxxxxxx. The dotted-decimal equivalent would be 255.0.0.0. A Class B network address uses the first 2 bytes and would therefore be represented as either FFFFxxxx or 255.255.0.0. A Class C network address uses the first 3

```
name      IPSUB      SNETMASK=n0.n1.n2.n3 I hhhhhhhh

                     ,NETADDR=n0.n1.n2.n3 I hhhhhhhh

                     [,INTFACE=name]
```

Figure 19.25 Format for the IPSUB definition statement.

bytes of the 4-byte address and is represented by FFFFFFxx and 255.255.255.0. The remaining bits in the address field may be used for subnetting or for host addresses. For example, if we have a Class A IP network address and want to provide 4096 host addresses for each subnet, we can define 12 bits of the host address area for the subnet value. This would be represented as a dotted-decimal 255.255.240.0 or in hexadecimal as FFFFF000.

The NETADDR operand specifies the IP network address to which the SNETMASK will be applied. All bits in the host area of the IP address must be specified as zero when coding the value for NETAD-DR. For Class A addresses, the last 24 bits must have all zeros. Class B addresses must have the last 16 bits all zeros, and Class C address-es must have the last 8 bits specified as zeros. As an example, a Class B network address may be defined as 128.255.0.0 in dotted decimal and 80FF0000 in hexadecimal.

The INTFACE operand is again optional and is associated with the INTFACE operand of the ESS LINE or NCST LU definition state-ments to which the subnet mask and network address apply. If the INTFACE operand is not coded, then the NETADDR and SNET-MASK values are used as the defaults for all interfaces that do not have a specific IPSUB definition statement. Only one IPSUB defini-tion statement may be defined with the same NETADDR-INTFACE pair.

19.5.3 IPGATE definition statement

The IPGATE definition statement is required when defining an Internet configuration. This definition statement specifies the routes in the IP route table. The IPGATE definition statement must follow the last line group and any IPLOCAL or IPSUB definition state-ments. Figure 19.26 lists the format for the IPGATE definition state-ment.

The DESTADDR operand is required and specifies the IP address of the final destination subnetwork. This value defines the subnet-work that is accessible through the interface specified on the INT-FACE operand. Basically it is defining the next IP subnetwork hop

name	IPGATE	DESTADDR=0	n0.n1.n2.n3	hhhhhhhh
		,INTFACE=name
		[,NEXTADDR=0	n0.n1.n2.n3	hhhhhhhh]

Figure 19.26 Format of the IPGATE definition statement.

along the way to the final IP subnetwork. The NEXTADDR operand defines the interface on the next NCP that can deliver the frame to the final IP subnet defined by the DESTADDR operand. A value of 0 for the NEXTADDR operand indicates that the route is direct or that the gateway to be used is an NCST logical unit. Again the INTFACE operand identifies the Internet interface to use for routing the IP datagrams. Figure 19.27 diagrams a simple IP routed network through NCP executing NCP V6R2.

19.6 SUMMARY

In this chapter, we learned about LINE, SERVICE, CLUSTER, and TERMINAL definition statements of NCP. A LINE definition statement defines the functional characteristics of a line, such as line speed, duplex, and recovery procedure from transmission errors. The SERVICE definition statement causes the generation procedure to

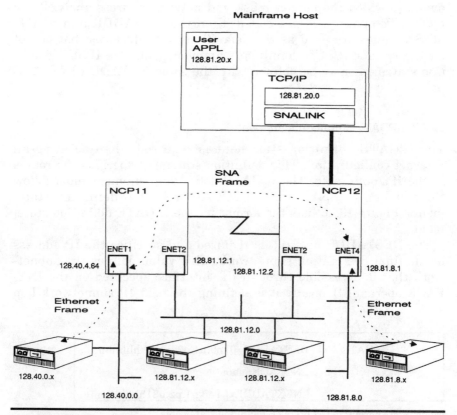

Figure 19.27 Ethernet workstation to workstation flow over an IBM SNA subarea connection.

create a service order table for the resources associated with a line. This statement must follow a LINE definition statement and is applicable to nonswitched links only. The CLUSTER definition statement defines BSC cluster controllers. The TERMINAL definition statement defines each terminal or printer attached to a BSC cluster controller. We also discussed the IP-related definition statements that provide the NCP with information on the IP network addressing scheme, IP subnetwork mask, and IP routing table definitions.

NCP Macros—PU, LU, SDLCST, and IPHOST

20.1   PU Definition Statement

The definition of an SNA resource must have a physical unit definition that defines several characteristics and functions. Among these are the name of the device, its address, the maximum amount of data the device can receive in one PIU or PIU segment, and the maximum number of PIUs the NCP will send to the device before servicing other devices in the service order table for this SDLC link. PUs are also coded for link stations when the SDLC link connects two NCPs. PU definitions for multipoint subarea links, switched subarea links, and T2.1 nodes are discussed in *Advanced SNA Networking*.

At least one PU definition statement must follow the SERVICE definition statement for an SDLC line. The PU definition statement may also follow an LU definition statement for a multipoint SDLC line configuration. When the line is an SDLC link between two NCPs, the PU definition statement must follow the LINE definition statement.

20.1.1.   Nonswitched PU definition statement

The name for the PU definition statement is required. The ADDR operand (Figure 20.1) of the PU definition statement assigns the SDLC station address of the cluster controller for use in polling by the NCP. This station address should match the station address specified on the cluster controller during the configuration process of the cluster controller. If the addresses do not match, the device will never be activated. The value for ADDR is an 8-bit address in hexadecimal representation ranging from 01 to FE. It is a required operand if the

```
N11SCC1    PU      ADDR=C1,                STATION ADDRESS         X
                   MAXDATA=265,            MAX PIU SIZE            X
                   MAXOUT=7,               MAX # OF FRAMES         X
                   PASSLIM=7,              MAX FRAMES / SERVICE    X
                   DLOGMOD=SNA3278,        DEFAULT LOGMODE ENTRY   X
                   MODETAB=MT01,           MODE TABLE NAME         X
                   USSTAB=USSSNA,          NON-SNA USSTAB          X
                   ISTATUS=ACTIVE
```

Figure 20.1   Nonswitched SDLC 3x74 cluster controller definition.

line is nonswitched and the PU type is PU Type 1 or PU Type 2 device. If LPDA-compatible modems are used on this SDLC line, the value FD for the ADDR operand cannot be coded.

An IBM 3274 cluster controller has a receive buffer size of 256 bytes in length. This is the maximum amount of data the cluster can receive in one data transfer. The data is the information that is destined for the end-user. This data plus the Transmission Header (TH) and Request/Response Header (RH) make up the total amount of data that can be transferred on one PIU or PIU segment. For a PU Type 2 device the TH and RH overhead is 9 bytes. The value for the MAXDATA operand is therefore $9 + 256 = 265$ bytes. This value is the largest PIU or PIU segment that the cluster represented by this PU definition statement will accept in one data transfer. This MAXDATA value is truly device-dependent, and you should consult the device publication for the appropriate value.

Transmission of SDLC frames to a device is governed by the MAXOUT operand (Figure 20.2) of the PU definition statement. The value specified on this operand determines the number of SDLC frames that will be sent to the device before requesting a response from it.

```
N11SCC1    PU      ADDR=C1,                STATION ADDRESS         X
                   MAXDATA=265,            MAX PIU SIZE            X
                   MAXOUT=7,               MAX # OF FRAMES         X
                   PASSLIM=7,              MAX FRAMES / SERVICE    X
                   DLOGMOD=SNA3278,        DEFAULT LOGMODE ENTRY   X
                   MODETAB=MT01,           MODE TABLE NAME         X
                   USSTAB=USSSNA,          NON-SNA USSTAB          X
                   ISTATUS=ACTIVE
```

Figure 20.2   Data transfer performance operands of the PU definition statement.

For IBM 3274 cluster controllers, the value coded is usually the maximum allowed for the cluster, which is MAXOUT=7. After sending seven SDLC frames to the PU, the NCP will send a request to the PU to ensure that the data has arrived intact. The value of 7 is based on the modulus of modulo 8 for the link. We have not coded the MODULO operand of the LINE definition statement because the default of 8 is most commonly used. Modulo 8 indicates to NCP that up to seven SDLC frames can be sent to a PU before a response from the PU is required. The other option for this operand is 128. This means that 127 SDLC frames will be transmitted to the PU before the NCP requests a response from the PU. This modulus is commonly used for satellite link connections.

The scan of the service order table is controlled by the SERVLIM operand. For each scan of the table, the NCP must know how many PIUs can be sent when servicing a PU. The PASSLIM operand provides this value. For a point-to-point line configuration, it is advisable to code the maximum value allowed for the PASSLIM operand. The maximum is 254 PIUs per service. If the line configuration is multipoint, a PASSLIM operand is the determining factor in the amount of service a PU can receive from the NCP.

The remaining operands in Figure 20.2 are VTAM-only operands. The DLOGMOD, MODETAB, USSTAB, and ISTATUS operands will sift down to the following LU definition statements. All of these operands are discussed in great detail in Chapter 13 and will not be reviewed here.

20.1.2 Switched PU definition statement

For the switched SDLC line, a PU definition statement is required (Figure 20.3) to specify the type of PUs allowed to connect to this dial-up line and the number of LUs that can be expected to have sessions. The PUTYPE operand specifies the PU type that can access this switched line. You may code PUTYPE=1 for PU Type 1 devices only; for PU Type 2 devices, code only PUTYPE=2. For PU Type 4 devices, code PUTYPE=4, and if both PU Type 1 and PU Type 2 devices can use this switched line, code PUTYPE=(1,2). When the PU definition is for a switched line, the MAXLU operand is required. The MAXLU operand specifies the number of LUs that can have active sessions at

```
N11SWS36   PU       PUTYPE=2,            PU TYPE ALLOWED        X
                    MAXLU=2              MAX # OF LUS
```

Figure 20.3 Switched PU definition statement.

any one time. We have coded MAXLU=2 to match the switched major node defined in Chapter 13.

Note that there is no service order table for a switched line since a switched connection can only be point to point.

20.1.3 CA PU definition statement

Previously we introduced the two types of CA links available in NCP V5R1/V5R2. The PU definitions required for these links are quite simple and represent the bus-and-tag technology for channel connection (Figure 20.4). When defining the channel-subarea link, the PUTYPE parameter must indicate a PU Type 5. No LU definition statements may follow the PU Type 5 definition. When defining a channel-peripheral link, the PUTYPE parameter is specified as PUTYPE=2. At least one LU definition statement must follow the PU definition statement. In both cases the PU definition statement is defining a host link station.

ESCON channel-adapter definitions require a physical PU definition for the physical line definition of the channel and a logical PU definition for each PU connection defined on the physical channel. Figure 20.5 shows the list of the PU definition for the ESCON physical PU definition. The PUTYPE parameter on the PU definition for the physical line is identified as a PUTYPE 1 device.

The logical line definition of an ESCON channel can have several PU definitions defined to it. In the example shown in Figure 20.6, the logical line definition has two PU definition statements. The first one represents the communications controller subarea definition. This is denoted by the PU Type 5 definition and the TGN parameter. The ADDR value of 01 matches the UNITADD parameter defined for the communications controller in the MVS IOCP. The second PU definition identifies a PU Type 2.1 device with a station address of 03, which matches the definition in the MVS IOCP. At least one LU definition statement follows the PU Type 2.1 PU definition statement.

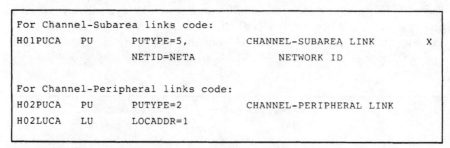

```
For Channel-Subarea links code:
H01PUCA    PU      PUTYPE=5,          CHANNEL-SUBAREA LINK        X
                   NETID=NETA         NETWORK ID

For Channel-Peripheral links code:
H02PUCA    PU      PUTYPE=2           CHANNEL-PERIPHERAL LINK
H02LUCA    LU      LOCADDR=1
```

Figure 20.4 Channel-adapter line PU definitions.

```
ESCONPGR    GROUP    LNCTL=CA,              Define channel adapter group

                     SRT=(32768,32768),

                     XMONLINK=YES

ESCONPLN    LINE     ADDRESS=2240,

                     SPEED=144000000

ESCONPPU    PU       ANS=CONTINUE,

                     PUTYPE=1
```

Figure 20.5 ESCON PU definition for the physical line definition.

```
ESCONLGR    GROUP    LNCTL=CA,              Define channel adapter group

                     PHYSRSC=ESCONPPU,      Name of physical ESCON PU

                     MONLINK=YES

ESCONLN1    LINE     ADDRESS=NONE,

                     HOSTLINK=1,            Logical line number

                     SPEED=144000000

ESCONPU1    PU       ADDR=01,              UNITADD=01

                     ANS=CONTINUE,

                     PUDR=NO,

                     PUTYPE=5,

                     TGN=1

ESCONPU2    PU       ADDR=03,              UNITADD=03

                     ANS=STOP,

                     PUDR=YES,

                     PUTYPE=2
```

Figure 20.6 ESCON PU definition for the logical line definition.

20.1.4   Token-ring PU definition statement

Each LINE definition needs a physical unit definition. Token-ring line definitions are defined as SNA physical unit Type 1 resources. The format of this physical unit definition is

```
name PU PUTYPE=1
```

Downstream Physical Units (DSPU) are SNA Type 2.0 or Type 2.1 nodes that are downstream from a LAN gateway. Figure 20.7 illustrates some possible token-ring network configurations for downstream devices off of an IBM 37x5. Notice that these downstream devices can be either establishment controllers or workstations.

20.1.5   Ethernet PU definition statements

Ethernet connectivity requires a PU definition statement for the NCST session support and a PU definition statement to support Ethernet-to-Ethernet communication through the SNA backbone. The PU definition for the NCST is needed to support NCST-to-NCST sessions. The VPACING parameter value of zero on the NCST PU definition maximizes the throughput over the virtual route between the NCST LUs. Figure 20.8 lists the PU definition statements for NCST and Ethernet.

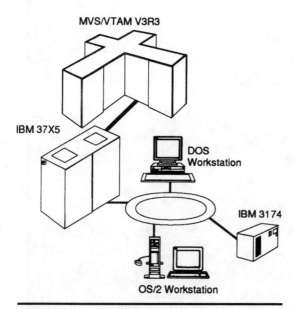

Figure 20.7   PU Type 2 and 2.1 down-stream physical unit connectivity through an IBM 37x5 Communications Controller.

```
NCSTGRP    GROUP    NCST=IP                    IP xport support

NCSTLINE   LINE

NCSTPU     PU       VPACING=0

ENETGRP    GROUP    ETHERNET=PHY               Enet support

ENETLINE   LINE     ADDRESS=(1056,FULL),       line number

                    FRAMECNT=(100000,5000),    xmit/receive cnt

                    INTFACE=(ETH1,1500),       Interface name

                    LOCADD=70001056FFFF        Enet MAC address

ENETPU01   PU       ANS=CONTINUE,

                    ARPTAB=2000,               2000 entries

                    LANTYPE=802.3,

                    PUTYPE=1
```

Figure 20.8 PU definitions required for Ethernet transport support through NCP
V6R1 and higher.

The Ethernet PU definition statement is defined as a PU Type 1 to represent the Ethernet connection as an SNA-type device. The ARPTAB parameter specifies the maximum number of entries allowed in the Address Resolution Protocol (ARP) table located in the NCP. The ARP table is used to convert the IP address of an Ethernet device into a physical MAC Layer network address. The value entered on the ARPTAB parameter should be large enough to support the number of temporary devices expected to be attached to the Ethernet LAN. This ARP table is equivalent to an ARP cache, found in many internet routers. The default value for ARPTAB is 1000, with a minimum of 1 and a maximum value of 65535. An ARPTAB value that is too large will waste valuable NCP memory, and a value that is too small to handle the number of Ethernet devices attached to the NCP will result in many ARP requests, causing poor performance.

The LANTYPE parameter identifies the type of Ethernet LAN that is being defined by this PU definition statement. The LANTYPE parameter has three possible values allowed. The ENETV2 value indicates that the Ethernet being attached to the LAN interface specifies that the PU being defined shall support only the Ethernet Version 2 frame format. A value of 802.3 indicates that the PU will support only the standard frame defined by IEEE 802.3. A value of DYNAMIC indicates that the PU will support both the IEEE 802.3

and the Ethernet Version 2 frame formats simultaneously. The DYNAMIC value lowers performance, as the NCP must determine the frame format for each frame that is received on the interface. Coding a value of DYNAMIC also requires the LANTYPE parameter on the IPHOST definition statement.

20.1.6 Frame relay PU definition statements

A PU definition statement is needed for each frame relay line definition. Figure 20.9 illustrates a sample definition for the physical and logical PU definitions of frame relay line groups. The first PU definition statement following the physical line definition statement for frame relay specifies support for the Local Management Interface (LMI) protocol standard and NCP echo detection for the subport. LMI is a protocol used in frame relay networks to provide information on Permanent Virtual Circuit (PVC) status and link integrity. LMI should be used if the adjacent subport also supports LMI. The LMI keyword indicates LMI support. If a value of NO is specified, then LMI is not supported. The first suboperand of the LMI keyword deter-

```
FRPGRP      GROUP    FRELAY=PHYS              Phys frame relay
FRPLINE     LINE     ADDRESS=(1036,FULL),
                     CLOCKNG=EXT,
                     MAXFRAME=2106,
                     NRZI=NO,
                     SPEED=1544000
FRPPU01     PU       ERRORT=(3,4),
                     LMI=(CCITT,SEC),
                     SPOLL=6,
                     TIMERS=(10,15)
FRPPU1E     PU       ADDR=1E                  FHSP DLCI x'1E'
FRPPU2E     PU       ADDR=2E                  FHSP DLCI x'2E'
FRLGRP      GROUP    FRELAY=LOG               Logical frame
                                              relay
FRLLINE     LINE     ISTATUS=ACTIVE
```

Figure 20.9 PU definition for frame relay physical group.

mines which LMI standard to support if LMI is to be used. The default for the LMI keyword is CCITT. CCITT instructs the NCP to use the CCITT Q.933 Annex A international standard. A value of ANSI may also be coded for the first suboperand of the LMI keyword. This will instruct NCP to use the ANSI T1.617 Annex D United States standard. Whatever value is defined here must also be defined on the other end of the permanent virtual circuit.

The second operand of the LMI keyword specifies support for NCP echo detection. Note that this is an NCP function and is in no way a frame relay standard. Echo detection is required when an NCP uses a satellite connection for the frame relay network. The transmission of frames using satellites sometimes results in echoed frames. To detect these frames, the NCP sets the Command/Response (C/R) bits in all outbound frames to 1 (PRIMARY) or 0 (SECONDARY). This allows the NCP to discern the difference between actual outbound frames and frames that are received because of echo. The NCP will reject any inbound frame with a C/R bit set the same as the outbound frame. The possible values for the second suboperand of the LMI keyword are PRIMARY, used to set the bit to a 1, and SECONDARY, used to set the bit to a 0. The C/R bit is usually not used and is therefore set to 0. When defining the subport partners, one side should be set to PRIMARY and the other set to SECONDARY.

The ERRORT keyword defines the error threshold by which NCP determines that an adjacent frame relay link station connection has been lost. The suboperands represent minimum and maximum thresholds. The default is ERRORT=(3,4), but the values may range from 1 to 10. The ERRORT keyword is valid only when the LMI keyword specifies either CCITT or ANSI.

The SPOLL keyword specifies the frequency at which LMI status requests are sent to the adjacent LMI subport. The default value for this is 6, and the range is from 1 to 255. This value is used in conjunction with the TIMERS keyword. The TIMERS keyword specifies the number of seconds between LMI Link Integrity Verification (LVI) status enquiry messages exchanged between adjacent subports. The range for each suboperand of the TIMERS keyword is 5 to 30, with 15 as the default. The first suboperand is the interval for sending LVI messages, and the second suboperand is the time in seconds for which an LVI message must be received before an error interval expires.

Each PU definition following the first PU definition on the physical frame relay line group has an address specified. The ADDR keyword is used to define the frame relay Data Link Connection Identifier (DLCI) value that defines the PVC associated with the frame relay connection. In NCP V6R1, the ADDR keyword is invalid but required for logical line PU definitions. In NCP V6R2, the ADDR is valid for any physical PU definition that follows the LMI PU definition. The minimum value for ADDR in NCP V6R1/V6R2 for logical lines is x'10'. The maximum value for ADDR is determined by the largest DLCI allowed for the line adapter type. If the frame relay line is defined on a TSS adapter, then the maximum DLCI allowed is x'D7'. A TSS adapter has a port address range of 0 to 895. If the line is attached to a HPTSS adapter, the maximum DLCI value is x'FE'. A HPTSS adapter has a line address range from 1024 to 1039.

20.2 LU Definition Statement

Devices attached to an SNA cluster controller gain access to the network through a Logical Unit (LU). Each LU is assigned a local address for use by the PU of the cluster controller and the NCP. The LU definition statement supplies the name of the logical unit, its logical address, control of data transmission to the LU, and the LU's dispatching priority in relation to the other LUs associated with the PU (Figure 20.10).

When defining LUs for an SNA 3x74 cluster controller, the LOCADDR operand specifies the local address for each LU. It is the only required operand of the LU definition statement. The sequence for defining the LUs is dependent upon the local address. You must code the LUs in ascending local address order. For an SNA 3274 cluster controller, the local address value begins at 2. The range for a PU Type 2 device is 1 to 255. A PU Type 1 device range for LOCADDR is 0 to 63. You should consult the cluster controller's manual for the valid local address range of that device.

The LU defined by this LU definition statement can be prioritized for dispatching to receive service from the NCP according to its primary use. The BATCH operand allows you to assign a low priority (BATCH=YES) or a high priority (BATCH=NO); the default value is NO. If an LU is representing a graphics printer, it may be worthwhile to assign a low priority to the printer. This will prohibit the large amount of data used for printing graphics from monopolizing the line and will allow interactive LUs to receive a greater portion of the service.

```
S11C1T00   LU      LOCADDR=02,   1ST AVAILABLE LOG. ADDR.        X
                   BATCH=NO,     HIGH DISPATCHING PRIORITY       X
                   PACING=0,     NO PACING TO LU                 X
                   VPACING=0     NO PACING TO NCP FOR THIS LU
```

Figure 20.10 LU definition statement.

The BATCH operand has been removed from NCP V4.3/V5. This is because these versions support virtual route transmission priority between the NCP and the peripheral node. This will prohibit print output from other data-intensive LUs from monopolizing the peripheral SDLC link as well as the INN links between subareas.

Performance of an LU-LU session is greatly influenced by the PACING and VPACING operands (Figure 20.11) of the LU definition statement. The PACING operand specifies the number of SDLC frames that the NCP may send to the LU before it must wait to receive a pacing response. The pacing response is used by the LU to signal to the NCP that it is ready to receive more data during this service time. The range of the pacing value is 0 to 255. The specification of 0 indicates that no pacing between the LU and the NCP occurs; the default value is 1. The VPACING operand defines the number of PIUs destined for this LU that VTAM can send to the NCP before requesting a pacing response from the NCP; the default value here is 2. The specification of 0 for this operand indicates that no pacing responses are necessary between the NCP and VTAM for the transmission of data destined to the LU defined by this LU definition statement.

These session-pacing indicators prohibit the higher-level node from sending data to the lower-level node until a pacing response has been received from the lower-level node. As you may have guessed, these parameters play an important role in session performance. Further information on performance can be found in *Advanced SNA Networking*. Figure 20.12 contains the full definition for all the LUs associated with PU N11SCC1.

20.3 GENEND Definition Statement

After all the resources attached to the NCP have been defined, the GENEND definition statement is coded. This statement must follow the last resource definition statement of the NCP and is required.

The GENEND definition statement allows you to define user-written code characteristics and options that are to be included with the NCP load module. The statement is used in conjunction with supple-

```
S11C1T00   LU        LOCADDR=02,     1ST AVAILABLE LOG. ADDR.       X
                     BATCH=NO,       HIGH DISPATCHING PRIORITY      X
                     PACING=0,       NO PACING TO LU                X
                     VPACING=0       NO PACING TO NCP FOR THIS LU
```

Figure 20.11   Flow control operands of the LU definition statement.

```
S11C1T00  LU  LOCADDR=02,PACING=0,VPACING=0
S11C1T01  LU  LOCADDR=03,PACING=0,VPACING=0
S11C1T02  LU  LOCADDR=04,PACING=0,VPACING=0
S11C1T03  LU  LOCADDR=05,PACING=0,VPACING=0
S11C1T04  LU  LOCADDR=06,PACING=0,VPACING=0
S11C1T05  LU  LOCADDR=07,PACING=0,VPACING=0
S11C1T06  LU  LOCADDR=08,PACING=0,VPACING=0
S11C1T07  LU  LOCADDR=09,PACING=0,VPACING=0
S11C1T08  LU  LOCADDR=10,PACING=0,VPACING=0
S11C1T09  LU  LOCADDR=11,PACING=0,VPACING=0
S11C1T10  LU  LOCADDR=12,PACING=0,VPACING=0
S11C1T11  LU  LOCADDR=13,PACING=0,VPACING=0
S11C1T12  LU  LOCADDR=14,PACING=0,VPACING=0
S11C1T13  LU  LOCADDR=15,PACING=0,VPACING=0
S11C1T14  LU  LOCADDR=16,PACING=0,VPACING=0
S11C1T15  LU  LOCADDR=17,PACING=0,VPACING=0
S11C1T16  LU  LOCADDR=18,PACING=0,VPACING=0
S11C1T17  LU  LOCADDR=19,PACING=0,VPACING=0
S11C1T18  LU  LOCADDR=20,PACING=0,VPACING=0
S11C1T19  LU  LOCADDR=21,PACING=0,VPACING=0
S11C1T20  LU  LOCADDR=22,PACING=0,VPACING=0
S11C1T21  LU  LOCADDR=23,PACING=0,VPACING=0
S11C1T22  LU  LOCADDR=24,PACING=0,VPACING=0
S11C1T23  LU  LOCADDR=25,PACING=0,VPACING=0
S11C1T24  LU  LOCADDR=26,PACING=0,VPACING=0
S11C1T25  LU  LOCADDR=27,PACING=0,VPACING=0
S11C1T26  LU  LOCADDR=28,PACING=0,VPACING=0
S11C1T27  LU  LOCADDR=29,PACING=0,VPACING=0
S11C1T28  LU  LOCADDR=30,PACING=0,VPACING=0
S11C1T29  LU  LOCADDR=31,PACING=0,VPACING=0
S11C1P30  LU  LOCADDR=32,PACING=1,VPACING=2,BATCH=YES,DLOGMOD=PRT
S11C1P31  LU  LOCADDR=33,PACING=1,VPACING=2,BATCH=YES,DLOGMOD=PRT
```

Figure 20.12   LU definitions for SDLC 3x74 cluster controller.

mentary programs such as IBM's Non-SNA Interconnection, NCP
Packet Switched Interface, and Network Terminal Operator, to name
a few.

20.4   Multiple NCPs in a Single Domain

Theoretically, in a single domain, 254 PU Type 4 nodes can be
attached to a single VTAM host in a pre-VTAM 3.2 network. This is

not advisable in practice; however, several single-domain networks do have more than one attached NCP. In this section we will discuss the NCP statements needed to define NCP-NCP communications.

20.4.1 SDLCST definition statement

In a single-domain network, it is not uncommon for NCP subareas to be connected by an SDLC link. In fact, most large networks have several SDLC subarea links between NCPs. This configuration allows for multiple paths between subareas. When these paths are activated, the NCPs must determine their roles and the subarea links' characteristics according to their roles. The role and link characteristics are determined by the SDLCST definition statement. The SDLCST definition statement must appear in the NCP source before any GROUP definition statement does.

As you can see from Figure 20.13, there are only two required operands of the SDLCST definition statement. The name operand is coded as any valid assembler-language symbol and is used by the SDLCST operand of the LINE definition statement that defines the SDLC subarea link.

The GROUP operand of the SDLCST definition statement is also required. The value coded here identifies the GROUP definition statement that defines the subarea link parameters associated with this SDLC selection table (SDLCST) definition.

The MAXOUT operand specifies the number of SDLC frames the NCP can receive on the line before issuing a response. This is specific to the NCP when operating in secondary mode. The MAXOUT value specified on the PU definition statement for this subarea link is the value that the primary-mode NCP will use for sending frames to the secondary-mode NCP before requesting an acknowledgment. The value for MAXOUT is determined by the modulus being enforced by the NCPs. If modulus 8 is being used, MAXOUT can range from 1 to 7. If modulus 128 is in use, MAXOUT can range from 8 to 127. The

```
name        SDLCST      GROUP=group name,
                        [,MAXOUT=n|7]
                        [,MODE=PRI|SEC]
                        [,PASSLIM=n|254]
                        [,RETRIES=NONE|(m[,t[,n]])]
                        [,SERVLIM=n|4]
                        [,TADDR=chars]
```

Figure 20.13 SDLCST definition statement format.

MODE operand specifies if this SDLCST definition describes functional characteristics for the NCP when it is in primary (polling responsibility and error recovery) or in secondary mode. The value coded here must be the same as the MODE operand of the associated GROUP definition statement pointed to by the GROUP operand of this SDLCST definition statement. Figure 20.14 outlines the procedure for determining which NCP of a subarea link is to act in primary or secondary mode.

The PASSLIM operand is functionally equivalent to the PASSLIM operand discussed for the SDLC LINE definition, except that here in the SDLCST definition the PASSLIM value can affect the mode of transmission on the subarea line (e.g., full or half duplex). Usually,

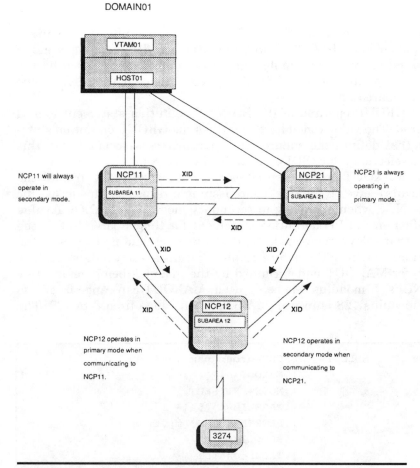

DOMAIN01

NCP11 will always operate in secondary mode.

NCP21 is always operating in primary mode.

NCP12 operates in primary mode when communicating to NCP11.

NCP12 operates in secondary mode when communicating to NCP21.

Figure 20.14 Primary and secondary mode is determined by the higher subarea number found in the exchange identification (XID) between NCPs in pre NCP 4.3 networks.

SDLC subarea links are defined with a send and receive address on the LINE definition statement to facilitate the transmission of data in both directions. If, however, the PASSLIM value specified for the SDLCST definition is less than the MAXOUT value specified for this SDLCST definition, the line will operate in half-duplex transmission mode. It is best to take the default for this operand to ensure full-duplex transmission when the LINE definition for the subarea link is assigned two addresses.

The RETRIES operand acts in the same manner here as the RETRIES operand for error recovery to peripheral nodes. However, ample time should be given to allow a link-attached NCP subarea to access its dump data sets on disk. Therefore, we suggest a minimum retry time for the RETRIES and REPLYTO values of 60.

The SERVLIM operand here also functions in the same manner as the SERVLIM operand of the LINE definition statement. Remember that the SERVLIM value determines the number of scans through the Service Order Table (SOT) to complete normal service of resources on the link before performing special services to the resources, that is, activation and deactivation or status requests from the SSCP for a resource on the link. The value for SERVLIM is dependent upon the number of status commands and activation and deactivation of devices on this link. If the commands are frequent, a low SERVLIM value is justifiable to avoid queuing of the status commands. However, for an SDLC subarea link, few status commands should traverse the link, and in most cases the subarea link is point-to-point and not multipoint, so the SERVLIM value can be set to 254 for optimal service to the INN link.

Finally, the TADDR operand of the SDLCST definition statement specifies a unique SDLC station address for the NCP operating in secondary mode and may be specified on the SDLCST definition describing secondary mode. The default is the hexadecimal representation of the NCPs's subarea address. Figure 20.15 shows the SDLCST definitions for NCP11.

20.4.2 SDLCST GROUP definition statements for single-domain NCP

Two GROUP definition statements must be defined for an SDLC subarea link when using the SDLCST definition statement. One group will specify the characteristics for the link during the primary mode of operation, and the second will define the characteristics of the link during the secondary mode of operation. A GROUP definition statement must still be defined just before the LINE definition statement that defines the actual SDLC subarea link. All in all, three GROUP definition statements must be defined for one SDLC subarea link.

```
PRISTE   SDLCST   GROUP=PRINCP,    NCP IN PRIMARY MODE             X
                  MAXOUT=127,      USING MODULO128                 X
                  MODE=PRI,        PRIMARY MODE                    X
                  PASSLIM=254,     ENSURE FULL DUPLEX              X
                  SERVLIM=254      MAX SERVICE FOR INN LINK
SECSTE   SDLCST   GROUP=SECNCP,    NCP IN SECONDARY MODE           X
                  MAXOUT=127,      USING MODULO128                 X
                  MODE=SEC,        PRIMARY MODE                    X
                  PASSLIM=254,     ENSURE FULL DUPLEX              X
                  SERVLIM=254      MAX SERVICE FOR INN LINK
```

Figure 20.15 SDLCST definitions for NCP NCP11.

```
PRINCP   GROUP    MODE=PRI,        NCP IN PRIMARY MODE             X
                  LNCTL=SDLC,      SDLC LINE CONTROL               X
                  TYPE=NCP,        NETWORK CONTROL MODE            X
                  DIAL=NO,         DEDICATED LINK                  X
                  REPLYTO=(,60),   60-S REPLY TIME OUT             X
                  TEXTTO=3         3-S TEXT TIME OUT
SECNCP   GROUP    MODE=SEC,        NCP IN SECONDARY MODE           X
                  LNCTL=SDLC,      SDLC LINE CONTROL               X
                  TYPE=NCP,        NETWORK CONTROL MODE            X
                  DIAL=NO,         DEDICATED LINK                  X
                  REPLYTO=(NONE,NONE),  NO REPLY TIME OUT          X
                  TEXTTO=NONE,     NO TEXT TIME OUT                X
                  ACTIVTO=420.0    MAX SERVICE FOR INN LINK
```

Figure 20.16 GROUP definition statements for primary and secondary mode.

However, the GROUP definition statements that reflect the mode of
operation identified by the SDLCST definition statement for primary
and secondary modes need be defined only once. Let's look at the
coded example of the GROUP definition statements for the SDLC
subarea links between NCP11 and NCP21 in Figure 20.16.

In Figure 20.16 we have defined the LINE characteristics that will
be enforced by the NCP when the SDLC subarea link is operating in
primary and secondary modes. It is these values and not those
defined on the GROUP, LINE, and PU definitions for the SDLC subarea link that will be used.

The first GROUP definition statement defines the characteristics of
the SDLC subarea link when the NCP is in the primary state. The

name PRINCP is the same name defined in the SDLCST definition statement operand GROUP of the statement that defines the primary mode of operation. The MODE operand of the GROUP statement in Figure 20.16 defines to NCP that the operands defined here are to be used when the NCP is in primary mode of operation with the SDLC subarea link.

The LNCTL, TYPE, and DIAL operand values are all typical for a dedicated SDLC link. The LNCTL operand specifies the usage of SDLC as the link protocol. The TYPE operand specifies that the lines in this group operate in network control mode. The DIAL operand identifies this link as a dedicated nonswitched line.

The REPLYTO operand of the PRINCP group specifies the number of seconds the primary NCP will wait for a response to a poll, selection, or message text that was sent to the secondary NCP before issuing a time-out error for the secondary NCP. The range is 0.1 to 60 for a link in network control mode. The comma in the value defined in Figure 20.16 indicates that the default for SDLC links in modulus 8 mode is taken. The default is 1. The 60 indicates that 60 must expire for SDLC links in modulus 128 mode between receipt of a response to a poll from the secondary NCP before issuing a time-out error.

The TEXTTO operand defines the amount of time in seconds the primary NCP will wait between receipt of text messages from the secondary NCP before issuing a text time-out error. Barring problems with the link, 3 is sufficient for this type of condition.

The group that defines the secondary-mode characteristics for SDLC subarea links in this NCP is labeled SECNCP. Again, this must match the GROUP operand of the SDLCST definition statement that defines the link characteristics for the NCP operating in secondary mode. The MODE operand of the GROUP definition statement SECNCP identifies this group for use in defining the link characteristics when the NCP is in secondary mode of operation.

The LNCTL, TYPE, and DIAL operands have the same meaning here as in the definition for the group PRINCP. The time-out operands are defined differently, and we have added a new time-out value, ACTIVTO.

The REPLYTO value for the SECNCP group specifies that the NCP will not keep elapsed time counts on this link. This prevents the link from becoming inoperative by the secondary NCP because of an error on the link from the primary NCP. The value REPLYTO=(NONE,NONE) is required when the group being defined describes the secondary mode of operation (e.g., MODE=SEC).

The same logic can be applied to the TEXTTO operand of the group SECNCP. In fact, if MODE=SEC for an SDLC subarea link group definition, the TEXTTO value must be equal to NONE. This makes sense, since the NCP in secondary mode is acting in a passive state and the primary NCP controls the error recovery on the link.

The ACTIVTO operand is pertinent to the NCP in secondary mode only. It specifies the time-out value, in tenths of a second, that the secondary NCP will wait for a response from the primary NCP before entering shutdown mode. The range is from 60 to 420. The default if ACTIVTO is not specified is 60 in an IBM 3725 or 3720 for both modulus 8 and modulus 128 links. The IBM 3705 defaults to 420. You do not want to have to go through the process of reactivating NCPs and subarea links if ACTIVTO has been reached and the secondary NCP goes into shutdown mode. Therefore, it is to your advantage to have the secondary NCP wait as long as possible for communication to be reestablished by the primary NCP. Hence we have coded the maximum value of 420.0.

20.4.3   GROUP, LINE, and PU definition statements for SDLC subarea links

The LINE definition statements for an NCP SDLC subarea link are the same for peripheral LINE definitions. There are, however, three operands that are specific to an SDLC subarea link: SDLCST, MONLINK, and MODULO. Figure 20.17 contains the GROUP, LINE, and PU definitions for the SDLC subarea links in NCP11.

On the LINE definition statement for each subarea link defined, we have specified the same values for the SDLCST, MONLINK, and MODULO operands. The SDLCST operand on the LINE definitions statement for subarea links tells NCP which SDLCST definition to use when the NCP is in primary or secondary mode. The first parameter specifies the name of the primary mode definitions. In this case we have NCP using the parameters specified on the SDLCST statement labeled PRISTE. The second parameter tells NCP to use the characteristics identified in the SDLCST statement defined for secondary mode of operation. In this instance the parameter points to the label SECSTE that identifies the SDLCST definitions for secondary mode. Used in this fashion, the link stations are considered "configurable."

NCP V4.3 and higher support predefined link station roles. This is accomplished by specifying either the primary or secondary SDLCST definition statement name on the LINE SDLCST operand. Used in this manner, the link station role is determined at NCP generation and is said to be a "predefined" link station. For predefined link stations, the XID sent indicates that this link station can be *only* primary or secondary. In this case, the receiving link station must be *either* configurable or predefined as the opposite link station role. For example, if in Figure 20.18 the SDLCST operand of the LINE definition statement named INNLINK were specified as

```
SALINKS    GROUP   LNCTL=SDLC,       SDLC IS LINE PROTOCOL      X
                   CLOCKNG=EXT,      MODEMS DO CLOCKING         X
                   NRZI=NO,          NO NONRETURN TO ZERO       X
                   PAUSE=0,          DO NOT WAIT TO POLL        X
                   SERVLIM=254,      MAXIMIZE SERVICE TO LINKS  X
                   SPEED=56000,      LINE SPEED IS 56 KPS       X
                   ISTATUS=ACTIVE
SA113221   LINE    ADDRESS=(32,FULL),  SEND/RECEIVE XMIT        X
                   ATTACH=MODEM,       MODEM ATTACHES LINK      X
                   SDLCST=(PRISTE,SECSTE), PRI & SEC SDLCST     X
                   MONLINK=NO,         MONITOR FOR ACTPU        X
                   MODULO=128          USE MODULUS 128
PU2132     PU      PUTYPE=4,         PU T.4 DEVICE (NCP)        X
                   MAXOUT=127,       USE MODULUS 128            X
                   TGN=21,           TG NUMBER FOR LINK         X
                   ANS=CONTINUE      KEEP X-DOMAIN SESSIONS
SA113421   LINE    ADDRESS=(34,FULL),  SEND/RECEIVE XMIT        X
                   ATTACH=MODEM,       MODEM ATTACHES LINK      X
                   SDLCST=(PRISTE,SECSTE), PRI & SEC SDLCST     X
                   MONLINK=NO,         MONITOR FOR ACTPU        X
                   MODULO=128          USE MODULUS 128
PU2134     PU      PUTYPE=4,         PU T.4 DEVICE (NCP)        X
                   MAXOUT=127,       USE MODULUS 128            X
                   TGN=21,           TG NUMBER FOR LINK         X
                   ANS=CONTINUE      KEEP X-DOMAIN SESSIONS
SA113612   LINE    ADDRESS=(36,FULL),  SEND/RECEIVE XMIT        X
                   ATTACH=MODEM,       MODEM ATTACHES LINK      X
                   SDLCST=(PRISTE,SECSTE), PRI & SEC SDLCST     X
                   MONLINK=NO,         MONITOR FOR ACTPU        X
                   MODULO=128          USE MODULUS 128
PU1236     PU      PUTYPE=4,         PU T.4 DEVICE (NCP)        X
                   MAXOUT=127,       USE MODULUS 128            X
                   TGN=21,           TG NUMBER FOR LINK         X
                   ANS=CONTINUE      KEEP X-DOMAIN SESSIONS
```

Figure 20.17   GROUP, LINE, and PU definitions for SDLC subarea links in NCP11.

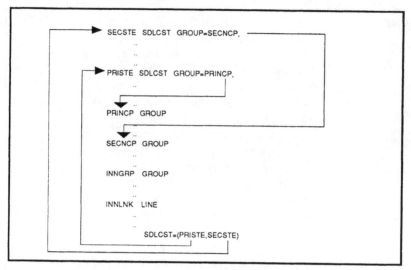

Figure 20.18   Diagram outlining the use of the SDLCST definition statement.

```
SDLCST=(,SECSTE)
```

then this link station would always operate in a secondary link station role. Hence, the opposite link station to which the line attaches would have to be predefined as a primary link station or be defined as a configurable link station. Likewise, if the SDLCST operand of the LINE definition statement named INNLINK in Figure 20.18 were specified as

```
SDLCST=(PRISTE)
```

then this link station would always operate in a primary link station role. The opposite link station to which this line attaches would have to be predefined as a secondary link station or be defined as a configurable link station. One more note on predefined link stations: If you predefine a link station as primary, then it is not necessary to define secondary SDLCST GROUP definitions for secondary. The opposite holds true if the link station is predefined as secondary.

The MONLINK operand is used on subarea link definitions by the NCP to actively monitor the link address for an ACTPU command from an SSCP when the NCP is not in session with an SSCP on this link. The code in Figure 20.17 is for the channel-attached NCP11, and therefore the SSCP-PU session is established via the channel and not the defined subarea links. For NCPs such as NCP12 in Figure 20.17, remotely loaded and activated NCPs should have MONLINK=YES specified on all their SDLC subarea links to turn the activation process around quickly. However, if NCP11 were defined to be activated by VTAM01 through NCP21, MONLINK=YES should also be specified. The default value for MONLINK is NO if the TYPGEN operand of the BUILD definition statement is equal to NCP (channel-attached). If the TYPGEN operand of the BUILD definition statement specifies NCP-R, the MONLINK operand defaults to YES (link-attached).

The MODULO operand of the LINE definition statement specifies the use of modulus 8 or modulus 128. We stated on the SDLCST definitions that MAXOUT=127; therefore, we must code MODULO=128 in order for the NCP to handle the amount of frames that the link-attached subarea can send or receive. Remember that the equivalent definition on the attached NCP must also specify MODULO=128.

The PU definitions under each subarea link define the link station in the attached NCP. PU2132 defines the characteristics of the link station for this subarea link that terminates in NCP21. The PUTYPE operand identifies the PU at the end of this link as a PU Type 4 device. Again, the MAXOUT operand defines the number of frames that can be sent before a response is requested. It is best for both documentation and implementation purposes that this value match the MAXOUT value specified on the MAXOUT operand of the SDLCST definition statement.

The TGN operand identifies the transmission group number assigned to the subarea link. This number is used in correlation with the PATH statements when defining explicit routes to the link-attached NCPs. The ANS operand tells the NCP whether to keep active cross-domain sessions enabled if the NCP loses contact with the owning SSCP. For the most part, ANS=CONTINUE is always coded to avoid a complete session outage.

20.4.4 Multipoint NCP subarea link definition

Multipoint line configurations are quite often used to reduce leased-line costs by allowing multiple devices to share the same physical line. NCP V5R2 introduced the ability to have subarea multipoint connection support. Figure 20.19 illustrates a multipoint line configuration between three NCPs as well as a cluster controller sharing the same physical line.

In the figure, NCP21, NCP12, and L24C1 are all attached to NCP11 through a multipoint SDLC leased line. Using a predefined configuration, the primary and secondary link station roles are predetermined. There can be only one primary link station on a multipoint

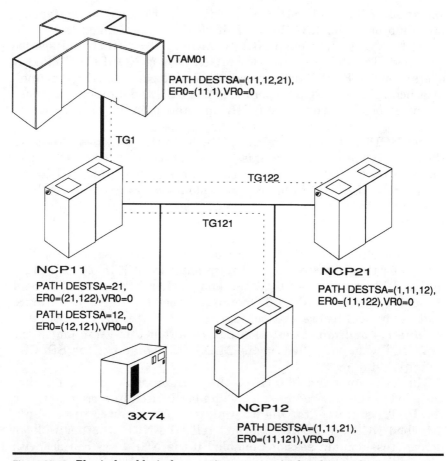

Figure 20.19   Physical and logical connections over a mixed multipoint line.

link, and it must reside in an NCP. We have selected NPC11 as the primary link station. Note that on the multipoint line we can mix PU types. All PU types, Node Types 1, 2, 2.1, 5, and 4, can be connected to the same multipoint line. The NCPs represent PU Type 4 and the cluster controller a PU Type 2. In this type of configuration, only the primary NCP (NCP11) can load or dump the secondary NCPs (NCP12 and NCP21). To support multipoint subarea links, VTAM V3R2 and NCP V4R3/V5R2 are required.

Figure 20.19 shows both the physical and the logical connections between the resources. All communication must pass through the primary link in a multipoint configuration. The primary NCP acts as an intermediate routing node. In our scenario, communications between any of the subareas must pass through NCP11. This is denoted by the PATH statements included in Figure 20.19. Each of the secondary

NCP11 definitions for a mixed multipoint line configuration		
MPTGROUP	GROUP	MOD=PRI,LNCTL=SDLC,DIAL=NO,TYPE=NCP
N11L24	LINE	ADDRESS=(24,FULL),DUPLEX=FULL,
		MONLINK=YES
	SERVICE	ORDER=(L24T421,L24T412,L24C1)
L24T421	PU	PUTYPE=4,ADDR=21,MODULO=128,TGN=122
L24T12	PU	PUTYPE=4,ADDR=12,MODULO=8,TGN=121
L24C1	PU	PUTYPE=2,MAXDATA=265
L24C1T00	LU	LOCADDR=2

Figure 20.20   Sample mixed multipoint LINE and PU definitions.

subareas must have an ER specified to the adjacent primary NCP (NCP11). Even though this is a multipoint configuration, PU Type 4 devices cannot be dynamically added or deleted. However, PU Type 2 or Type 1 devices may be dynamically reconfigured.

Figure 20.20 illustrates the definitions for NCP11 to support a multipoint subarea line configuration. The GROUP definition statement must specify GROUP=PRI for this NCP to act as the primary link station on this multipoint link. This operand does not refer back to a previously defined SDLCST definition. For multipoint subarea links, the SDLCST definition and its corresponding SDLCST operand on the LINE definition statement are not coded. There are no special operands to a LINE definition for a multipoint subarea link. However, it is recommended that you specify FULL on the ADDRESS and DUPLEX parameters.

The SERVICE statement does not support the MAXLST operand for a multipoint subarea link. Instead, the NCP takes the default for MAXLST, which is the number of PUs defined in the service order table.

The ADDR operand specifies the polling address of the PU. In point-to-point subarea links, the NCP issues a broadcast station address (x'FF'). Here, in a multipoint configuration, the primary NCP will issue unique polling addresses for each PU defined on the link. This is no different from multipoint polling for multipoint links that have multiple PU Type 2 devices attached.

The MODULO operand can now be coded on the PU definition-statement rather than on the LINE definition statement. For NCP-NCP communication, the modulus can be 8 or 128. Communication between the primary NCP and the PU Type 2 is dependent on PU Type 2 support for modulus 8 or 128.

The DATMODE operand of the PU definition statement is not applicable to multipoint subarea links. Full-duplex versus half-duplex data transmission is specified in the XID. In the NCP, the specification of ADDRESS=(nn,FULL) on the LINE definition statement sets

```
NCP12 definitions for a mixed multipoint line configuration

N12MPT              GROUP       MOD=SEC,LNCTL=SDLC,DIAL=NO,TYPE=NCP
N12L18              LINE        ADDRESS=(18,FULL),DUPLEX=FULL,
                                TADDR=12,IPL=YES
L18T411             PU          PUTYPE=4,MODULO=128,TGN=121

NCP21 definitions for a mixed multipoint line configuration

N21MPT              GROUP       MOD=SEC,LNCTL=SDLC,DIAL=NO,TYPE=NCP
N21L48              LINE        ADDRESS=(48,FULL),DUPLEX=FULL,
                                TADDR=21,IPL=YES
L48T411             PU          PUTYPE=4,MODULO=128,TGN=ANY
```

Figure 20.21   Mixed multipoint LINE and PU definitions in the down-stream PU Type 4 devices.

the two-way simultaneous indicator in the XID. IF the parameter is defined as ADDRESS=(nn,HALF), the NCP sets the two-way alternate indicator in the XID. The full-duplex mode is used only if both NCPs have ADDRESS=(nn,FULL) specified. Otherwise half-duplex is used. The determination of transmission modes allows for mixed modes on the multipoint subarea link, as shown in Figure 20.21.

The TGN operand specifies which transmission group number will be used for communicating to its respective PU. In a multipoint subarea link, one line can be associated with several transmission groups.

For the secondary NCPs on the multipoint subarea link, the definitions are quite similar to those for a normal point-to-point subarea link. In the GROUP definition statement, the MODE operand must specify SEC to indicate that the NCP will operate in secondary mode. Again, the SDLCST operand of the LINE definition statement is not coded. The key assignments are the DUPLEX and TADDR operands.

The DUPLEX operand of a secondary NCP must specify HALF. This disables the Ready-To-Send (RTS) signal on the link from this NCP. This must be done so that other PUs on the link can send to the primary NCP when they are polled. If DUPLEX=FULL is coded, indicating that RTS is permanently on, only one of the secondary NCPs will be allowed to send, since the others will never be able to turn RTS on.

The value of the TADDR operand for this NCP must be the same as that of the ADDR operand of the PU definition statement in the primary NCP. Recall that the ADDR operand of the PU statement identifies the polling address of the secondary NCP. The TADDR value can range from x'00' to x'FF'.

The IPL operand of the LINE definition statement indicates to the secondary NCP that it may load or dump over the address specified in the TADDR operand. The IPL ports table in MOSS on the communi-

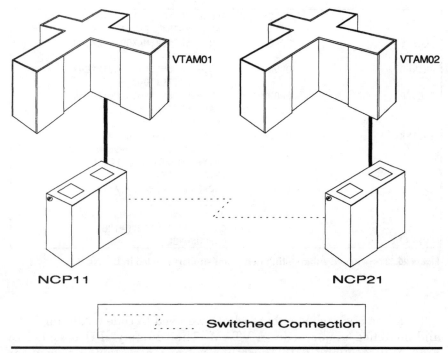

Figure 20.22 Configuration depicting a switched NCP subarea link connection.

cations controller must have an SDLC controller address equal to the TADDR value.

20.4.5 Switched NCP subarea link definition

The use of switched subarea support in Figure 20.22 provides increased connectivity between subareas in the network. The support can be used either for subarea link backup or for frequent access to other subareas. Switched subarea support allows for dial-up connection between NCP and NCP, NCP and VTAM, or VTAM and VTAM. Multiple switched subarea links may be combined into a single transmission group. They can also be included with leased lines to form a multilink transmission group. These two configurations are applicable only to NCP-NCP subarea links.

To make the connection, the calling subarea node requires an Auto Calling Unit (ACU); the call cannot be manual. Once the connection is established, the primary and secondary link station modes are determined through the XID. You can use either configurable or predefined. It is recommended that you use configurable, since this gives you the most flexibility.

PRISTE	SDLCST	GROUP=PRINCP,MAXOUT=127,MODE=PRI,
		PASSLIM=254,SERVLIM=254
SECSTE	SDLCST	GROUP=SECNCP,MAXOUT=127,MODE=SEC,
		PASSLIM=254
PRINCP	GROUP	MODE=PRI,LNCTL=SDLC,TYPE=NCP,
		DIAL=YES,TEXTO=3
SECNCP	GROUP	MODE=SEC,LNCTL=SDLC,TYPE=NCP,
		DIAL=YES,TEXTO=NONE,ACTIVTO=420.0,
		SERVLIM=254
SALINK	GROUP	LNCTL=SDLC,NRZI=NO,PAUSE=0,
		SERVLIM=254,SPEED=9600,DIAL=YES,
		PUTYPE=4,ACTIVTO=120.0,
		ISTATUS=ACTIVE
SA113221	LINE	ADDRESS=(32,HALF),ATTACH=MODEM,
		SDLCST(PRISTE,SECSTE),
		MONLINK=NO,ANSER=NO,AUTO=33,
		CALL=INOUT,MODULO=128
PU2132	PU	BRKCON=NONE

Figure 20.23 Switched subarea link definition support needed in the NCP.

To define a switched subarea link, we need to code a "dummy" PU and switched major node in VTAM. The "dummy" PU is coded in Figure 20.23 along with the other definition statements used for point-to-point subarea links. In the group labeled SALINK, the DIAL, PUTYPE, and ACTIVTO parameters are highlighted. The DIAL parameter is specified as YES to indicate that the following line in this group requires switched line control procedures. The PUTYPE parameter identifies to the NCP that this group is defining a switched subarea link. It must be in the GROUP definition statement in order for the NCP to prepare for the switched subarea support. Finally, the ACTIVTO parameter specifies the number of seconds allowed for receipt of I-frames before disconnecting the link with an error.

In the LINE definition, the ANSWER operand tells VTAM to accept dial-in PUs from this NCP. The AUTO operand specifies the address of the auto call unit in the NCP. This address must be different from the address specified for the LINE ADDRESS operand. The CALL operand indicates that incoming and outgoing calls may be made through this switched link.

The PU definition statement specifies the BRKCON operand. This operand tells the NCP how to break the connection when ACTIVTO has been reached. The default is NONE, meaning that the connection will stay alive. Two other options are CONNECTO and NOWNERTO. CONNECTO tells the NCP to start checking for time ACTIVTO at

```
VTAM02 Switched Major Node:

SWNODE1            VBUILD      TYPE=SWNET
SWNCP11            PU          PUTYPE=4,TGN=ANY,ANS=CONT,
                               NETID=NETA,IDNUM=FFE00,SUBAREA=11
                   PATH        DIALNO=201|555|1212,GID=1,PID=1,
                               GRPNM=SALINK,REDIAL=3,USE=YES
SWVTAM02           PU          PUTYPE=5,TGN=ANY,ANS=CONT,
                               NETID=NETA,IDNUM=FFE00,SUBAREA=02
                   PATH        DIALNO=201|555|1212,GID=1,PID=1,
                               GRPNM=SALINK,REDIAL=3,USE=YES
```

Figure 20.24 Switched major node definition required to support switched subarea links.

connect time. Coding NOWNERTO starts time-out activity when the owning SSCP begins Automatic Network Shutdown (ANS).

A switched major node in VTAM is used to merge with the dummy PU definition found in the calling NCP or VTAM. The definition in Figure 20.24 is your basic switched major node. The differences are the inclusion of the PUTYPE, TGN, and SUBAREA operands. These are needed to tell VTAM that this switched major node is supporting switched subarea links. This example shows the definition for a PU Type 4 dial connection by the specification of PUTYPE=4. The TGN operand indicates which transmission group is to be used. The specification of ANY makes this connection totally flexible in this regard. The SUBAREA operand value must match the subarea of the switched connected subarea. In this example, the calling NCP has a subarea address of 11. The PU labeled SWVTAM02 is the code for accepting a dial connection with VTAM02 in Figure 20.24.

For VTAM to perform a switched subarea link connection, it too needs a dummy PU definition. This type of connection is currently supported by the IBM 4361 with the Integrated Communications Adapter (ICA) and by the IBM 9370 with the Transmission Subsystems Controller (TSC). The definitions in Figure 20.25 can be used to dial both PU Type 4 and PU Type 5 subareas. The SUBDIAL

```
VTAM02 "dummy" Definitions:

                   VBUILD      TYPE=CA
SWGRP              GROUP       LNCTL=SDLC,DIAL=YES,SUBDIAL=YES
SWLINE             LINE        ADDRESS=420,AUTO=420,CALL=INOUT,
                               ANSWER=ON
SWDMY              PU
```

Figure 20.25 The "dummy" PU definition required by VTAM to support switched subarea link connections.

operand identifies this channel-attachment major node as supporting switched subarea links. The AUTO operand specifies the same address as the ADDRESS operand on the LINE definition statement. This is required for making outgoing calls. The function of calling is actually provided by microcode of the ICA and TSC.

20.4.6 Token-ring NCP subarea link definition

Logical definitions are also required for specifying INN traffic over token-ring. The ECLTYPE keyword in the logical GROUP definition for INN communications specifies the connection as being logical, and the type of connection is set to subarea (Figure 20.26). The PHYPORT keyword relates this logical definition to the physical definition.

The logical LINE definition statement specifies the transmission group number associated with the logical line. Prior to NCP V6R2, token-ring links between NCP subareas must be single-link transmission groups. That is, each token-ring INN link defined between two NCP subareas must have a unique transmission group number. NCP V6R2 allows multimedia transmission groups and hence multilink token-ring INN connections, so this single-link restriction is void with this new release.

The logical physical unit definition defines the type of station on the remote TIC. For subarea INN links, the PUTYPE keyword on the PU definition statement is always 4. The ADDR keyword specifies the destination MAC address for which the NCP TIC will establish a session. The ADDR value follows the format ss4000abbbbbbb. The ss value is the SAP of the token-ring defined by this PU definition statement. The ss value is always X'04' when the INN link is attached to another NCP. If the INN link is attached to an IBM 9370 host computer, the ss value must be a multiple of X'04'. The 4000abbbbbbb value must match the LOCADD keyword value of the physical line definition found in the attached NCP subarea. Figure 20.27 lists a completed token-ring definition for both BNN and INN TICs in two NCP subareas.

20.4.7 Frame relay NCP subarea link definition

The frame relay support found in NCP V6 allows INN links to traverse frame relay PVCs. Figure 20.28 details the definition for frame

```
TRINNGRP                GROUP      ECLTYPE=(LOGICAL,SUBAREA),
                                   PHYPORT=1,SDLCST=(PRISTE,SECSTE)
```

Figure 20.26 Token-ring GROUP definition statement needed for token-ring subarea link connectivity.

```
NCP11:

S11TRGR      GROUP     ECLTYPE=(PHYSICAL,ANY),ADAPTER=TIC2,
                       TRSPEED=16
S11TRLNE     LINE      ADDRESS=(1088,FULL),
                       LOCADDR=400041000111,PORTADD=1,
                       MAXSTL=4096,RCVBUFC=4096
S11TRPU1     PU        PUTYPE=1
S11TRLOG     GROUP     ECLTYPE=(LOGICAL,SUBAREA),PHYPORT=1,
                       SDLCST=(PRIGRP,SECGRP)
S11S12LN     LINE      TGN=11
S11S12PU     PU        PUTYPE=4,ADDR=04400041000121
S11BNN       GROUP     ECLTYPE=(LOGICAL,PERIPHERAL),
                       PHYPORT=1,CALL=INOUT,AUTOGEN=140
NCP12:

S12TRGR      GROUP     ECLTYPE=(PHYSICAL,ANY),ADAPTER=TIC2,
                       TRSPEED=16
S12TRLNE     LINE      ADDRESS=(1088,FULL),
                       LOCADDR=400041000121,PORTADD=1,
                       MAXSTL=4096,RCVBUFC=4096
S12TRPU      PU        PUTYPE=1
S12TRLOG     GROUP     ECLTYPE=(LOGICAL,SUBAREA),PHYPORT=1,
                       SDLCST=(PRIGRP,SECGRP)
S12S11LN     LINE      TGN=11
S12S11PU     PU        PUTYPE=4,ADDR=04400041000111
S12BNN       GROUP     ECLTYPE=(LOGICAL,PERIPHERAL),
                       PHYPORT=1,CALL=INOUT,AUTOGEN=140
```

Figure 20.27 Sample NCP definition for BNN and INN definitions over token-ring.

```
FRPGRP      GROUP   FRELAY=PHYS              Phys frame relay
FRPLINE     LINE    ADDRESS=(1036,FULL),
                    CLOCKNG=EXT,
                    MAXFRAME=2106,
                    NRZI=NO,
                    SPEED=1544000
FRPPU01     PU      ERROR=(3,4),
                    LMI=(CCITT,SEC),
                    SPOLL=6,
                    TIMERS=(10,15)
FRPPU1E     PU      ADDR=1E                  FHSP DLCI x'1E'
FRPPU2E     PU      ADDR=2E                  FHSP DLCI x'2E'
FRLGRP      GROUP   FRELAY=LOG,              Logical frame
                                             relay
                    ISTATUS=ACTIVE,
                    MAXOUT=127,
                    MODULO=128,
                    PHYSRSC=FRPPU01,
                    RETRIES=(5,5,5),
                    SDLCST=(FRPRI,FRSEC)
FRLLINE     LINE    ISTATUS=ACTIVE,
                    MONLINK=YES,
                    IPL=YES
FRLPU01     PU      ADDR=60,
                    ANS=CONTINUE,
                    PUTYPE=4,
                    TGN=1
```

Figure 20.28 PU definition for frame relay subarea logical group.

relay INN links. Note that the definitions are quite similar to those already defined for SDLC INN links. The GROUP definition statement has two parameters unique to frame relay. These are the FRELAY and PHYSRSC keywords.

The FRELAY keyword indicates that the group being defined is a logical frame relay definition. The PHYSRSC keyword identifies the name of the LMI PU on the physical line definition that is to be used for the logical line being defined.

The logical line definition resembles the line definition for any other INN link definition. The PU definition for the INN subarea link uses the frame relay DLCI address on the ADDR keyword. The remaining keywords for a PU definition statement that defines an INN link station are used in the same manner as for SDLC INN link stations.

20.5 Summary

In this chapter we explored the use and definition of Ethernet, token-ring, and frame relay for the transmission of information through a communications controller executing NCP V6R1 and NCP V6R2. The functions discussed for these NCP versions provide enhanced functional support for Ethernet-Ethernet communications, BNN and INN communications over token-ring, and INN communications using frame relay.

Operations and Network Management

Network Operations

Management of an SNA network through VTAM is accomplished by using three operational commands. These are the DISPLAY, MODIFY, and VARY VTAM operator commands. The DISPLAY command is used to obtain the current status of the network resources and to request route test information for display on the network operator's console. The MODIFY command can be used to alter VTAM start list options while VTAM is executing. The VARY command is used to alter the state of a network resource.

21.1 VTAM Operator Commands

VTAM allows operating system console operators to modify network resource status and to obtain this status through operator commands. These commands are found in VTAM's USS table named ISTNCNO. This table is found in VTAMLIB and is modifiable. The actual table used is dependent on the USSTAB parameter value specified on the USSTAB= start option in ATCSTR00 of VTAMLST. The operator commands can also be modified to provide a unique set of commands for a network management application, such as NetView, by specifying the USSTAB= parameter on the VTAM APPL definition statement that defines the NetView application control block.

21.1.1 The DISPLAY ID command

The DISPLAY ID command is by far the most commonly used VTAM command when operating an SNA network. The results of the command provide the operator with the current status of the resource selected, its desired status as a result of a VARY command, the name of the major node that this resource is associated with if it is a minor

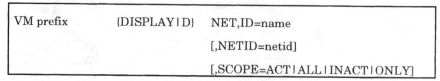

VM prefix	{DISPLAY	D}	NET,ID=name
			[,NETID=netid]
			[,SCOPE=ACT	ALL	INACT	ONLY]

Figure 21.1 VTAM DISPLAY ID operator command format.

node, and LU-LU session data if the resource is so capable. Figure 21.1 outlines the format of the DISPLAY ID command.

The DISPLAY ID command is commonly used to view the current status of resources. For instance, if the command entered is

```
D NET,ID=NCP11
```

only information specific to the node name NCP11 will be displayed. In this example, only the status of NCP11 itself is displayed. This can be seen in Figure 21.2. If more information about the NCP were needed, we could have entered

```
D NET,ID=NCP11,E
```

to obtain the status of the links attached to the NCP as well as the status of the NCP itself. We could have entered

```
D NET,LINES
```

to obtain the status of all lines in the domain.

Use the SCOPE operand to limit the resource display to a specific status. An example is using the INACT operand to quickly determine which subordinate resources of the requested resource are currently inactive.

```
DISPLAY NET,ID=NCP11
IST075I NAME = NCP11, TYPE = PU T4/5
IST486I STATUS= ACTIV, DESIRED STATE= ACTIV
IST247I LOAD/DUMP PROCEDURE STATUS = RESET
IST484I SUBAREA = 11
IST391I ADJ LINK STATION = TR11L121, LINE = TR11L12, NODE = NCP12
IST391I ADJ LINK STATION = 0A0-S, LINE = 0A0-L, NODE = VTAM01
IST391I ADJ LINK STATION = SD21L111, LINE = SD21111, NODE = NCP21
IST391I ADJ LINK STATION = SD21L112, LINE = SD21112, NODE = NCP21
IST391I ADJ LINK STATION = TR21L111, LINE = TR21L11, NODE = NCP21
IST654I I/O TRACE = OFF, BUFFER TRACE = OFF
IST077I SIO = 07718 CUA = 0A0
IST675I VR = 0, TP = 2
```

Figure 21.2 Sample output from DISPLAY ID command.

So, as you can see, by using the DISPLAY ID command we can determine the current state of any network resource. The resource can be a major or minor node, an application name, or an NCP, PU, LU, or LINE name.

Let's look at some DISPLAY commands that have built-in functions for specific types of resources. Figure 21.3 contains a list of VTAM DISPLAY commands frequently used to gather status information on related resources. The formats listed can be used as written.

21.1.2 DISPLAY MAJNODES command

The D NET,MAJNODES command will display the current status of all major nodes known to this VTAM—that is, major nodes that were activated at start-up or activated by the operator. The display, as shown in Figure 21.4, shows only those major nodes that are currently active and those that were active but are now in another state.

The D NET,APPLS command displays the state of all application minor node names of active application major nodes, including appli-

VM prefix	{DISPLAY	D}	NET,MAJNODES
			NET,APPLS
			NET,PENDING
			NET,PATHTAB

Figure 21.3 VTAM operator command formats with built-in functions.

```
D NET,MAJNODES
IST350I DISPLAY   TYPE = MAJOR NODES
IST089I VTAMSEG   TYPE = APPL SEGMENT      , ACTIV
IST089I VTAM01    TYPE = PU T4/5 MAJ NODE , ACTIV
IST089I ISTPDILU  TYPE = CDRSC SEGMENT     , ACTIV
IST089I ISTCDRDY  TYPE = CDRSC SEGMENT     , ACTIV
IST089I ISTDSWMN  TYPE = SW SNA MAJ NODE   ; ACTIV
IST089I CAVTM01   TYPE = CA MAJ NODE       , ACTIV
IST089I JES201    TYPE = APPL SEGMENT      , ACTIV
IST089I TSO01     TYPE = APPL SEGMENT      , ACTIV
IST089I CICS01    TYPE = APPL SEGMENT      , ACTIV
IST089I IMS01     TYPE = APPL SEGMENT      , ACTIV
IST089I DB201     TYPE = APPL SEGMENT      , ACTIV
IST089I SN52010   TYPE = SW SNA MAJ NODE   , ACTIV
IST089I LCLSNA01  TYPE = LCL SNA MAJ NODE , ACTIV
IST089I LCLNSNA1  TYPE = LCL 3270 MAJ NODE, ACTIV
IST089I NCP11     TYPE = PU T4/5 MAJ NODE , ACTIV
IST314I END
```

Figure 21.4 Sample output from DISPLAY MAJNODES command.

cation minor node names that were never activated when the application major node was activated. This display supplies session information for the application minor node only. If the status is CONCT, the application is in a connectable state waiting for the application program to issue an OPEN for the application ACB name. In most cases this means that the application program has not been started by the operator.

There are some 69 pending states for resources in an SNA network, far too many to include in this book. The display from the VTAM command D NET,PENDING is useful in obtaining the status of all resources that are pending some type of action. Two of the more common pending states are Pending Activate PU (2) (PAPU2) and Pending Contacted (2) (PCTD2). You may see these states during activation of a peripheral PU. PAPU2 specifies that the ACTPU has been sent to the device, but the device has not yet responded. The PCTD2 indicates that VTAM has started the activation of the PU, but VTAM has not received the CONTACTED reply from the device. The PU address may not be initialized. The (2) on these states refers to the state of the secondary link station.

21.1.3   DISPLAY PATHTAB command

Throughout the book we have referred to the PATH table used by VTAM to route data through the network. The D NET,PATHTAB command requests VTAM to display the current PATH table on the operator's console. The output of this command is shown in Figure 21.5. The display will identify the destination subarea (DESTSUB), the adjacent subarea (ADJSUB) to reach the DESTSUB, the Explicit Route (ER) used to reach the ADJSUB, the status of the ER, and the Virtual Routes (VRs) that have the ER mapped. Note that this display does not list the Transmission Group Number (TGN). A common status for the ER STATUS column of the display is INOP. This indi-

```
D NET,PATHTAB
IST350I DISPLAY TYPE = PATH TABLE CONTENTS
IST516I DESTSUB    ADJSUB TGN   ER    ER STATUS    VR(S)
IST517I      12        12   1    4    ACTIV3       1
IST517I      12        12   2    2    ACTIV3       2
IST517I      12        12   2    0    ACTIV3       0
IST517I      21        11   1    3    PDEFO
IST517I      21        11   1    0    PDEFO
IST517I      21        12   2    1    PDEFO
IST314I END
```

Figure 21.5   Sample output from the DISPLAY PATHTAB command.

cates that the ER is defined to VTAM, but there is no physical connection between VTAM and the adjacent subarea.

21.1.4 DISPLAY ROUTE command

There is one more DISPLAY command that we need to discuss, especially when debugging a route problem. The command is the DISPLAY ROUTE command. Figure 21.6 outlines the format.

Using this command, we can determine the routes defined to specific subareas. For example, we can request the display of routes from NCP to NCP, or from SSCP to SSCP, or SSCP to NCP, or NCP to SSCP. This is accomplished by using the DESTSUB and ORIGIN operands of the DISPLAY ROUTE command. To verify routes from an NCP named NCP11 to an NCP named NCP21, we enter the command

```
D NET,ROUTE,DESTSUB=21,ORIGIN=NCP11
```

This will display, as shown in Figure 21.7, the routes defined for the NCP NCP11 to communicate with the NCP21. The display will identi-

```
VM prefix    {DISPLAY|D}    NET,ROUTE

                            ,DESTSUB=subarea number

                            [,COSNAME=name|ER=n|ER=ALL|VR=n]

                            [,NETID=netid]

                            [,ORIGIN=subarea PU name]

                            [,TEST=YES|NO]]
```

Figure 21.6 Format for VTAM DISPLAY ROUTE command.

```
D NET,ROUTE,DESTSUB=21,ORIGIN=NCP11
IST535I ROUTE DISPLAY 77 FROM SA 11 TO SA 21
IST808I ORIGIN PU = NCP11 DEST PU = NCP21 NETID = NETA
IST535I ROUTE DISPLAY 77 FROM SA 11 TO SA 21
IST808I ORIGIN PU = NCP11 DEST PU = NCP21 NETID = NETA
IST536I VR  TP    STATUS    ER          ADJSUB  TGN  STATUS
IST537I  0   0    ACT        0           21      2   ACT
IST537I  0   1    INACT      0           21      2   ACT
IST537I  0   2    INACT      0           21      2   ACT
IST537I  1   0    ACT        4           12      2   ACT
IST537I  1   1    INACT      4           12      2   INACT
IST537I  1   2    INACT      4           12      2   INACT
IST314I END
```

Figure 21.7 Sample output from DISPLAY ROUTE command.

fy the origin and destination PU names. It also displays their VRs, their Transmission Priority (TP), and their status along with the ER status. This display will assist you in mapping the exact paths that are active and available for use. If the TEST=YES operand is added, VTAM performs explicit route tests for each explicit route defined in the path tables. This is because we defaulted to ER=ALL. The results of this test are displayed and provide the operator with the status of the ER, the number of subareas traversed to perform the test of the ER, and the TG used for the test. The results also indicate whether the test succeeded or failed. An all-too-common result is that the ER tested was nonreversible—that is, a reverse ER through the same subareas was not defined properly. Remember, data can traverse different forward and reverse ERs, but the ERs must pass through the same subareas.

Now that you have a feel for how you can display the status of network resources, let's look at how we can alter the state of these resources.

21.1.5 VARY ACTivate command

The VARY ACTivate command is used to activate an inactive resource that is defined to VTAM. The format of the command is found in Figure 21.8. There are several parameters associated with loading an NCP into a communications controller. These will not be discussed here but are shown to identify the defaults. For more information on these NCP-specific parameters, consult the IBM VTAM Operation manual.

The ID parameter allows the operator to specify the name of the VTAM resource that is to be activated. Any valid name associated with a network resource defined to VTAM, such as an NCP name, line name, link station name, PU name, LU name, application name, or path major node, can be specified as the value of the ID parameter. The LOGON parameter allows the operator to activate a resource and pass its control to an application for logon processing. This is typically done while activating a 3270 type of resource. The LOGMODE parameter specifies the logon mode name to use in concert with the LOGON parameter. This will be the logon mode name used until the resource is recycled or is modified through dynamic modification. The SCOPE parameter indicates the number of resources that will be activated by this command if the resource being activated is a major node. The ALL value specifies that all resources subordinate to the resource being activated will begin the activation process regardless of the ISTATUS value defined for the resource. The COMP value specifies that activation processing for the subordinate resources will

VM	{VARY I V}	NET,ACT
prefix		
		,**ID**=name
		[,ANS={ON I <u>OFF</u>}]]
		[,DUMPLOAD={YES I <u>NO</u>}]*
		[,DUMPSTA=name]*
		[,LOAD={YES I NO I <u>U</u>}]*
		[,LOADFROM={<u>HOST</u> I EXTERNAL}]*
		[,LOADMOD=load module name]*
		[,LOADSTA=name]*
		[,LOGMODE=logon mode name]
		[,LOGON=[pluname I cdrscname I uservarname}}
		[,NEWPATH={name I (name1,...,name3)}]
		[,RNAME={name I (name1,...name13)}]*
		[,SAVEMOD={YES I <u>NO</u>}]*
		[,SCOPE={ALL I <u>COMP</u> I ONLY I U}]
		[,U=channel device name]
		[,WARM]
		* NCP specific parameters

Figure 21.8   Format for VTAM VARY ACTivate command.

```
VARY NET,ACT,ID=BT10111
IST093I BT10111 ACTIVE
```

Figure 21.9   Output from VARY ACT command.

begin in accordance with the ISTATUS value of the resource. The ONLY value tells VTAM to activate only the named resource and not its subordinate resources. Figure 21.9 contains a sample output from a successful activate command.

21.1.6   VARY INACTivate command

The VARY INACTivate command is used to change the status of a resource from active to inactive. In the inactive state, the resource is

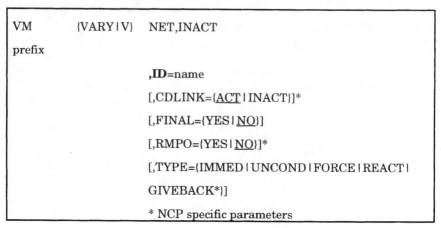

Figure 21.10 Format of the VARY INACTivate command.

not communicating with VTAM. Figure 21.10 details the format of the VARY INACTivate command.

The CDLINK parameter is used to indicate to an NCP that cross-domain links with active cross-domain sessions are to be left active even though VTAM is inactivating the NCP. Specifying INACT will cause even cross-domain links to go inactive. The FINAL parameter indicates to the PU whether it will be reactivated after inactivation. The RMPO parameter is used to indicate to a communications controller through the NCP that at completion of NCP inactivation, the communications controller will be powered off. The TYPE parameter specifies a deactivation process other than normal. Normal processing will wait for the end of LU-LU sessions of subordinate resources before inactivation occurs. The IMMED and UNCOND values will disrupt the LU-LU sessions and force VTAM to issue the CLSDST command to the application. The FORCE command is used when a resource is not responding to an immediate or unconditional inactivation request. The difference is that with the IMMED or UNCOND values, communication with the PU resource still exists. The FORCE command is used when VTAM is not communicating with the resource but still believes it has active sessions with the resource. The REACT value is used to recycle a resource. This command causes VTAM to issue the inactivate command and then an activate command to the resource and its subordinates. The GIVEBACK value is used in multidomain configurations where a backup VTAM is to return control and ownership of the resource to the original VTAM owner of the resource without disrupting any LU-LU sessions. Figure 21.11 gives an example of the VTAM display after a successful inactivation of a logical unit.

```
VARY NET,INACT,ID=BT10111
IST105I BT10111 NODE NOW INACTIVE
```

Figure 21.11 VTAM output from inactivating a logical unit.

```
VM        {VARY|V}   NET,DRDS

prefix

                     ,ID=dynamic reconfiguration member name
```

Figure 21.12 Format of the VARY DRDS command.

21.1.7 VARY DRDS command

The VARY DRDS command is used to dynamically add, change, and move resources without the need to perform NCP generations, loading of communications controllers, or recycling VTAM. The format of the command is found in Figure 21.12. The ID value is the name of a member in the dynamic reconfiguration data set (DRDS), which is normally VTAMLST.

When used for NCP changes, the command will allow the following without regenerating the NCP and loading the communications controller:

- Adding a nonswitched PU and its LUs to a defined line
- Adding an LU to a defined nonswitched PU
- Moving a PU and its LUs to another line in the same NCP
- Moving an LU from one PU to another PU in the same NCP
- Changing the station address of nonswitched PUs
- Changing the VTAM-only parameters of PUs and LUs
- Deleting a nonswitched PU and its LUs
- Deleting an LU from a nonswitched PU

For channel-attached devices, only LUs can be dynamically deleted and added using the VARY DRDS command. The resources affected by the command must be in a certain state for the command to successfully execute. The major node for the resource must be active. If the DRDS affects a LINE definition, then the status of the line can be either active or inactive. For the addition or deletion of an LU, the PU can be active or inactive. When a PU is being added, deleted, or

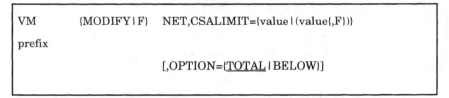

```
VM        {MODIFY I F}   NET,CSALIMIT={value I (value{,F})}

prefix

                        [,OPTION={TOTAL I BELOW}]
```

Figure 21.13   Format of the MODIFY CSALIMIT command.

moved, the PU must have an inactive status. For all instances of adding, moving, or deleting, the LUs must have an inactive status.

21.1.8   The MODIFY command

Of all the MODIFY commands available, the MODIFY CSALIMIT command (Figure 21.13) could make your day. If the values selected for CSALIMIT during VTAM initialization prove to be insufficient for production, VTAM will let you know. In fact, no new requests will be handled by VTAM until the resource shortage has been recovered.

The MODIFY CSALIMIT command allows you to alter the address space available for VTAM's region in the host computer. You can increase or decrease the storage available to VTAM. Depending on the operating system, the value can go as high as 2 GB. The value is specified on the CSALIMIT operand in kilobytes. The OPTION operand determines whether the value specified is for the CSA24 start option (OPTION=BELOW) or for the CSALIMIT start option (OPTION=TOTAL).

The MODIFY TNSTAT command (Figure 21.14) can be used to turn off tuning statistics that were started at initialization time. The TNSTAT start option must have been specified in the start options list in order for this MODIFY command to take effect. If NO is specified, the tuning statistics are written to the MVS SMF data set, the VSE Trace file, and the VM "FILE TUNSTATS A" CMS file, or an alternative file with the DDname TUNSTATS defined with the GCS FILEDEF command. Specifying YES will cause the tuning statistics

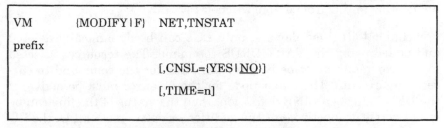

```
VM        {MODIFY I F}   NET,TNSTAT

prefix

                        [,CNSL={YES I NO}]

                        [,TIME=n]
```

Figure 21.14   Format of the MODIFY TNSTAT command.

to also write to the system console. The TIME operand specifies the number of minutes between tuning statistics recording events.

21.2 SUMMARY

In this chapter we discussed various VTAM commands to display, modify, and change resources and VTAM parameters. We reviewed the usage of the VTAM DISPLAY command and how it can be used to determine the status of network resources. The VARY command was discussed, some of the more commonly used VARY commands for managing the SNA network were illustrated. The MODIFY command was reviewed, and we discussed its ability to modify VTAM start options. There are several more variations of these commands. For further details on these, consult the IBM VTAM operators manual. The following chapter will discuss a network management architecture that augments VTAM's abilities to manage an SNA network.

22

Network Management

The importance of managing an SNA network became evident after IBM introduced the concept of multidomain networks in the late 1970s with the Multi-System Networking Facility (MSNF). MSNF provided VTAM hosts with the ability to communicate and share resources between them. SNA networks have since migrated from single-domain to multiple-domain, and in 1984 to multiple network configurations with SNA Network Interconnection (SNI). The complexity of managing these networks has increased even further as the management of the network facility equipment has been included. A comprehensive network management system is needed to manage and control all of the variables that make up today's complex networks.

NetView is IBM's strategic tool for managing these highly complex SNA networks. NetView resides under SNA's Open Network Management (ONM) architecture as the cornerstone of a full, comprehensive network management system that incorporates five major network management functions:

1. Configuration management

2. Problem management

3. Performance management

4. Accounting and availability management

5. Distribution management

To support a multivendor environment, non-IBM equipment and non-SNA resources must be included for true end-to-end centralized network management. Before we describe the functions and facilities of NetView, we must introduce you to the architecture and message flows utilized by them.

22.1 Open Network Management Architecture

Open Network Management (ONM), through published network management architectures, allows users and vendors to incorporate non-IBM and non-SNA resource management under SNA. An Application Program Interface (API) is provided under ONM that allows users and vendors to access network management data and commands. This facilitates the notion of centralized network management which includes both voice and data. Finally, network management products can interpret the architecture and utilize the APIs which are supported. These network management products have three distinct roles, which are the focal point, entry point, and service point.

22.1.1 Focal point

A focal point provides central network management for the domain. It resides in the host and can be a product or a set of products that supplies comprehensive support for managing the network. The network operator, along with the focal point, determines the actions necessary for managing the network. In Figure 22.1, we see that a focal point is

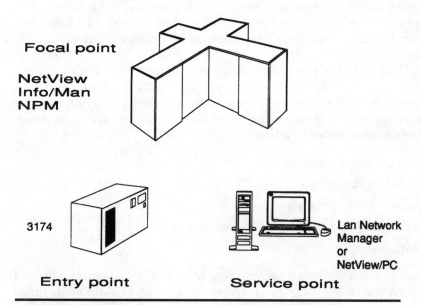

Figure 22.1 Focal point, Entry point, and Service point device types and applications support of each.

residing on a host computer that is running VTAM. The figure indicates that NetView is the focal point for this network. However, any CNM application that has the AUTH=CNM operand specified on the APPL definition statement possesses the ability to be a focal point. More than one focal-point application can reside on a host. Examples of other focal points are Sterling Software's Net/Master, IBM's NetView Performance Monitor (NPM), and IBM Information Management (Info/Man). Each of these focal-point applications can coincide with NetView or operate without NetView.

22.1.2   Entry point

An entry point transports network management data and session data to a host over the same link. An SNA PU is an entry point. The PU performs the functions of network management as well as those functions concerned with transporting session data for its peripheral resources. The entry point is in the same domain as the focal point. This is because the focal point works in conjunction with VTAM, and VTAM owns any entry point that it activates. The entry point supports the SNA management formats and protocols to the focal point. Examples of an entry point are IBM 3274 and 3174 cluster controllers.

22.1.3   Service point

The service point rounds out the ONM architecture roles by providing SNA network management for non-SNA products. Non-SNA products do not have SNA addressability and do not implement the SNA network management services formats and protocols. The service point converts native vendor protocols to SNA formats and then transmits them to the focal point. Like the entry point, the service point must be in the same domain as the focal point. The service point communicates with the focal point on an SSCP-PU session. NetView/PC, Lan Network Manager, and applications that can execute with NetView/PC are examples of a service point.

22.2   SNA Network Services Flow

Prior to ONM, information pertaining to non-SNA resources (e.g., modems, multiplexers, matrix switches) was not consolidated by the SNA CNM application. A different management system outside of the main processor of the computer center handled the management of non-SNA resources. ONM now eliminates the need for two separate management systems to ascertain and diagnose network problems. The consolidation of network fault messages under one single application can greatly decrease the time needed to resolve network problems. However, ONM does not eliminate the need for the management system that passes the network fault to NetView/PC. But it does impose a requirement on the non-SNA resource vendor: to write a NetView/PC application that can translate non-SNA alerts to the ONM NMVT format and to allow NetView commands to initiate and request information from the non-SNA resource vendor management system.

Take a look at the diagram in Figure 22.2. This figure shows a typical ONM flow using SNA management services. In this diagram, a focal point CNM application resides in an MVS/XA operating system. The NetView application named NPDA, the hardware facility, receives all unsolicited and solicited NMVT alerts from the network.

Notice that we have two forms of Network Management Service (NS) Request Units (RU) that can flow to the focal point. Prior to NMVTs, an NS RU named Record Formatted Maintenance Statistics (RECFMS) was used for unsolicited alerts. In fact, this format is still used by many devices for both unsolicited and solicited RUs. RECFMS is sent to the focal point as a solicited reply in response to the Request Maintenance Statistic (REQMS) NS RU.

Once the NetView/PC application has received and translated the proprietary protocol of the vendor's alarms into an NMVT, the service-point and entry-point flows are the same. Each transmits the

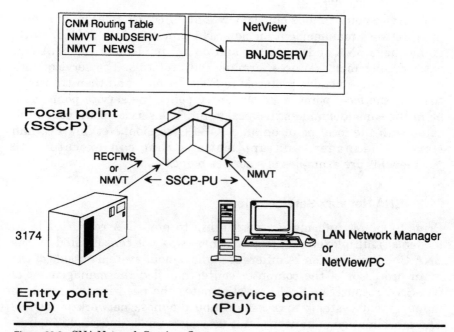

Figure 22.2 SNA Network Services flows.

alarm to the focal point via the SSCP-PU session. The SSCP receives the NS RU and must determine the recipient CNM application for this NS RU. It does this by scanning the CNM routing table and comparing the NS RU header received from the network with values defined in the CNM routing table. The SSCP can then deliver the NS RU to the associated CNM application.

In our example, the CNM application is NPDA, which has an ACB named BNJDSERV. The BNJDSERV ACB is defined to receive both RECFMS and NMVTs. After determining the receiving application, the SSCP delivers the NS RU to the named ACB, which will then process the alarms. Notice that we have also defined an ACB named NEWS for receiving NMVTs. A stipulation for SSCP routing of NS RUs is that only one CNM application can receive an NS RU. In this case, the ACB for NPDA (BNJDSERV) was opened and participating in an SSCP-LU session with VTAM before the NEWS ACB was opened. This can be accomplished by using the NetView Command Facility operator command STOP TASK=BNJDSERV to close the BNJDSERV ACB and using the VARY ACTivate command to open the NEWS ACB. The NEWS CNM application is not a task of NetView, but rather an independent CNM application. Therefore, the NetView command START TASK=NEWS will not activate the NEWS ACB. Once the NEWS ACB is successfully opened by VTAM, it will process all NMVT NS RUs that are received by the SSCP. However, the RECFMS NS RUs are lost because the ACB for BNJDSERV was closed to allow NEWS to receive NMVTs. We can overcome this by issuing the NetView command START TASK=BNJDSERV to open the BNJDSERV ACB once again. In this case, BNJDSERV will not receive the NMVTs, since NEWS is already receiving them, but it will receive the RECFMS NS RUs.

22.3   NetView Release 1 Overview

In May 1986, IBM launched its long-range plan for centralized network management. At the core of the plan is NetView. In this initial release, NetView was a conglomeration of previously independent Communications Network Management (CNM) program products. This repackaging of CNM program products allowed IBM to deliver a comprehensive network management package at a reasonable price. There are five main functions provided with NetView R1: Network Command Control Facility (NCCF), which was released in 1979 as a program product along with the Network Problem Determination Application (NPDA); Network Logical Data Manager (NLDM), which was released in 1984; and the VTAM Node Control Application (VNCA) and Network Management Productivity Facility (NMPF), which were both originally offered as field-developed programs (FDP).

All five applications were released as supported CNM program products in 1986.

22.3.1 Network Command Control Facility (NCCF)

At the heart of the NetView CNM programs is the NCCF, now known as the NetView Command Facility (Figure 22.3). This program encompasses the role of the VTAM Primary Program Operator (PPO). The PPO is allowed to issue VTAM operator commands and receive solicited and unsolicited VTAM operator messages. These messages are not the same as solicited and unsolicited NS RUs. The operator messages are of the VTAM IST type found in the *VTAM Messages and Codes* manual. The Command Facility provides points of entry into the CNM interface so that end users can capture and modify network management data. These customizable points of entry are known as exits. The exit routines must be coded in IBM assembler language. The facility also provides an interpretive language called Command LIST (CLIST). The CLIST language provides a rudimenta-

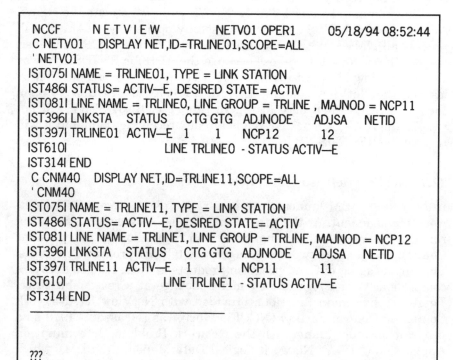

Figure 22.3   Sample display of the NetView Command Services screen.

ry means of simplifying and automating the network operator's responsibilities.

22.3.2 NetView Hardware Monitor

The NetView Hardware Monitor [previously known as Network Problem Determination Application (NPDA)] receives SNA network management services data that concerns hardware faults for resources in an SNA network. In addition to SNA resources, the Hardware Monitor can receive link-level diagnostic data from modems that support IBM's Link Problem Determination Aid (LPDA) facility. This capability marks IBM's entrance into managing network facilities as well as SNA resources. The Hardware Monitor application notifies a network operator of resource outages and their probable cause, and recommends actions to rectify the problem (Figures 22.4 and 22.5). The Hardware Monitor receives solicited and unsolicited network management services data from SNA resources (e.g.,

```
N E T V I E W          SESSION DOMAIN:  NETV01      OPER1              05/18/94 09:16:02

NPDA-31A                    * ALERTS-HISTORY *                        PAGE  1 OF  1


SEL# DOMAIN RESNAME TYPE TIME  ALERT DESCRIPTION:PROBABLE CAUSE
( 1)  NETV01  LANMGR01  LAN  09:08 COMMUNICATIONS OVERRUN:TOKEN-RING ADPT INTF
( 2)  NETV01  LANMGR01  LAN  09:08 COMMUNICATIONS OVERRUN:TOKEN-RING ADPT INTF
( 3)  NETV01  LANMGR02  LAN  09:08 COMMUNICATIONS OVERRUN:TOKEN-RING ADPT INTF
( 4)  NETV01  LANMGR02  LAN  09:08 COMMUNICATIONS OVERRUN:TOKEN-RING ADPT INTF


















ENTER SEL# (ACTION),OR SEL# PLUS M (MOST RECENT), P (PROBLEM), DEL (DELETE)

???
CMD==>
```

Figure 22.4 Sample display of the NetView Hardware Monitor Alerts Dynamic screen.

```
N E T V I E W        SESSION DOMAIN: NETV01  OPER1       05/18/94 09:27:26

NPDA-45A        * RECOMMENDED ACTION FOR SELECTED EVENT *    PAGE 1 OF 1

NETV01    UA6R1NH    LANMGR    LANMGR01


DOMAIN    | SP  | --- | TP  | --  (LAN )



USER   CAUSED - NONE


INSTALL CAUSED - NONE


FAILURE CAUSED - COMMUNICATIONS PROGRAM

                    TOKEN-RING ADAPTER

   ACTIONS - I120 - REVIEW LINK DETAILED DATA

            I132 - CONTACT TOKEN-RING ADMINISTRATOR RESPONSIBLE FOR THIS
                LAN




ENTER ST (MOST RECENT STATISTICS), DM (DETAIL MENU), OR D (EVENT DETAIL)

???

CMD==>
```

Figure 22.5 Sample display of the NetView Hardware Monitor Recommended Action screen.

PUs and LUs). This data makes up a formatted Request Unit (RU) that contains code points and is known as the Network Management Vector Transport (NMVT) or Record Formatted Maintenance Statistics (RECFMS). The code points are used to display predefined alert display messages and accompanying recommended actions that reside in files on the host processor's peripheral storage devices. These files can be customized by end users to suit their network management needs. The resulting alert errors are logged to the Hardware Monitor alert database and to an external logging file, such as IBM's System Management Facility (SMF), for further processing and analysis at a later time.

22.3.3 NetView Session Monitor

To assist in troubleshooting SNA session errors, NetView has incorporated the Network Logical Data Manager (NLDM) program product under the guise of the NetView Session Monitor. Depending on the NLDM initialization parameters, all SNA sessions, including SSCP-SSCP, SSCP-PU, SSCP-LU, and LU-LU sessions, may be traced to gather session information. This session information includes session partners, explicit and virtual routes, and specific error and reason codes for session failures. This facility also provides the capability of tracing SNA Path Information Units (PIUs) that travel between the session partners. This data is logged and recalled by the operator for further in-depth analysis of the session error. The Session Monitor is also used in conjunction with the Response Time Monitor (RTM). RTM provides solicited and unsolicited response time reporting to the Session Monitor. Again, this data is logged and can be used by the operator for further analysis.

22.3.4   NetView Status Monitor (STATMON)

STATMON provides the network operator with a "quick glance at the network" for the status of network resources. STATMON is made up of functions from the VTAM Node Control Application (VNCA). This application uses a hierarchical display of the SNA resources in the operator's network. If the status of a resource has changed, STAT-MON will update its display appropriately. NCCF command lists can be incorporated into STATMON's monitoring capability to automate the recovery procedure for resources that have become inactive. To enhance a network operator's control of the network, STATMON can be used with a light pen for issuing VTAM commands on SNA resources, simplifying the operator's interface with VTAM. No longer does the network operator have to memorize the format of every VTAM command. Data obtained from STATMON can be logged to an external logging file system (e.g., IBM's SMF) for further report processing on resource availability.

22.3.5   Communications network
management router function

In VTAM, the CNM routing table is used to direct the unsolicited network services alerts (NMVTs, RECFMS, RECMSs) to the unsolicited command processor responsible for handling unsolicited alerts. VTAM requires that only one unsolicited command processor can be active at one time for each network service request unit. With this facility, the CNM Router Task (DSICRTR) receives all network service alerts. This allows NetView to interpret the alert in finer detail. Using this facility, the various NetView functions can each receive the same network service RU. For instance, the NMVT has several different types of major vectors within it. CNM Router Task can scan the NMVT received from the network and then reroute the alert to the appropriate NetView task. For example, an Alert Major Vector (x'0000') within an NMVT is routed to the Hardware Monitor, and the Response Time Monitor (RTM) Major Vector (x'0080') is routed to the Session Monitor.

22.3.6   Generic alerts

Under NetView V1R2, the Hardware Monitor display panels and messages have been enhanced by the new generic alert format of the NMVT. The implementation of this new format eliminates the customization of NetView Hardware Monitor display panels and messages for product-specific alerts. Instead, the alerts' dynamic message, event detail, and recommended action screens are built upon the text identified by the generic alert code points. The code points are used as indices for tables that reside at the host on disk. This provides individualized alert text for each type of alert received, as opposed to NetView R1 predefined alert messages and displays, which were indexed by a component identifier and an alert descriptor code.

In keeping with IBM's ONM architecture, NetView R2 provides a facility for vendors to add their own generic alert code points. A code-point text table can be created to provide product-specific code-point text that can be used with IBM's code-point text to create the Hardware Monitor alert messages and screen displays. Prior to NetView V1R2, only IBM's text or the product-specific text could be displayed. Using generic alerts, product-specific and IBM code-point text can be combined on the same message and/or screen display.

NetView will, however, continue to support nongeneric alert formats that were supported under NetView R1 and the individual CNM program products.

22.3.7   Service Point Command Service (SPCS)

Management of non-IBM equipment by NetView is now possible through a newly added function of the NetView Command Facility. This new function is called Service Point Command Service (SPCS). Using the command facility, a NetView operator can enter four supported commands that are to be executed by an application program of a service point. The supported commands are

RUNCMD

LINKDATA

LINKTEST

LINKPD

RUNCMD command. RUNCMD is a generic command that is recognized by the NetView Command Facility and transported to the service point for processing. For example, RUNCMD can be used by the network operator to obtain the status of the vendor's equipment. Command syntax checking is not performed by NetView and must be in agreement with the vendor's application executing on NetView/PC. The vendor's application must be coded to handle command errors, as well as to execute the requested command. The response from the service point is displayed under the command facility function of NetView.

LINKDATA command. This command is used to request control, error, and statistical data maintained by the service-point application. The service-point application responds to this command with an architected record. The information supplied in this record is displayed by NetView in full-screen mode.

LINKTEST command. Using this command from NetView R2, an application executing on the service point will request DCE tests. The results of the tests will then be sent to NetView in an architected record for presentation by NetView in full-screen mode.

LINKPD command. Problem determination analysis can be performed by the service-point application for a given element. Using the LINKPD command, the network operator can request the service-point application to perform alert correlation for a specific problem. The response to this command is a generic alert. The generic alert is then processed by NetView R2 in the same manner as solicited and unsolicited NMVTs. In this case, the NMVT is solicited.

22.4   NetView V2R3

Since NetView's introduction, the product has gone through some major enhancements. NetView V2R3 is the first release of NetView to support IBM's SystemView architectural blueprint for managing network resources and adhering to application and database standards for application presentation and database format and storage.

22.4.1   SystemView overview

ONM opened the doors to consolidating and integrating systems and network management data. In doing so, different vendors that adhered to ONM's focal-point, entry-point, and service-point functions implemented them using different user interfaces, data definitions, communications, and application services. This led to further confusion for operations personnel and longer training periods for new staff. A strategy was needed for defining all the various disciplines of systems and network management (i.e., information systems management). Enter SystemView from IBM.

SystemView was announced September 5, 1990, and like its counterpart for application standardization, Systems Application Architecture (SAA), SystemView identifies a clear structure for defining the standards for information systems management. SystemView is a strategy for managing information systems while providing a business solution that follows the standards set by SAA and Open Systems Interconnection (OSI). These include a consistent user interface, a common communications interface, and a standardized definition of resources and data.

The SystemView structure addresses a set of guidelines, standards, and interfaces that will create a seamless view of the integration of information management systems across the enterprise. These are

- End-use dimension
- Application dimension
- Data dimension

Through the implementation of these three dimensions, it is hoped that SystemView will provide a coherent information systems management solution across heterogeneous systems.

End-use dimension.   The end-use dimension outlines the guidelines and standards for presentation of a SystemView application to an end user. These guidelines call for the presentation to be either graphic, textual, or command language. Once the end user selects his or her interface, no matter which SystemView application the end user interfaces with, that is the presentation shown. The end user could be an operator, a system administrator, a business analyst, or a systems programmer. The end user can switch from one SystemView application to another and not be aware of the application switch. This is accomplished by utilizing the common user access interface defined in SAA. The initial conformance to this is the announcement of NetView Version 2 Release 1 and the NetView Graphic Monitor Facility (GMF). These two offerings work in conjunction to provide the beginnings of a graphical user interface adhering to the SystemView end-use dimension. The information systems enterprise network is depicted graphically on an OS/2-based personal computer. Each graphical object depicted is a "managed object." These display objects will have a defined appearance and characteristics. The appearance and characteristics of each display object follow them between SystemView applications. This reduces possible end user errors resulting from misinterpretation of the presentation. The display objects correspond to data objects defined in the SystemView data dimension.

The initial offering of NetView GMF is used for problem management by notifying the end user of resource status changes graphically. In the future, the graphical interface will be used for configuration management. Modifications to the configuration of the network will be done with a simple "point and click." For instance, the network administrator will add, delete, move, or modify network resources using the graphical display objects. Coding of VTAM and NCP definition statements for SNA resources will be a thing of the past. This method of system definition requires a comprehensive set of disciplines. These are defined by the application dimension.

Application dimension. Management applications used today are inherently dedicated to a specific function, system, or resource. This leaves information systems managers with the task of correlating the data and events recorded by various information management systems. The objective of the application dimension is to provide a comprehensive set of management applications and tools that will not only facilitate the integration of different management data but also automate various information systems management tasks.

It has become quite evident that the main purpose of MIS is to provide a product to the business end of a company. That product is information. The information is provided by means of a service. That service is information systems. The success of the MIS department in providing the product and service in a timely and consistent manner is achieved through well-orchestrated information systems management disciplines.

Under the application dimension there are six disciplines, as shown in Figure 22.6.

Business management offers inventory management, registration, financial administration, business planning, and management services for computer-related facilities that can affect the business (e.g., environmental management). Under this discipline, purchases, leases, and maintenance contracts for information systems hardware and software can be kept under tight control. As a business unit expands, plans for procurement of office space, furniture, voice and data facilities, and equipment can be utilized by change management.

Change management can be viewed as the key SystemView application. Nothing affects the stability of an information system quite as much as an uncoordinated change. Under change management,

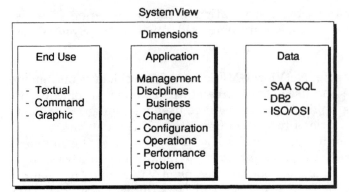

Figure 22.6   The three dimensions of IBM's System View architecture.

information from the business management application can automatically be introduced into the planning and scheduling of enterprisewide changes. Conflicts that may or will cause problems if the changes are implemented as planned can be flagged. This feature greatly enhances the MIS goal of providing product and service in a timely and consistent manner.

Configuration management can use information from business and change management to define the physical and logical connections and dependencies of enterprisewide information system resources. Configuration templates provided by the data dimension can be used to generate configuration standard definitions for these resources.

Operations management is responsible for managing the use of the information system's resources and supporting the processing workloads. Through information from the previous management applications, the operations management application can dynamically create automated tasks for the new configuration implemented by the configuration management application. Automation of operations will play a major role in SystemView, changing the status of operations personnel. With the automation of repetitive and mundane operational tasks, highly skilled operations personnel will be required to handle those instances where automation has not been or cannot be implemented. True, these highly skilled operations people will require a higher salary than many companies are used to paying; however, fewer people will be needed because of the enhanced use of automated techniques.

Performance management under the application dimension will still execute its normal functions of capacity planning and collection of performance data. However, under SystemView we will see the performance application make configuration changes and tuning suggestions through the change management application. This may be done directly or through the problem management application.

Problem management has the tasks of detecting, analyzing, correcting, and tracking incidents and problems reported by information systems across the enterprise. It is possible that change, configuration, and performance management applications can report change or configuration conflicts and performance thresholds exceeded, thus providing a type of checks-and-balances procedure.

The disciplines described here use generic definitions of resources and adhere to compatibility with open standards such as OSI. The cooperative processing between the disciplines is made possible by the sharing of commonly defined data. This data is defined in the data dimension.

Data dimension. The data dimension describes the long-awaited data repository. The database for the data dimension will initially use the SAA Structured Query Language (SQL) database interface. The data dimension defines the data models utilized by the end-use and application dimensions. This common repository will be consistent with OSI standards (ISO/IEC 10165-4, *Guidelines for the Definition of Managed Objects,* and ISO/IEC 10165-1, *Management Information Model*). The SystemView Data Model, as previously stated, will describe the characteristics of and relationships among enterprisewide resources. It is this standardization of resource characteristics and relationships that will facilitate the seamless switch between SystemView applications.

22.4.2 Resource Object Data Management (RODM)

The RODM function of NetView V2R3 automates the configuration discipline of SystemView. The configuration discipline is composed of several functional pieces of NetView and three other IBM offerings that work together with RODM to build the configuration database for SNA.

IBM controller product offerings contain Vital Product Data (VPD) information. The newer offerings send this VPD information to VTAM during activation, either through an SSCP-PU session using NMVT or through an LU6.2 session with NetView. Figure 22.7 depicts the flow of this VPD data and its usage.

VPD flows up to the mainframe, where VTAM sends the data to NetView's RODM. RODM translates the VPD data format into OSI standard format for object definitions and enters the data into a data

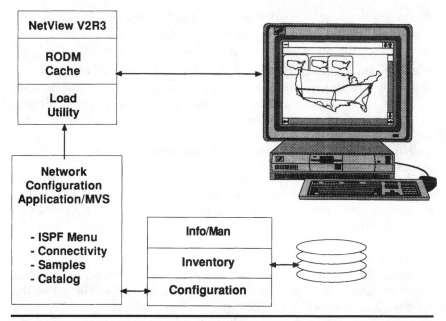

Figure 22.7 Flow of Vital Product Data (VPD); the RODM function of NetView V2R3.

cache in memory. This is done through RODM's Load Utility. Information on how the resource is to be defined is provided through the Network Configuration Application/MVS (NCA/MVS) offering. This offering allows you to define templates for defining objects and their resources. IBM's Information/Manager (Info/Man) is used to create an object-oriented relational database which is kept on disk and in storage. SNA/SDLC resources provide VPD data, including connectivity data, to VTAM. Token-ring LAN resources can provide VPD data by using IBM's LAN Station Manager V1R1.

22.4.3 OS/2 Graphic Monitor Facility

IBM's NetView Graphic Monitor Facility graphically represents a network, a portion of a network, or a group of networks in various levels of detail. It offers interactive views (pictures) of the SNA resources, both the nodes and the links between the nodes, that are being monitored. Large portions of complex communications networks can be monitored, providing information similar to that provided by the NetView status monitor. The difference lies in the graphical representation of the status of resources across several domains, the connectivity, and the history of resource status updates offered by the NetView Graphic Monitor Facility. The overall status as well as the status of individual resources can be monitored through the various

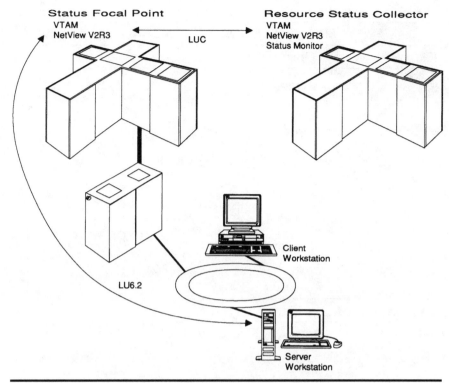

Figure 22.8   NetView Graphic Monitor Facility flows and client/server configuration.

views. The NetView Graphic Monitor Facility supports a client-server configuration such as that depicted in Figure 22.8.

The NetView Graphic Monitor Facility presents the networks using two types of views: hierarchical and nonhierarchical. Hierarchical views, which allow you to view the network in detail, come in four flavors:

1. *Cluster.*   This is the highest view level, showing a clustered view of aggregate resources. Figure 22.9 shows an example of a cluster view.

2. *Backbone.*   This provides different views of the host and the communications controllers which make up the network backbone. It shows PU Types 4 and 5, transmission groups, links, subarea links, and gateways.

3. *Composite.*   This level displays grouped peripheral nodes when they are too numerous to monitor in a single peripheral view or when there are several categories of peripheral resources, such as switched, LAN, or general.

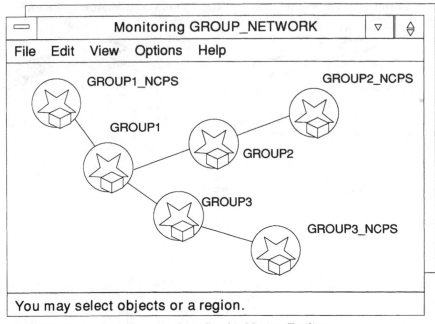

Figure 22.9   A sample cluster view from Graphic Monitor Facility.

4. *Peripheral.*   This is the lowest level of views containing real resources. It presents separate views for each category of peripheral resources.

Nonhierarchical views are selected by name. There are three types:

1. *Retained.*   This view displays important real resources in different portions of a network.

2. *Snapshot.*   This is a view created by the user to display the network at a given time.

3. *Background.*   This is a view containing only background lines, i.e., a map.

22.5 Summary

In this chapter IBM's NetView and SystemView were reviewed. Their capabilities and their roles in IBM's network management strategy were discussed. NetView serves as a focal point, and LAN Network Manager plays the role of a service point. Together they can assist you in managing an enterprisewide network.

SystemView is not a product. It is not something tangible that can be touched. It is an information systems management strategy that utilizes SAA's CUA and SAA's communications interface guidelines for LU6.2. Not only is SystemView using IBM standards, it is also complying with and supporting the ISO/IEC 9595 architected interface Common Management of Information Services (CMIS) and the ISO/IEC 9596 protocol Common Management Information Protocol (CMIP) for the exchange of management information with OSI networks. By using consistent user interfaces, a common repository definition of data, true integration, and automation, SystemView can and will provide management functions across SNA and non-SNA networks.

The following table is the IBM-supplied default logon mode table for VTAM V3R2. You may use the entries in the table for your denfinitions. If you elect to create your own logon mode table, be sure to link-edit your table under a name other than ISTINCLM.

```
ISTINCLM MODETAB
IBM3767  MODEENT LOGMODE=INTERACT,FMPROF=X'03',TSPROF=X'03',PRIPROT=X'B*
              1',SECPROT=X'A0',COMPROT=X'3040'

         TITLE 'TWXDEVPT'
***********************************************************************
*                                                                     *
*             TWX DEVICE WITH THE DCODE SET TO KEYBOARD               *
*             AND PRINTER.  THIS IS THE DEFAULT SETTING.              *
*                                                                     *
***********************************************************************
TWXDEVPT MODEENT LOGMODE=TWXDECPT,FMPROF=X'03',TSPROF=X'03',PRIPROT=X'B*
              1',SECPROT=X'A0',COMPROT=X'3040',DCODE=X'00'

         TITLE 'TWXDEVDP'
***********************************************************************
*                                                                     *
*             TWX DEVICE WITH THE DCODE SET TO KEYBOARD               *
*             AND DISPLAY.                                            *
*                                                                     *
***********************************************************************
TWXDEVDP MODEENT LOGMODE=TWXDEVDP,FMPROF=X'03',TSPROF=X'03',PRIPROT=X'B*
              1',SECPROT=X'A0',COMPROT=X'3040',DCODE=X'80'
```

```
IBM3770   MODEENT LOGMODE=BATCH,FMPROF=X'03',TSPROF=X'03',PRIPROT=X'A3',*
                SECPROT=X'A3',COMPROT=X'7080'


IBMS3270  MODEENT LOGMODE=S3270,FMPROF=X'02',TSPROF=X'02',PRIPROT=X'71',*
                SECPROT=X'40',COMPROT=X'2000'


IBM3600   MODEENT LOGMODE=IBM3600,FMPROF=X'04',TSPROF=X'04',PRIPROT=X'F1*
                ',SECPROT=X'F1',COMPROT=X'7000'


IBM3650I  MODEENT LOGMODE=INTRACT,FMPROF=X'04',TSPROF=X'04',PRIPROT=X'B1*
                ',SECPROT=X'90',COMPROT=X'6000'


IBM3650U  MODEENT LOGMODE=INTRUSER,FMPROF=X'04',TSPROF=X'04',PRIPROT=X'3*
                1',SECPROT=X'30',COMPROT=X'6000'


IBMS3650  MODEENT LOGMODE=IBMS3650,FMPROF=X'04',TSPROF=X'04',PRIPROT=X'B*
                0',SECPROT=X'B0',COMPROT=X'4000'


IBM3650P  MODEENT LOGMODE=PIPELINE,FMPROF=X'04',TSPROF=X'04',PRIPROT=X'3*
                0',SECPROT=X'10',COMPROT=X'0000'


IBM3660   MODEENT LOGMODE=SMAPPL,FMPROF=X'03',TSPROF=X'03',PRIPROT=X'A0'*
                ,SECPROT=X'A0',COMPROT=X'0081'


IBM3660A  MODEENT LOGMODE=SMSNA100,FMPROF=X'00',TSPROF=X'00',PRIPROT=X'0*
                0',SECPROT=X'00',COMPROT=X'0000'
          TITLE 'D6327801'
*******************************************************************
*                                                                 *
*              3276 SNA WITH 3278 MODEL 1 SCREEN                   *
*              PRIMARY SCREEN 12 X 40 (480)                        *
*              ALTERNATE SCREEN 12 X 80 (960)                      *
*                                                                 *
*******************************************************************
D6327801  MODEENT LOGMODE=D6327801,FMPROF=X'03',TSPROF=X'03',PRIPROT=X'B*
                1',SECPROT=X'90',COMPROT=X'3080',RUSIZES=X'88F8',PSERVIC*
                =X'0200000000000C280C507F00'
          TITLE 'D6327802'
*******************************************************************
*                                                                 *
*              3276 SNA WITH 3278 MODEL 2 SCREEN                   *
*              PRIMARY SCREEN 24 X 80 (1920)                       *
*              NO ALTERNATE SCREEN DEFINED                         *
*                                                                 *
*******************************************************************
D6327802  MODEENT LOGMODE=D6327802,FMPROF=X'03',TSPROF=X'03',PRIPROT=X'B*
                1',SECPROT=X'90',COMPROT=X'3080',RUSIZES=X'88F8',PSERVIC*
                =X'0200000000000185000007E00'
          TITLE 'D6327803'
*******************************************************************
*                                                                 *
```

```
*              3276 SNA WITH       MODEL 3 SCREEN               *
*              PRIMARY SCREEN 24 X 80 (1920)                    *
*              ALTERNATE SCREEN 32 X 80 (2560)                  *
*                                                               *
*****************************************************************
D6327803 MODEENT LOGMODE=D6327803,FMPROF=X'03',TSPROF=X'03',PRIPROT=X'B*
             1',SECPROT=X'90',COMPROT=X'3080',RUSIZES=X'88F8',PSERVIC*
             =X'020000000000185020507F00'
         TITLE 'D6327804'
*****************************************************************
*                                                               *
*              3276 SNA WITH       MODEL 4 SCREEN               *
*              PRIMARY SCREEN 24 X 80 (1920)                    *
*              ALTERNATE SCREEN 43 X 80 (3440)                  *
*                                                               *
*****************************************************************
D6327804 MODEENT LOGMODE=D6327804,FMPROF=X'03',TSPROF=X'03',PRIPROT=X'B*
             1',SECPROT=X'90',COMPROT=X'3080',RUSIZES=X'88F8',PSERVIC*
             =X'02000000000018502B507F00'
         TITLE 'D6327805'
*****************************************************************
*                                                               *
*              3276 SNA WITH       MODEL 5 SCREEN               *
*              PRIMARY SCREEN 24 X 80 (1920)                    *
*              ALTERNATE SCREEN 27 X 132 (3564)                 *
*                                                               *
*****************************************************************
D6327805 MODEENT LOGMODE=D6327805,FMPROF=X'03',TSPROF=X'03',PRIPROT=X'B*
             1',SECPROT=X'90',COMPROT=X'3080',RUSIZES=X'88F8',PSERVIC*
             =X'02000000000018501B847F00'
         TITLE 'D6328904'
*****************************************************************
*                                                               *
*              3276 SNA WITH 3289 MODEL 4 PRINTER              *
*                                                               *
*****************************************************************
D6328904 MODEENT LOGMODE=D6328904,FMPROF=X'03',TSPROF=X'03',PRIPROT=X'B*
             1',SECPROT=X'90',COMPROT=X'3080',RUSIZES=X'8787',PSERVIC*
             =X'03000000000018502B507F00'
         TITLE 'D6328902'
*****************************************************************
*                                                               *
*              3276 SNA WITH 3289 MODEL 2 PRINTER              *
*                                                               *
*****************************************************************
D6328902 MODEENT LOGMODE=D6328902,FMPROF=X'03',TSPROF=X'03',PRIPROT=X'B*
             1',SECPROT=X'90',COMPROT=X'3080',RUSIZES=X'8787',PSERVIC*
             =X'030000000000185018507F00'
         TITLE 'D4A32781'
*****************************************************************
*                                                               *
```

```
*                  3274 MODEL 1A WITH MODEL 1 SCREEN (LOCAL SNA)        *
*                  PRIMARY SCREEN 12 X 40 (480)                         *
*                  ALTERNATE SCREEN 12 X 80 (960)                       *
*                                                                       *
************************************************************************
D4A32781 MODEENT LOGMODE=D4A32781,FMPROF=X'03',TSPROF=X'03',PRIPROT=X'B*
               1',SECPROT=X'90',COMPROT=X'3080',RUSIZES=X'87C7',PSERVIC*
               =X'0200000000000C280C507F00'
         TITLE 'D4A32782'
************************************************************************
*                                                                       *
*                  3274 MODEL 1A WITH MODEL 2 SCREEN (LOCAL SNA)        *
*                  PRIMARY SCREEN 24 X 80 (1920)                        *
*                  NO ALTERNATE SCREEN DEFINED                          *
*                                                                       *
************************************************************************
D4A32782 MODEENT LOGMODE=D4A32782,FMPROF=X'03',TSPROF=X'03',PRIPROT=X'B*
               1',SECPROT=X'90',COMPROT=X'3080',RUSIZES=X'87C7',PSERVIC*
               =X'020000000000185000007E00'
         TITLE 'LSK32782'
************************************************************************
*                                                                       *
*                  3274 MODEL 1A WITH MODEL 2 SCREEN (LOCAL SNA)        *
*                  PRIMARY SCREEN 24 X 80 (1920)                        *
*                  NO ALTERNATE SCREEN DEFINED                          *
*                  KATAKANA                                             *
*                                                                       *
************************************************************************
LSK32782 MODEENT LOGMODE=LSK32782,FMPROF=X'03',TSPROF=X'03',PRIPROT=X'B*
               1',SECPROT=X'90',COMPROT=X'3080',RUSIZES=X'87C7',PSERVIC*
               =X'020000000000185000007E00',LANG=X'11'
         TITLE 'D4A32783'
************************************************************************
*                                                                       *
*                  3274 MODEL 1A WITH MODEL 3 SCREEN (LOCAL SNA)        *
*                  PRIMARY SCREEN 24 X 80 (1920)                        *
*                  ALTERNATE SCREEN 32 X 80 (2560)                      *
*                                                                       *
************************************************************************
D4A32783 MODEENT LOGMODE=D4A32783,FMPROF=X'03',TSPROF=X'03',PRIPROT=X'B*
               1',SECPROT=X'90',COMPROT=X'3080',RUSIZES=X'87C7',PSERVIC*
               =X'020000000000185020507F00'
         TITLE 'D4A32784'
************************************************************************
*                                                                       *
*                  3274 MODEL 1A WITH MODEL 4 SCREEN (LOCAL SNA)        *
*                  PRIMARY SCREEN 24 X 80 (1920)                        *
*                  ALTERNATE SCREEN 43 X 80 (3440)                      *
*                                                                       *
************************************************************************
D4A32784 MODEENT LOGMODE=D4A32784,FMPROF=X'03',TSPROF=X'03',PRIPROT=X'B*
```

```
                     1',SECPROT=X'90',COMPROT=X'3080',RUSIZES=X'87C7',PSERVIC*
                     =X'02000000000018502B507F00'
         TITLE 'D4A32785'
*********************************************************************
*                                                                   *
*               3274 MODEL 1A WITH MODEL 5 SCREEN (LOCAL SNA)        *
*               PRIMARY SCREEN 24 X 80 (1920)                        *
*               ALTERNATE SCREEN 27 X 132 (3564)                     *
*                                                                   *
*********************************************************************
D4A32785 MODEENT LOGMODE=D4A32785,FMPROF=X'03',TSPROF=X'03',PRIPROT=X'B*
                     1',SECPROT=X'90',COMPROT=X'3080',RUSIZES=X'87C7',PSERVIC*
                     =X'02000000000018501B847F00'
         TITLE 'D4A32XX3'
*********************************************************************
*                                                                   *
*               3274 MODEL 1A (LOCAL SNA)                            *
*               PRIMARY SCREEN 24 X 80 (1920)                        *
*               ALTERNATE SCREEN TO BE DETERMINED BY APPLICATION     *
*                                                                   *
*********************************************************************
D4A32XX3 MODEENT LOGMODE=D4A32XX3,FMPROF=X'03',TSPROF=X'03',PRIPROT=X'B*
                     1',SECPROT=X'90',COMPROT=X'3080',RUSIZES=X'87C7',PSERVIC*
                     =X'02800000000000000000000300'
         TITLE 'D4A32771'
*********************************************************************
*                                                                   *
*               3274 MODEL 1A WITH 3277 MODEL 1 SCREEN              *
*                                                                   *
*********************************************************************
D4A32771 MODEENT LOGMODE=D4A32771,FMPROF=X'03',TSPROF=X'03',PRIPROT=X'B*
                     1',SECPROT=X'90',COMPROT=X'3080',RUSIZES=X'87C7',PSERVIC*
                     =X'02000000000000000000000100'
         TITLE 'D4A32772'
*********************************************************************
*                                                                   *
*               3274 MODEL 1A WITH 3277 MODEL 2 SCREEN              *
*                                                                   *
*********************************************************************
D4A32772 MODEENT LOGMODE=D4A32772,FMPROF=X'03',TSPROF=X'03',PRIPROT=X'B*
                     1',SECPROT=X'90',COMPROT=X'3080',RUSIZES=X'87C7',PSERVIC*
                     =X'02000000000000000000000200'
         TITLE 'D4C32781'
*********************************************************************
*                                                                   *
*               3274 MODEL 1C WITH MODEL 1 SCREEN(REMOTE SNA)       *
*               PRIMARY SCREEN 12 X 40 (480)                         *
*               ALTERNATE SCREEN 12 X 80 (960)                       *
*                                                                   *
*********************************************************************
D4C32781 MODEENT LOGMODE=D4C32781,FMPROF=X'03',TSPROF=X'03',PRIPROT=X'B*
```

```
                 1',SECPROT=X'90',COMPROT=X'3080',RUSIZES=X'87F8',PSERVIC*
                 =X'0200000000000C280C507F00'
         TITLE 'D4C32782'
***********************************************************************
*                                                                     *
*              3274 MODEL 1C WITH MODEL 2 SCREEN(REMOTE SNA)          *
*              PRIMARY SCREEN 24 X 80 (1920)                          *
*              NO ALTERNATE SCREEN DEFINED                            *
*                                                                     *
***********************************************************************
D4C32782 MODEENT LOGMODE=D4C32782,FMPROF=X'03',TSPROF=X'03',PRIPROT=X'B*
                 1',SECPROT=X'90',COMPROT=X'3080',RUSIZES=X'87F8',PSERVIC*
                 =X'020000000000185000007E00'
         TITLE 'RSK32782'
***********************************************************************
*                                                                     *
*              3274 MODEL 1C WITH MODEL 2 SCREEN(REMOTE SNA)          *
*              PRIMARY SCREEN 24 X 80 (1920)                          *
*              NO ALTERNATE SCREEN DEFINED                            *
*              KATAKANA                                                *
*                                                                     *
***********************************************************************
RSK32782 MODEENT LOGMODE=RSK32782,FMPROF=X'03',TSPROF=X'03',PRIPROT=X'B*
                 1',SECPROT=X'90',COMPROT=X'3080',RUSIZES=X'87F8',PSERVIC*
                 =X'020000000000185000007E00',LANG=X'11'
         TITLE 'D4C32783'
***********************************************************************
*                                                                     *
*              3274 MODEL 1C WITH MODEL 3 SCREEN(REMOTE SNA)          *
*              PRIMARY SCREEN 24 X 80 (1920)                          *
*              ALTERNATE SCREEN 32 X 80 (2560)                        *
*                                                                     *
***********************************************************************
D4C32783 MODEENT LOGMODE=D4C32783,FMPROF=X'03',TSPROF=X'03',PRIPROT=X'B*
                 1',SECPROT=X'90',COMPROT=X'3080',RUSIZES=X'87F8',PSERVIC*
                 =X'020000000000185020507F00'
         TITLE 'D4C32784'
***********************************************************************
*                                                                     *
*              3274 MODEL 1C WITH MODEL 4 SCREEN(REMOTE SNA)          *
*              PRIMARY SCREEN 24 X 80 (1920)                          *
*              ALTERNATE SCREEN 43 X 80 (3440)                        *
*                                                                     *
***********************************************************************
D4C32784 MODEENT LOGMODE=D4C32784,FMPROF=X'03',TSPROF=X'03',PRIPROT=X'B*
                 1',SECPROT=X'90',COMPROT=X'3080',RUSIZES=X'87F8',PSERVIC*
                 =X'02000000000018502B507F00'
         TITLE 'D4C32785'
***********************************************************************
*                                                                     *
*              PRIMARY SCREEN 24 X 80 (1920)                          *
```

```
*                 NO ALTERNATE SCREEN DEFINED                          *
*                                                                      *
**********************************************************************
D4C32782 MODEENT LOGMODE=D4C32782,FMPROF=X'03',TSPROF=X'03',PRIPROT=X'B*
              1',SECPROT=X'90',COMPROT=X'3080',RUSIZES=X'87F8',PSERVIC*
              =X'020000000000185000007E00'
         TITLE 'RSK32782'
**********************************************************************
*                                                                      *
*              3274 MODEL 1C WITH MODEL 2 SCREEN(REMOTE SNA)           *
*              PRIMARY SCREEN 24 X 80 (1920)                           *
*              NO ALTERNATE SCREEN DEFINED                             *
*              KATAKANA                                                 *
*                                                                      *
**********************************************************************
RSK32782 MODEENT LOGMODE=RSK32782,FMPROF=X'03',TSPROF=X'03',PRIPROT=X'B*
              1',SECPROT=X'90',COMPROT=X'3080',RUSIZES=X'87F8',PSERVIC*
              =X'020000000000185000007E00',LANG=X'11'
         TITLE 'D4C32783'
**********************************************************************
*                                                                      *
*              3274 MODEL 1C WITH MODEL 3 SCREEN(REMOTE SNA)           *
*              PRIMARY SCREEN 24 X 80 (1920)                           *
*              ALTERNATE SCREEN 32 X 80 (2560)                         *
*                                                                      *
**********************************************************************
D4C32783 MODEENT LOGMODE=D4C32783,FMPROF=X'03',TSPROF=X'03',PRIPROT=X'B*
              1',SECPROT=X'90',COMPROT=X'3080',RUSIZES=X'87F8',PSERVIC*
              =X'020000000000185020507F00'
         TITLE 'D4C32784'
**********************************************************************
*                                                                      *
*              3274 MODEL 1C WITH MODEL 4 SCREEN(REMOTE SNA)           *
*              PRIMARY SCREEN 24 X 80 (1920)                           *
*              ALTERNATE SCREEN 43 X 80 (3440)                         *
*                                                                      *
**********************************************************************
D4C32784 MODEENT LOGMODE=D4C32784,FMPROF=X'03',TSPROF=X'03',PRIPROT=X'B*
              1',SECPROT=X'90',COMPROT=X'3080',RUSIZES=X'87F8',PSERVIC*
              =X'02000000000018502B507F00'
         TITLE 'D4C32785'
**********************************************************************
*                                                                      *
*              3274 MODEL 1C WITH MODEL 5 SCREEN(REMOTE SNA)           *
*              PRIMARY SCREEN 24 X 80 (1920)                           *
*              ALTERNATE SCREEN 27 X 132 (3564)                        *
*                                                                      *
**********************************************************************
D4C32785 MODEENT LOGMODE=D4C32785,FMPROF=X'03',TSPROF=X'03',PRIPROT=X'B*
              1',SECPROT=X'90',COMPROT=X'3080',RUSIZES=X'87F8',PSERVIC*
              =X'02000000000018501B847F00'
```

```
          TITLE 'D4C32XX3'
***********************************************************************
*                                                                     *
*              3274 MODEL 1C (REMOTE SNA)                             *
*              PRIMARY SCREEN 24 X 80 (1920)                          *
*              ALTERNATE SCREEN TO BE DETERMINED BY APPLICATION       *
*                                                                     *
***********************************************************************
D4C32XX3 MODEENT LOGMODE=D4C32XX3,FMPROF=X'03',TSPROF=X'03',PRIPROT=X'B*
              1',SECPROT=X'90',COMPROT=X'3080',RUSIZES=X'87F8',PSERVIC*
              =X'028000000000000000000300'
          TITLE 'D4C32771'
***********************************************************************
*                                                                     *
*              3274 MODEL 1C WITH 3277 MODEL 1 SCREEN                 *
*                                                                     *
***********************************************************************
D4C32771 MODEENT LOGMODE=D4C32771,FMPROF=X'03',TSPROF=X'03',PRIPROT=X'B*
              1',SECPROT=X'90',COMPROT=X'3080',RUSIZES=X'87F8',PSERVIC*
              =X'020000000000000000000100'
          TITLE 'D4C32772'
***********************************************************************
*                                                                     *
*              3274 MODEL 1C WITH 3277 MODEL 2 SCREEN                 *
*                                                                     *
***********************************************************************
D4C32772 MODEENT LOGMODE=D4C32772,FMPROF=X'03',TSPROF=X'03',PRIPROT=X'B*
              1',SECPROT=X'90',COMPROT=X'3080',RUSIZES=X'87F8',PSERVIC*
              =X'020000000000000000000200'
          TITLE 'D4B32781'
***********************************************************************
*                                                                     *
*      3274 MODEL 1B/1D WITH MODEL 1 SCREEN (LOCAL NON-SNA)           *
*      3274 1C BSC WITH MODEL 1 SCREEN                                *
*      3276 BSC WITH MODEL 1 SCREEN                                   *
*      PRIMARY SCREEN 12 X 40 (480)                                  *
*      ALTERNATE SCREEN 12 X 80 (960)                                *
*                                                                     *
***********************************************************************
D4B32781 MODEENT LOGMODE=D4B32781,FMPROF=X'02',TSPROF=X'02',PRIPROT=X'7*
              1',SECPROT=X'40',COMPROT=X'2000',RUSIZES=X'0000',PSERVIC*
              =X'0000000000000C280C507F00'
          TITLE 'D4B32782'
***********************************************************************
*                                                                     *
*      3274 MODEL 1B/1D WITH MODEL 2 SCREEN (LOCAL NON-SNA)           *
*      3274 1C BSC WITH MODEL 2 SCREEN                                *
*      3276 BSC WITH MODEL 2 SCREEN                                   *
*      PRIMARY SCREEN 24 X 80 (1920)                                 *
*      NO ALTERNATE SCREEN DEFINED                                   *
*                                                                     *
***********************************************************************
```

```
**************************************************************
D4B32782 MODEENT LOGMODE=D4B32782,FMPROF=X'02',TSPROF=X'02',PRIPROT=X'7*
               1',SECPROT=X'40',COMPROT=X'2000',RUSIZES=X'0000',PSERVIC*
               =X'000000000000185000007E00'
         TITLE 'LNK32782'
**************************************************************
*                                                            *
*      3274 MODEL 1B/1D WITH MODEL 2 SCREEN (LOCAL NON-SNA)   *
*      3274 1C BSC WITH MODEL 2 SCREEN                        *
*      3276 BSC WITH MODEL 2 SCREEN                           *
*      PRIMARY SCREEN 24 X 80 (1920)                          *
*      NO ALTERNATE SCREEN DEFINED                            *
*      KATAKANA                                               *
*                                                            *
**************************************************************
LNK32782 MODEENT LOGMODE=LNK32782,FMPROF=X'02',TSPROF=X'02',PRIPROT=X'7*
               1',SECPROT=X'40',COMPROT=X'2000',RUSIZES=X'0000',PSERVIC*
               =X'000000000000185000007E00',LANG=X'11'
         TITLE 'D4B32783'
**************************************************************
*                                                            *
*      3274 MODEL 1B/1D WITH MODEL 3 SCREEN (LOCAL NON-SNA)   *
*      3274 1C BSC WITH MODEL 3 SCREEN                        *
*      3276 BSC WITH MODEL 3 SCREEN                           *
*      PRIMARY SCREEN 24 X 80 (1920)                          *
*      ALTERNATE SCREEN 32 X 80 (2560)                        *
*                                                            *
**************************************************************
D4B32783 MODEENT LOGMODE=D4B32783,FMPROF=X'02',TSPROF=X'02',PRIPROT=X'7*
               1',SECPROT=X'40',COMPROT=X'2000',RUSIZES=X'0000',PSERVIC*
               =X'000000000000185020507F00'
         TITLE 'D4B32784'
**************************************************************
*                                                            *
*      3274 MODEL 1B/1D WITH MODEL 4 SCREEN (LOCAL NON-SNA)   *
*      3274 1C BSC WITH MODEL 4 SCREEN                        *
*      3276 BSC WITH MODEL 4 SCREEN                           *
*      PRIMARY SCREEN 24 X 80 (1920)                          *
*      ALTERNATE SCREEN 43 X 80 (3440)                        *
*                                                            *
**************************************************************
D4B32784 MODEENT LOGMODE=D4B32784,FMPROF=X'02',TSPROF=X'02',PRIPROT=X'7*
               1',SECPROT=X'40',COMPROT=X'2000',RUSIZES=X'0000',PSERVIC*
               =X'00000000000018502B507F00'
         TITLE 'D4B32785'
**************************************************************
*                                                            *
*      3274 MODEL 1B/1D WITH MODEL 5 SCREEN (LOCAL NON-SNA)   *
*      3274 1C BSC WITH MODEL 5 SCREEN                        *
*      3276 BSC WITH MODEL 5 SCREEN                           *
*      PRIMARY SCREEN 24 X 80 (1920)                          *
```

```
*       ALTERNATE SCREEN 27 X 132 (3564)                               *
*                                                                      *
**********************************************************************
D4B32785 MODEENT LOGMODE=D4B32785,FMPROF=X'02',TSPROF=X'02',PRIPROT=X'7*
             1',SECPROT=X'40',COMPROT=X'2000',RUSIZES=X'0000',PSERVIC*
             =X'00000000000018501B847F00'
         TITLE 'D4B32XX3'
**********************************************************************
*                                                                      *
*       3274 MODEL 1B/1D (LOCAL NON-SNA)                               *
*       3274 1C BSC                                                    *
*       3276 BSC                                                       *
*       PRIMARY SCREEN 24 X 80 (1920)                                 *
*       ALTERNATE SCREEN TO BE DETERMINED BY APPLICATION              *
*                                                                      *
**********************************************************************
D4B32XX3 MODEENT LOGMODE=D4B32XX3,FMPROF=X'02',TSPROF=X'02',PRIPROT=X'7*
             1',SECPROT=X'40',COMPROT=X'2000',RUSIZES=X'0000',PSERVIC*
             =X'00800000000000000000000300'
      TITLE 'SCS'
**********************************************************************
*                                                                      *
*       PRINTER WITH SNA CHARACTER SET                                *
*                                                                      *
**********************************************************************
SCS      MODEENT LOGMODE=SCS,FMPROF=X'03',TSPROF=X'03',PRIPROT=X'B1',  *
             SECPROT=X'90',COMPROT=X'3080',RUSIZES=X'87C6',            *
             PSERVIC=X'01000000E1000000000000000',                    *
             PSNDPAC=X'01',SRCVPAC=X'01'
         TITLE 'DSC4K'
**********************************************************************
*                                                                      *
*       PRINTER WITH 4K BUFFER                                        *
*                                                                      *
**********************************************************************
DSC4K    MODEENT LOGMODE=DSC4K,FMPROF=X'03',TSPROF=X'03',PRIPROT=X'B1',*
             SECPROT=X'90',COMPROT=X'3080',RUSIZES=X'8787',            *
             PSERVIC=X'03000000000018502B507F00'
         TITLE 'DSC2K'
**********************************************************************
*                                                                      *
*       PRINTER WITH 2K BUFFER                                        *
*                                                                      *
**********************************************************************
DSC2K    MODEENT LOGMODE=DSC2K,FMPROF=X'03',TSPROF=X'03',PRIPROT=X'B1',*
             SECPROT=X'90',COMPROT=X'3080',RUSIZES=X'8787',            *
             PSERVIC=X'03000000000018501850BF00'
         TITLE 'BAT13790'
**********************************************************************
*                                                                      *
*       3790 BATCH                                                    *
```

```
*                                                                        *
*************************************************************************
BAT13790 MODEENT LOGMODE=BAT13790,FMPROF=X'03',TSPROF=X'03',             *
             PRIPROT=X'00',SECPROT=X'00',COMPROT=X'0000',                *
             RUSIZES=X'0000'
         TITLE 'EMU3790'
*************************************************************************
*                                                                        *
*     3790 IN DATA STREAM COMPATIBILITY MODE                             *
*                                                                        *
*************************************************************************
EMU3790  MODEENT LOGMODE=EMU3790,FMPROF=X'03',TSPROF=X'03',              *
             PRIPROT=X'B1',SECPROT=X'B0',COMPROT=X'3080',                *
             RUSIZES=X'85C7',PSERVIC=X'02000000000000000000000200'
         TITLE 'RJE3790A'
*************************************************************************
*                                                                        *
*     3790 RJE                                                           *
*                                                                        *
*************************************************************************
RJE3790A MODEENT LOGMODE=RJE3790A,FMPROF=X'03',TSPROF=X'03',             *
             PRIPROT=X'A3',SECPROT=X'A1',COMPROT=X'7080',                *
             RUSIZES=X'8585',PSERVIC=X'01106000F100800000010040'
         TITLE 'RJE3790B'
*************************************************************************
*                                                                        *
*     3790 RJE                                                           *
*                                                                        *
*************************************************************************
RJE3790B MODEENT LOGMODE=RJE3790B,FMPROF=X'03',TSPROF=X'03',             *
             PRIPROT=X'A3',SECPROT=X'A1',COMPROT=X'7080',                *
             RUSIZES=X'8585',PSERVIC=X'01102000F100800000010040'
         TITLE 'BAT23790'
*************************************************************************
*                                                                        *
*     3790 BATCH                                                         *
*                                                                        *
*************************************************************************
BAT23790 MODEENT LOGMODE=BAT23790,FMPROF=X'03',TSPROF=X'04',             *
             PRIPROT=X'B1',SECPROT=X'B0',COMPROT=X'7080',                *
             RUSIZES=X'8585',PSERVIC=X'01310000000000000000000000'
         TITLE 'BLK3790'
*************************************************************************
*                                                                        *
*     3790 BULK PRINT                                                    *
*                                                                        *
*************************************************************************
BLK3790  MODEENT LOGMODE=BLK3790,FMPROF=X'03',TSPROF=X'03',              *
             PRIPROT=X'B1',SECPROT=X'B0',COMPROT=X'3080',                *
             RUSIZES=X'8585',PSERVIC=X'03000000000000000000000000'
         TITLE 'SCS3790'
```

```
***********************************************************************
*                                                                     *
*       3790 WITH SNA CHARACTER SET                                   *
*                                                                     *
***********************************************************************
SCS3790   MODEENT LOGMODE=SCS3790,FMPROF=X'03',TSPROF=X'03',          *
              PRIPROT=X'B1',SECPROT=X'B0',COMPROT=X'3080',            *
              RUSIZES=X'8585',PSERVIC=X'010000000000000000000000'
          TITLE 'EMUDPCX'
***********************************************************************
*                                                                     *
*       3790 IN DPCX EMULATION MODE                                   *
*                                                                     *
***********************************************************************
EMUDPCX   MODEENT LOGMODE=EMUDPCX,FMPROF=X'03',TSPROF=X'03',          *
              PRIPROT=X'B1',SECPROT=X'B0',COMPROT=X'3080',            *
              RUSIZES=X'85C7',PSERVIC=X'020000000000000000000200'
          TITLE 'DSILGMOD'
***********************************************************************
*                                                                     *
* DSILGMOD  LOGMODE TABLE FOR BSC,LOCAL,SDLC 3275,3277,3278,3279      *
*           MODEL 2 OR 12, 24 X 80 SCREEN. MAY BE USED TO RUN         *
*           MODELS 3, 4, 5, 2C OR 3C AS MODEL 2                       *
*           ALSO FOR 3284, 3286, 3287, 3288, 3289 PRINTERS            *
*           THROUGH A 3271, 3272, 3274, 3275, OR 3276 CONTROLLER      *
*                                                                     *
***********************************************************************
DSILGMOD MODEENT LOGMODE=DSILGMOD,FMPROF=X'02',TSPROF=X'02',          *
              PRIPROT=X'71',SECPROT=X'40',COMPROT=X'2000',            *
              RUSIZES=X'0000',PSERVIC=X'000000000000000000000200'
          TITLE 'ISTNLDM'
***********************************************************************
*         NLDM LOGMODE FOR LU - LU SESSION WITH NCCF                  *
***********************************************************************
ISTNLDM   MODEENT LOGMODE=ISTNLDM,FMPROF=X'02',TSPROF=X'03',          *
              PRIPROT=X'30',SECPROT=X'40',COMPROT=X'0000',            *
              SSNDPAC=X'02',RUSIZES=X'0000',                          *
              PSERVIC=X'000000000000000000000000'
          TITLE 'D329001'
***********************************************************************
*         LOGMODE TABLE ENTRY FOR THE 3290 TERMINAL                  *
*             OR EXTENDED DATA STREAM TERMINAL OFF 3274-1A            *
*             PRIMARY SCREEN SIZE 24 X 80                             *
*             ALTERNATE SCREEN SIZE 62 X 160                          *
***********************************************************************
D329001   MODEENT LOGMODE=D329001,FMPROF=X'03',TSPROF=X'03',          *
              PRIPROT=X'B1',SECPROT=X'90',COMPROT=X'3080',            *
              RUSIZES=X'8787',                                        *
              PSERVIC=X'02800000000018503EA07F00'
          TITLE 'NSX32702'
***********************************************************************
```

```
*         LOGMODE TABLE ENTRY FOR NON-SNA 3270 DEVICES WITH          *
*           EXTENDED DATA STREAMS (3278 OR 3279).                    *
*           SCREEN SIZE IS 24 X 80.                                  *
*********************************************************************
NSX32702 MODEENT LOGMODE=NSX32702,FMPROF=X'02',TSPROF=X'02',         *
              PRIPROT=X'71',SECPROT=X'40',COMPROT=X'2000',           *
              RUSIZES=X'0000',                                       *
              PSERVIC=X'008000000000185000007E00'
         TITLE 'NSX32703'
*********************************************************************
*         LOGMODE TABLE ENTRY FOR NON-SNA 3270 DEVICES WITH          *
*           EXTENDED DATA STREAMS (3278 OR 3279).                    *
*           PRIMARY SCREEN 24 X 80                                   *
*           ALTERNATE SCREEN 32 X 80                                 *
*********************************************************************
NSX32703 MODEENT LOGMODE=NSX32703,FMPROF=X'02',TSPROF=X'02',         *
              PRIPROT=X'71',SECPROT=X'40',COMPROT=X'2000',           *
              RUSIZES=X'0000',                                       *
              PSERVIC=X'008000000000185020507F00'
         TITLE 'NSX32704'
*********************************************************************
*         LOGMODE TABLE ENTRY FOR NON-SNA 3270 DEVICES WITH          *
*           EXTENDED DATA STREAMS (3278 OR 3279).                    *
*           PRIMARY SCREEN 24 X 80                                   *
*           ALTERNATE SCREEN 43 X 80                                 *
*********************************************************************
NSX32704 MODEENT LOGMODE=NSX32704,FMPROF=X'02',TSPROF=X'02',         *
              PRIPROT=X'71',SECPROT=X'40',COMPROT=X'2000',           *
              RUSIZES=X'0000',                                       *
              PSERVIC=X'00800000000018502B507F00'
         TITLE 'NSX32705'
*********************************************************************
*         LOGMODE TABLE ENTRY FOR NON-SNA 3270 DEVICES WITH          *
*           EXTENDED DATA STREAMS (3278 OR 3279).                    *
*           PRIMARY SCREEN 24 X 80                                   *
*           ALTERNATE SCREEN 27 X 132                                *
*********************************************************************
NSX32705 MODEENT LOGMODE=NSX32705,FMPROF=X'02',TSPROF=X'02',         *
              PRIPROT=X'71',SECPROT=X'40',COMPROT=X'2000',           *
              RUSIZES=X'0000',                                       *
              PSERVIC=X'008000000000018501B847F00'
         TITLE 'SNX32702'
*********************************************************************
*         LOGMODE TABLE ENTRY FOR REMOTE SNA 3270 DEVICES            *
*           WITH EXTENDED DATA STREAMS (3278 OR 3279).               *
*           SCREEN SIZE IS 24 X 80.                                  *
*********************************************************************
SNX32702 MODEENT LOGMODE=SNX32702,FMPROF=X'03',TSPROF=X'03',         *
              PRIPROT=X'B1',SECPROT=X'90',COMPROT=X'3080',           *
              RUSIZES=X'87F8',                                       *
              PSERVIC=X'028000000000185000007E00'
```

```
          TITLE 'SNX32703'
*****************************************************************
*         LOGMODE TABLE ENTRY FOR REMOTE SNA 3270 DEVICES       *
*            WITH EXTENDED DATA STREAMS (MOD3).                 *
*               PRIMARY SCREEN 24 X 80 (1920)                   *
*               ALTERNATE SCREEN 32 X 80 (2560)                 *
*****************************************************************
SNX32703 MODEENT LOGMODE=SNX32703,FMPROF=X'03',TSPROF=X'03',    *
               PRIPROT=X'B1',SECPROT=X'90',COMPROT=X'3080',     *
               RUSIZES=X'87F8',                                 *
               PSERVIC=X'028000000000185020507F00'
          TITLE 'SNX32704'
*****************************************************************
*         LOGMODE TABLE ENTRY FOR REMOTE SNA 3270 DEVICES       *
*            WITH EXTENDED DATA STREAMS (MOD4).                 *
*               PRIMARY SCREEN 24 X 80 (1920)                   *
*               ALTERNATE SCREEN 43 X 80 (3440)                 *
*****************************************************************
SNX32704 MODEENT LOGMODE=SNX32704,FMPROF=X'03',TSPROF=X'03',    *
               PRIPROT=X'B1',SECPROT=X'90',COMPROT=X'3080',     *
               RUSIZES=X'87F8',                                 *
               PSERVIC=X'02800000000018502B507F00'
          TITLE 'SNX32705'
*****************************************************************
*                                                               *
*         LOGMODE TABLE ENTRY FOR REMOTE SNA 3270 DEVICES       *
*            WITH EXTENDED DATA STREAMS (MOD5).                 *
*               PRIMARY SCREEN 24 X 80 (1920)                   *
*               ALTERNATE SCREEN 27 X 132 (3564)                *
*                                                               *
*****************************************************************
SNX32705 MODEENT LOGMODE=SNX32705,FMPROF=X'03',TSPROF=X'03',    *
               PRIPROT=X'B1',SECPROT=X'90',COMPROT=X'3080',     *
               RUSIZES=X'87F8',                                 *
               PSERVIC=X'02800000000018501B847F00'
          TITLE 'LSX32702'
*****************************************************************
*                                                               *
*            3274 MODEL 1A WITH MODEL 2 SCREEN (LOCAL SNA)       *
*            WITH EXTENDED DATA STREAMS (MOD2)                  *
*            PRIMARY SCREEN 24 X 80 (1920)                      *
*            NO ALTERNATE SCREEN DEFINED                        *
*                                                               *
*****************************************************************
LSX32702 MODEENT LOGMODE=LSX32702,FMPROF=X'03',TSPROF=X'03',PRIPROT=X'B*
               1',SECPROT=X'90',COMPROT=X'3080',RUSIZES=X'87C7',PSERVIC*
               =X'028000000000185000007E00'
          TITLE 'LSX32703'
*****************************************************************
*                                                               *
*            3274 MODEL 1A WITH MODEL 3 SCREEN (LOCAL SNA)       *
```

```
*                    WITH EXTENDED DATA STREAMS (MOD3)            *
*                    PRIMARY SCREEN 24 X 80 (1920)               *
*                    ALTERNATE SCREEN 32 X 80 (2560)             *
*                                                               *
*****************************************************************
LSX32703 MODEENT LOGMODE=LSX32703,FMPROF=X'03',TSPROF=X'03',PRIPROT=X'B*
               1',SECPROT=X'90',COMPROT=X'3080',RUSIZES=X'87C7',PSERVIC*
               =X'028000000000185020507F00'
         TITLE 'LSX32704'
*****************************************************************
*                                                               *
*             3274 MODEL 1A WITH MODEL 4 SCREEN (LOCAL SNA)      *
*             WITH EXTENDED DATA STREAMS (MOD4)                  *
*             PRIMARY SCREEN 24 X 80 (1920)                      *
*             ALTERNATE SCREEN 43 X 80 (3440)                    *
*                                                               *
*****************************************************************
LSX32704 MODEENT LOGMODE=LSX32704,FMPROF=X'03',TSPROF=X'03',PRIPROT=X'B*
               1',SECPROT=X'90',COMPROT=X'3080',RUSIZES=X'87C7',PSERVIC*
               =X'02800000000018502B507F00'
         TITLE 'LSX32705'
*****************************************************************
*                                                               *
*             3274 MODEL 1A WITH MODEL 5 SCREEN (LOCAL SNA)      *
*             WITH EXTENDED DATA STREAMS (MOD5)                  *
*             PRIMARY SCREEN 24 X 80 (1920)                      *
*             ALTERNATE SCREEN 27 X 132 (3564)                  *
*                                                               *
*****************************************************************
LSX32705 MODEENT LOGMODE=LSX32705,FMPROF=X'03',TSPROF=X'03',PRIPROT=X'B*
               1',SECPROT=X'90',COMPROT=X'3080',RUSIZES=X'87C7',PSERVIC*
               =X'02800000000018501B847F00'
         TITLE 'LNK32782'
*****************************************************************
*                                                               *
*     3274 MODEL 1B/1D WITH MODEL 2 SCREEN (LOCAL NON-SNA)       *
*     3274 1C BSC WITH MODEL 2 SCREEN                            *
*     3276 BSC WITH MODEL 2 SCREEN                               *
*     PRIMARY SCREEN 24 X 80 (1920)                             *
*     NO ALTERNATE SCREEN DEFINED                               *
*     KATAKANA                                                  *
*                                                               *
*****************************************************************
LNK32782 MODEENT LOGMODE=LNK32782,FMPROF=X'02',TSPROF=X'02',PRIPROT=X'7*
               1',SECPROT=X'40',COMPROT=X'2000',RUSIZES=X'0000',PSERVIC*
               =X'000000000000185000007E00',LANG=X'11'
         TITLE 'NED32702'
*****************************************************************
*        LOGMODE TABLE ENTRY FOR NON-SNA 3270 DEVICES WITH      *
*           EXTENDED DATA STREAMS (3278 OR 3279).               *
*           SCREEN SIZE IS 24 X 80.                             *
```

```
*           LANGUAGE IS ENGLISH.                                      *
*           QUERY FOR DOUBLE BYTE CAPABILITY.                         *
***********************************************************************
NED32702 MODEENT LOGMODE=NED32702,FMPROF=X'02',TSPROF=X'02',          *
               PRIPROT=X'71',SECPROT=X'40',COMPROT=X'2000',           *
               RUSIZES=X'0000',                                       *
               PSERVIC=X'008000000000185000007E00',LANG=X'81'
         TITLE 'NKD32702'
***********************************************************************
* NAME:  NON-SNA KATAKANA, DOUBLE BYTE CAPABLE, 3270-2                *
*           LOGMODE TABLE ENTRY FOR NON-SNA 3270 DEVICES WITH         *
*              EXTENDED DATA STREAMS (3278 OR 3279).                  *
*           SCREEN SIZE IS 24 X 80.                                   *
*           LANGUAGE IS KATAKANA.                                     *
*           QUERY FOR DOUBLE BYTE CAPABILITY.                         *
***********************************************************************
NKD32702 MODEENT LOGMODE=NKD32702,FMPROF=X'02',TSPROF=X'02',          *
               PRIPROT=X'71',SECPROT=X'40',COMPROT=X'2000',           *
               RUSIZES=X'0000',                                       *
               PSERVIC=X'008000000000185000007E00',LANG=X'91'
         TITLE 'LED32702'
***********************************************************************
*                                                                    *
*           LOGMODE TABLE ENTRY FOR LOCAL SNA                         *
*              3274 MODEL 1A WITH MODEL 2 SCREEN (LOCAL SNA)          *
*              PRIMARY SCREEN 24 X 80 (1920)                          *
*              NO ALTERNATE SCREEN DEFINED                            *
*              EXTENDED DATA STREAMS                                  *
*              ENGLISH LANGUAGE                                       *
*              QUERY FOR DOUBLE BYTE CAPABILITY                       *
***********************************************************************
LED32702 MODEENT LOGMODE=LED32702,FMPROF=X'03',TSPROF=X'03',PRIPROT=X'B*
               1',SECPROT=X'90',COMPROT=X'3080',RUSIZES=X'87C7',PSERVIC*
               =X'028000000000185000007E00',LANG=X'81'
         TITLE 'LKD32702'
***********************************************************************
*                                                                    *
*           LOGMODE TABLE ENTRY FOR LOCAL SNA                         *
*              3274 MODEL 1A WITH MODEL 2 SCREEN (LOCAL SNA)          *
*              PRIMARY SCREEN 24 X 80 (1920)                          *
*              NO ALTERNATE SCREEN DEFINED                            *
*              EXTENDED DATA STREAMS                                  *
*              KATAKANA LANGUAGE                                      *
*              QUERY FOR DOUBLE BYTE CAPABILITY                       *
***********************************************************************
LKD32702 MODEENT LOGMODE=LKD32702,FMPROF=X'03',TSPROF=X'03',PRIPROT=X'B*
               1',SECPROT=X'90',COMPROT=X'3080',RUSIZES=X'87C7',PSERVIC*
               =X'028000000000185000007E00',LANG=X'91'
         TITLE 'SED32702'
***********************************************************************
*           LOGMODE TABLE ENTRY FOR REMOTE SNA 3270 DEVICES          *
```

```
*           WITH EXTENDED DATA STREAMS (3278 OR 3279).          *
*           SCREEN SIZE IS 24 X 80.                             *
*           LANGUAGE IS ENGLISH.                                *
*           QUERY FOR DOUBLE BYTE CAPABILITY                    *
***************************************************************************
SED32702 MODEENT LOGMODE=SED32702,FMPROF=X'03',TSPROF=X'03',          *
             PRIPROT=X'B1',SECPROT=X'90',COMPROT=X'3080',             *
             RUSIZES=X'87F8',                                         *
             PSERVIC=X'028000000000185000007E00',LANG=X'81'
         TITLE 'SKD32702'
***************************************************************************
*           LOGMODE TABLE ENTRY FOR REMOTE SNA 3270 DEVICES     *
*           WITH EXTENDED DATA STREAMS (3278 OR 3279).          *
*           SCREEN SIZE IS 24 X 80.                             *
*           LANGUAGE IS KATAKANA.                               *
*           QUERY FOR DOUBLE BYTE CAPABILITY                    *
***************************************************************************
SKD32702 MODEENT LOGMODE=SKD32702,FMPROF=X'03',TSPROF=X'03',          *
             PRIPROT=X'B1',SECPROT=X'90',COMPROT=X'3080',             *
             RUSIZES=X'87F8',                                         *
             PSERVIC=X'028000000000185000007E00',LANG=X'91'
         TITLE 'SNASVCMG'
***************************************************************************
*           LOGMODE TABLE ENTRY FOR RESOURCES CAPABLE OF ACTING *
*                          AS LU 6.2 DEVICES                    *
***************************************************************************
SNASVCMG MODEENT LOGMODE=SNASVCMG,FMPROF=X'13',TSPROF=X'07',          *
             PRIPROT=X'B0',SECPROT=X'B0',COMPROT=X'D0B1',             *
             RUSIZES=X'8585',ENCR=B'0000',                            *
             PSERVIC=X'060200000000000000000300'
         MODEEND
```

IBM Default USS Table

The following table is an IBM-supplied default USS table for VTAM V3R2. If you plan to customize your own USS table, be sure to link-edit your table with a name other than ISTINCDT.

```
ISTINCDT USSTAB    TABLE=STDTRANS
LOGON    USSCMD    CMD=LOGON,FORMAT=PL1
         USSPARM   PARM=APPLID
         USSPARM   PARM=LOGMODE
         USSPARM   PARM=DATA
LOGOFF   USSCMD    CMD=LOGOFF,FORMAT=PL1
         USSPARM   PARM=APPLID
         USSPARM   PARM=TYPE,DEFAULT=UNCOND
         USSPARM   PARM=HOLD,DEFAULT=YES
IBMTEST  USSCMD    CMD=IBMTEST,FORMAT=BAL
         USSPARM   PARM=P1,DEFAULT=10
         USSPARM   PARM=P2,DEFAULT=ABCDEFGHIJKLMNOPQRSTUVWXYZ0123456789
MESSAGES USSMSG    MSG=1,TEXT='INVALID COMMAND SYNTAX'
         USSMSG    MSG=2,TEXT='% COMMAND UNRECOGNIZED'
         USSMSG    MSG=3,TEXT='% PARAMETER EXTRANEOUS'
         USSMSG    MSG=4,TEXT='% PARAMETER VALUE INVALID'
         USSMSG    MSG=5,TEXT='UNSUPPORTED FUNCTION'
         USSMSG    MSG=6,TEXT='SEQUENCE ERROR'
         USSMSG    MSG=7,TEXT='%(1) UNABLE TO ESTABLISH SESSION - %(2) F-
              AILED WITH SENSE %(3)'
         USSMSG    MSG=8,TEXT='INSUFFICIENT STORAGE'
         USSMSG    MSG=9,TEXT='MAGNETIC CARD DATA ERROR'
         USSMSG    MSG=11,TEXT='% SESSIONS ENDED'
         USSMSG    MSG=12,TEXT='REQUIRED PARAMETER OMITTED'
         USSMSG    MSG=13,TEXT='IBMECHO % '
         EJECT
```

```
STDTRANS DC     X'000102030440060708090A0B0C0D0E0F'
         DC     X'101112131415161718191A1B1C1D1E1F'
         DC     X'202122232425262728292A2B2C2D2E2F'
         DC     X'303132333435363738393A3B3C3D3E3F'
         DC     X'404142434445464748494A4B4C4D4E4F'
         DC     X'505152535455565758595A5B5C5D5E5F'
         DC     X'606162636465666768696A6B6C6D6E6F'
         DC     X'707172737475767778797A7B7C7D7E7F'
         DC     X'80C1C2C3C4C5C6C7C8C98A8B8C8D8E8F'
         DC     X'90D1D2D3D4D5D6D7D8D99A9B9C9D9E9F'
         DC     X'A0A1E2E3E4E5E6E7E8E9AAABACADAEAF'
         DC     X'B0B1B2B3B4B5B6B7B8B9BABBBCBDBEBF'
         DC     X'C0C1C2C3C4C5C6C7C8C9CACBCCCDCECF'
         DC     X'D0D1D2D3D4D5D6D7D8D9DADBDCDDDEDF'
         DC     X'E0E1E2E3E4E5E6E7E8E9EAEBECEDEEEF'
         DC     X'F0F1F2F3F4F5F6F7F8F9FAFBFCFDFEFF'
         END
```

Product Support of SNA Network Addressable Unit Types

The two tables provided in this appendix correlate IBM product types to SNA network addressable unit types. This is not a complete list, but rather a list of the most widely used products that support the various SNA network addressable units.

NODE TYPES				
PU 1	**PU 2**	**PU 2.1**	**PU 4**	**PU 5**
3271	3174	3174	NCP	VTAM
6670	3274	S/36	37X5	4300
3767	3276	S/38	3720	308X
	PC	AS/400		3090
	3770	PC		
	AS/400	TPF		
		VTAM V4R1		

IBM 3270 and 3745 Line and Channel-Adapter Considerations

These tables provide you with the line and channel-adapter definition values that require special attention for the IBM 3720 and 3745 communications controllers. This is partly because of the differences between the IBM 37x5 communications controllers and the added enhanced features of the IBM 3720 and 3745 communications controllers.

HISPEED=YES for 3720 LINE Definition		
Line Adapter Base	*lnbr* value for one scanner	*lnbr* value for two scanners
1	0	*
2	32	32 or 48

* Only one scanner is permitted on this line adapter base.
For highspeed lines on the IBM 3720, the lnbr value must be the lowest position on that scanner

lnbr value for IBM 3745		
Adapter Type	*lnbr* values	USGTIER Notes
LSS	0 - 511	
HSS	1024 - 1039	USGTIER1 & 2 do not support high speed scanners.
TRA	1088 - 1095	Only 1088 & 1089 are valid for USGTIER1 & 2. Only first TRA is supported.

channl adapt addr value for IBM 3745										
One CCU CCU A or B		Two CCUs CCU A and B			USGTIER Notes					
Physical Adapter Position	*channl adapt addr* value	CCU	Physical Adapter Position	*channl adapt addr* value	1	2	3	4	5	
1	8	A	1	8	*	*	*	*	*	
2	9	A	2	9			*	*	*	
3	10	A	3	10			*	*	*	
4	11	A	4	11			*	*	*	
5	0	B	5	0	*	*	*	*	*	
6	1	B	6	1			*	*	*	
7	2	B	7	2			*	*	*	
8	3	B	8	3			*	*	*	
9	12	A	9	12			*	*	*	
10	13	A	10	13			*	*	*	
11	14	A	11	14			*	*	*	
12	15	A	12	15			*	*	*	
13	4	B	13	4			*	*	*	
14	5	B	14	5			*	*	*	
15	6	B	15	3			*	*	*	
16	7	B	16	7			*	*	*	

VTAM and NCP Performance Tuning Tables

The following tables have been supplied for your reference. They contain information on performance factors for VTAM, NCP, LU-LU flows, Network Data Flows, and communications links.

TUNING TABLE		
NETWORK COMPONENT	**PERFORMANCE FACTOR**	**SPECIFIED IN**
VTAM	Buffer Expansion	VTAM Start List
	Buffer Sizes	VTAM Start List
	ITLIM (Transient)	VTAM Start List
	EAS	VTAM Start List
	VTAMEAS	VTAM Start List
	VTAM Internal Trace (VIT)	VTAM Start List
	SONLIM (Transient)	VTAM Start List
	CSALIMIT/CSA24	VTAM Start List
	MAXPVT	Application Minor Node
	UNITSZ, MAXBFRU	NCP HOST Definition Statement
	DELAY, MAXBFRU	VTAM CTC Definition Statement
NCP	PAUSE	NCP Source LINE Definition
	NEGPOLP	NCP Source LINE Definition
	RETRIES	NCP Source LINE Definition
	ADDRESS=(n,FULL)	NCP Source LINE Definition
	NCP BUFFER size	NCP Source BUILD Definition
	MAXOUT (Modulus8|128)	NCP Source PU Definition
	MAXDATA (Segmentation)	NCP Source PU Definition
	PASSLIM	NCP Source PU Definition
	SERVLIM	NCP Source GROUP Definiton
	REPLYTO	NCP Source LINE Definition
	Service Order Table (SOT)	NCP Source SERVICE Definition
	DELAY	NCP Source BUILD Definition
	TG Activation Sequence	NCP Source TG Definitions

TUNING TABLE		
NETWORK COMPONENT	**PERFORMANCE FACTOR**	**SPECIFIED IN**
LU-LU Flows	Chaining	Application Definitions
	PACING, VPACING	LOGMODE Table, NCP Source, Appl. Definition
	Type of Response (Defintie or Exception)	LOGMODE Table, Appl. Definition
	Compaction and Compression	Application Definitions
	Flip-Flop vs. Contention	Appl. Definitions, LOGMODE Table
Network Data Flows	INN Congestion Thresholds	NCP Source PATH Definition
	VR Window Sizes	VTAM/NCP Source PATH Definition
	VR Priority	LOGMODE Table COS Table
	Message Size	Application Definitions
	Alternate Routing	COS Table, PATH Tables
Links	Line Speed	Modem clocking, NCP clocking
	BSC vs. SDLC	NCP Source LINE Definition
	DUPLEX=FULL	NCP Source LINE Definition
	Point to point vs. Multi-point	NCP Source LINE and PU Definitions

Subsystem and Device Performance Considerations

This appendix contains several tables relating to network performance for subsystems and devices. The information in the tables does not represent definitive values. These tables are provided as a reference. For accurate values, consult the respective subsystem and device manuals.

SUBSYSTEM CONSIDERATIONS

CONSIDERATION	CICS	IMS	JES2	JES2/NJE
LU TYPE	DFHTCT TYPE = TERMINAL TRMTYPE= SESTYPE=	TYPE UNITYPE= SLUTYPEP SLUTYPE1 SLUTYPE2 SLUTYPE4 LUTYPE6	RMTnn LUTYPE1	INHERENTLY LU TYPE 0 i.e., NOT DEFINED USES FDX
TYPE OF RSP (PLU-TO-SLU)	DFHPCT TYPE=OPTGRP MSGOPT= (MSGINTEG)	NONREC NONREC - RQE RECOV - RQD	RQD	RQE
TYPE OF RSP (SLU-TO-PLU)	SESSION PARM. CODED RQD/RQE	(SEE IMS TABLE)	RQD	RQE
MAXRU (PLU-TO-SLU)	DFHTCT TYPE=TERMINAL BUFFER=	TERMINAL OUTBUF=	RMTnn BUFSIZE=	INIT. STATE & TPBUFSZ=nn i.e., NO LIMIT
MAXRU (SLU-TP-PLU)	DFHTCT TYPE=TERMINAL RUSIZE=	COMM RECANY=	RMTnn BUFSIZE=	see above
BID	ALWAYS	TERMINAL OPTION (SLUP)		NO BRACKETS
COMPRESSION	N/A	N/A	COMP / NOCOMP	STANDARD
COMPACTION	N/A	N/A	CMPCT / NOCMPCT	APPL COMPACT=
CHAIN SIZE	DEPENDS ON MSG SIZE & BUFFER	DEPENDS ON MSG SIZE & OUTBUF	Rnn.PRnn CKPTPGS= CKPTLNS=	USES SINGLE ELEMENT CHAINS

CICS LU TYPE DEFINITION EXAMPLES

DEVICE	LOGICAL UNIT	TRMTYPE=	SESTYPE=
3270	3270 · DS	3275 3270	
	3270 - PRT	3270P	
	LU 2	LUTYPE2	
	LU 3	LUTYPE3	
	SCS PR	SCSPRT	
3770	INTER-FLIP FLOP	3770I	
	INTER-CONTENTION	3770C	
	BATCH FLIP FLOP	3770 3770B BCHLU	
	FULL FUNCTION	3770 3770B	USERPROG
	BATCH DATA INTER	3770	BATCHDI

IMS RESPONSE TYPES

DATA TYPE	UNITYPE=				
	SLUTYPE, 3600		3790	3767, 3770	SLUTYPE2
	ACK	OPTACK		SLUTYPE1	
UPDATE TRANS	DRX	DRX/EXCP DRX	N/A	DR1/EXCP DR1	EXCP DR1
RECOV INQ TRANS	DRX	DRX/EXCP DRX	N/A	DR1/EXCP DR1	EXCP DR1
NONRECOV INQ TRANS	DRX/EXCP DRX	DRX/EXCP DRX	EXCP DRX	DR1/EXCP DR1	EXCP DR1
IMS MSG SWITCH	DRX	DRX/EXCP DRX	EXCP DRX	DR1/EXCP DR1	EXCP DR1
IMS COMMAND	DRX/EXCP DRX	DRX/EXCP DRX	EXCP DRX	DR1/EXCP DR1	EXCP DR1
SNA COMMAND	DR1	DR1	DR1	DR1	DR1
MFS CTL REQUEST	DRX/EXCP DRX	DRX/EXCP DRX	N/A	N/A	EXCP DR1
BROADCAST OUTPUT MSG SWITCH REPLIES (RECOV) /FOR./DIS./RDIS LAST MFS PGED OUTPUT	DR2	DR2	DR2	DR2	DR2
ALL OTHER IMS CMDS REPLIES (NONRECOV) MFS PAGED OUTPUT	EXCP DR2	EXCP DR2	EXCP DR2	EXCP DR2	EXCP DR2
SNA COMMANDS	DR1	DR1	DR1	DR1	DR1

I
N
P
U
T

O
U
T
P
U
T

SUBSYSTEM CONSIDERATIONS

CONSIDERATION	POWER	TSO	RES	BDT(MVS)
LU TYPE	PRMT TYPE = LUT1	BIND	TDESCR=	LU TYPE 0
TYPE OF RSP (PLU-TO-SLU)	RQD	RQD	RQD	RQE
TYPE OF RSP (SLU-TO-PLU)	RQD	BIND	RQD	RQE
MAXRU (PLU-TO-SLU)	FIXED 256	BIND	TERMINAL BUFSIZE=	4K MAX INIT. DECK
MAXRU (SLU-TP-PLU)	FIXED 256	BIND	TERMINAL BUFSIZE=	4K MAX
BID	N/A			
COMPRESSION	OUT - BIND IN - NO	N/A	RTAM SNACOMP=	YES
COMPACTION	PRMT CMPACT=name	N/A	RTAM CAPCT=	- NO
CHAIN SIZE	PRINT DATA SET	APPLICATION & MAXRU DEPENDENT	TERMINAL VBUF=	NO CHAINING

DEVICE CONSIDERATIONS

CONSIDERATION	3174	3274/3276	MLU 3770	3770
PU TYPE	2/2.1	2	2	2
NUMBER OF PUs	1	1	1	1
FDX or HDX PU	HDX	HDX	FDX	HDX
TYPE OF LOG ON	UNFORMATTED	UNFORMATTED	UNFORMATTED	UNFORMATTED
LU TYPES	1 - SCS PRTR. 2 - 3270 D.S. 3 - 3270 PRTR. 6.2 - DLU/ILU	1 - SCS PRTR 2 - 3270 D.S. 3 - 3270 PRTR.	1	1
RJE (SINGLE OR MULTIPLE LUs)	N/A	N/A	MULTIPLE (1 - 6)	SINGLE
COMPRESSION	NO	NO	YES	YES
COMPACTION	NO	NO	YES	YES
CONTENTION or FLIP-FLOP	FLIP-FLOP	FLIP-FLOP	BOTH	BOTH
ENCRYPTION	YES	YES	YES	NO
NUMBER OF LUs	MULTIPLE	MULTIPLE	DEPENDS ON MAXRU & PACING	1
TS PROFILE	3	3	3	3
MAX. RUSIZE OUTBOUND FROM HOST	REMOTE - ANY LOCAL - 1536	REMOTE - ANY LOCAL - 1536	512	512
MAX. RUSIZE INBOUND TO HOST	1K or 2K	3274 - 1K 3276 - 2K	512	512

DEVICE CONSIDERATIONS

CONSIDERATION	3174	3274/3276	MLU 3770	3770
MAXDATA	265 or 521	265	265 or 521	265 or 521
SEGMENTATION	YES	YES	NO	NO
MAX. # SEGMENTS	NO LIMIT	NO LIMIT	N/A	N/A
MAXOUT	7	7	7	1
PACING TO DEVICE	LU 1 - ANY LU 2 - ANY LU 3 - RQD LU 6.2 - ANY	LU 1 - ANY LU 2 - ANY LU 3 - RQD	VARIABLE	1
PACING FROM DEVICE	ANY	ANY	VARIABLE	VARIABLE
LINE SPEED	T1 or 56 kbps	T1 or 56 kbps	19.2 bps	4.8 bps 3770 19.2 bps 3776, 3777
LU LOCADDR	2 - 0	2 - 0	1 - 6	1
ACTPU/ACTLU ERP	YES	YES	YES	YES
MULTIPLE ELEMENT CHAINS	YES	YES	YES	YES
SLU-PLU RSP DEFINITION	BIND	BIND	BIND	BIND
SLU-PLU RSP DEFAULT	RQE	RQE	RQD	RQD

RU Size Table Values for MAXRUSIZE Parameter

Exponent (b)	MANTISSA (a)							
	8	9	A (10)	B (11)	C (12)	D (13)	E (14)	F (15)
0	8	9	10	11	12	13	14	15
1	16	18	20	22	24	26	28	30
2	32	36	40	44	48	52	56	60
3	64	72	80	88	96	104	112	120
4	128	144	160	176	192	208	224	240
5	256	288	320	352	384	416	448	480
6	512	576	640	704	768	832	896	960
7	1024	1152	1280	1408	1536	1664	1792	1920
8	2048	2304	2560	2816	3072	3328	3584	3840
9	4096	4608	5120	5632	6144	6656	7168	7680
A (10)	8192	9216	10240	11264	12288	13312	14336	15360
B (11)	16384	18432	20480	22528	24576	26624	28672	30720
C (12)	32768	36864	40960	45056	49152	53248	57344	61440
D (13)	65536	73728	81920	90112	98304	106496	114688	122880
E (14)	131072	147456	163840	180224	196608	212992	229376	245760
F (15)	262144	294912	327680	360448	393216	425984	458752	491520

Use the following formula: $a * 2^{**b}$

Bibliography of Suggested IBM Manuals

SC31-6438	VTAM Resource Definition Reference, Version 3 Release 4.1
SC31-6209	Network Control Program Resource Definition Guide, Version 6 Release 2
SC31-6210	Network Control Program Resource Definition Reference, Version 6 Release 2
GG24-3379	IBM 3745 Communication Controller Guide
	Mark A. Miller, P.E., *LAN Protocol Handbook*, ISBN 1-55851-099-0
GG24-3420	IBM AS/400, S/38 and PS/2 as T2.1 Nodes in a Subarea Network
GG24-3669	APPN Architecture and Product Implementations Tutorial
GG24-1562	IBM 3745 Communication Controller Guide for x10 and x1A Models
GG24-3785	NCP V6 Planning and Implementation Guide
SG22-1050	A Technical Overview: VTAM V3R4 and V3R4.1, NCP V5R3.1, V5R4 and V6R1, SSP V3R5.1, V3R6 and V3R7
	Sidnie Feit, *TCP/IP Architecture, Protocols, and Implementation*, ISBN 0-07-020346-6
GG24-2266	IBM 3174 Remote Token-Ring Gateway
GG24-3110	IBM 3725 Network Control Program Token-Ring Interface Planning and Implementation
GG24-4011	VTAM Version 4 Release 1 for MVS/ESA Implementation Guide
GG24-4012	NCP V6R2 Planning and Implementation Guide
GG24-3941	VTAM Version 4 Release 1 for MVS/ESA Planning Guide

List of Abbreviations

ACB	Access method control block or application control blocks.
ACF	Advanced Communications Function.
ACTLU	Activate logical unit.
ACTPU	Activate physical unit.
API	Application program interface.
APPL	Application program.
BIU	Basic information unit.
BSC	Binary synchronous communications.
BTU	Basic transmission unit.
CA	Channel adapter.
CDRM	Cross-domain resource manager.
CDRSC	Cross-domain resource.
CLSDST	Close destination.
CNM	Communications network management.
COS	Class of service.
CP	Control program.
CSP	Communications scanner processor.
CUA	Channel unit address.
CVT	Communications vector table.
DAF	Destination address field.
DFC	Data flow control.
DLU	Destination logical unit.
EBCDIC	Extended binary-coded decimal interchange code.

ER	Explicit route.
EREP	Environmental recording, editing, and printing.
FID	Format identification.
GTF	Generalized trace facility.
INN	Intermediate networking node.
I/O	Input/Output.
IRN	Intermediate routing node.
LU	Logical unit.
MOSS	Maintenance and operator subsystem.
MVS	Multiple Virtual Storage.
MVS/XA	MVS for Extended Architecture.
MVS/370	MVS for System/370.
NAU	Network addressable unit.
NCB	Node control block.
NCCF	Network Communications Control Facility.
NCP	Network control program.
NIB	Node Identification Block.
NPDA	Network Problem Determination Application.
OAF	Origin address field.
OPNDST	Open destination.
OS	Operating system.
PEP	Partition emulation program.
PIU	Path information unit.
PLU	Primary logical unit.
PTF	Program temporary fix.
PU	Physical unit.
PUT	Program update tape.
RDT	Resource definition table.
RH	Request/response header.
RPL	Request parameter list.
RU	Request/response unit.
SBA	Set buffer address.
SDLC	Synchronous data link control.
SIO	Start I/O.
SLU	Secondary logical unit.
SMF	System management facilities.
SMP	System Modification Program.

SNA	Systems Network Architecture.
SNI	SNA network interconnection.
SSCP	System services control point.
SVA	Shared virtual area.
SVC	Supervisor call.
TAP	Trace Analysis Program.
TCB	Task control block.
TG	Transmission group.
TH	Transmission header.
TSO	Time-sharing option.
USS	Unformatted system services.
VM	Virtual machine.
VM/SNA	Virtual machine with SNA function.
VM/SP	Virtual Machine System Product.
VR	Virtual route.
VS	Virtual storage.
VSCS	VM SNA console support.
VSE	Virtual Storage Extended.
VTAM	Virtual Telecommunications Access Method.

Glossary

ACB. In the context of VTAM, it refers to application control block or access method control block. In the context of NCP, it refers to adapter control block.

ACB name. Name specified in the ACBNAME parameter of VTAM's APPL statement.

Access method. Software responsible for moving data between the main storage and I/O devices (e.g., disk drives, tapes, etc.). Process by which a VTAM application program (e.g., CICS) initiates and establishes a session with another LU.

adapter control block. A control block of NCP having control information and the current state of operation for SDLC, BSC, and start/stop lines.

Advanced Communications Function. A group of SNA-compliant IBM program products such as ACF/VTAM, ACF/TCAM, ACF/NCP, and ACF/SSP.

alert. Occurrence of a very high priority event that requires immediate attention and response.

alias name. A name defined in the name translation program when alias name does not match the real name. It is primarily used for an LU name, Logon mode table name, and class of service name in a different SNA network.

API. See application program interface.

application control block. A control block linking a VTAM application (e.g., CICS) to VTAM.

application program. A program (e.g., CICS) using the services of VTAM to communicate with different LUs and providing a platform for users to perform business-oriented activities.

application program identification. A name specified in the APPLID parameter of ACB macro. VTAM identifies an application program by this name.

application program interface. Interface through which an application program interacts with VTAM.

application program major node.   A group or collection of application program minor nodes. It is a partitioned data set (PDS) member of VMS containing one or more APPL statements.

ASCII.   American Standard Code for Information Interchange.

automatic logon.   A process by which VTAM automatically starts a session request between PLU and SLU.

begin bracket.   An indicator in the request header (RH) indicating the first request in the first chain of a bracket. (See also end bracket.)

binary synchronous communications.   A non-SNA link-level protocol for synchronous communications.

bind.   Request to activate session between a PLU and SLU.

BIU.   A request header (RH) followed by all or part of a request/response unit (RU).

boundary function.   Capability of a subarea node to provide support for adjacent peripheral nodes.

bracket.   One or more RUs exchanged between two LU-to-LU half-sessions which must be completed before another bracket can be started.

BSC.   See binary synchronous communications.

buffer.   A portion of main storage for holding I/O data temporarily.

CCP.   Configuration control program.

CDRSC.   Cross-domain resource.

chain.   See RU chain.

channel-attached.   Attachment of a device directly to the computer's byte or block multiplexer channel.

CICS.   Customer Information Control System.

class of service.   Designation of transmission priority, bandwidth, and path security to a particular session.

cluster controller.   A channel-attached or link-attached communications device (e.g., 3174) which acts as an interface between a cluster of terminal devices and the CPU or communications controller.

CNM.   Communications network management.

communications adapter.   Optional hardware available on IBM 9370 and IBM 4331 that allows communications lines to be directly attached to it, thus alleviating the need for a communications controller.

communications controller.   Communications hardware that operates under the control of NCP and manages communications lines, cluster controllers, workstations, and routing of data through a network.

Configuration control program (CCP).   An interactive application program used to define and modify the configuration for an IBM 3710.

COS.   See class of service.

cross-domain. Pertaining to more than one domain.

cross-domain resource (CDRSC). A resource owned and controlled by a cross-domain resource manager (CDRM) of another domain.

cross-network. Resources involving more than one SNA network.

Customer Information Control System (CICS). A database/data communications teleprocessing and transaction management system which runs as a VTAM application.

data link control (DLC) layer. A layer of SNA implemented in the SDLC protocol which schedules data transfer over a pair of links and performs error checking.

deactivate. To render a network resource inoperable by taking it out of service.

definite response. A value in the RH directing the receiver to respond unconditionally.

DFC. Data flow control.

disabled. An indication to SSCP that a particular LU is unable to establish an LU-LU session.

disconnected. Loss of physical connection.

domain. An SSCP and the PUs, LUs, links, and other resources that are controlled by that SSCP.

duplex. Capability to transmit in both directions simultaneously.

EBCDIC. Extended Binary-Coded Decimal Interchange Code.

element. A resource in a subarea.

enabled. An indication to SSCP that a particular LU is ready to establish an LU-LU session.

end bracket. A value in the RH indicating an end of the bracket.

ER. See explicit route.

explicit route (ER). A set of one or more TGs that connect two subarea nodes.

FID. See format identification.

FMH. Function management header.

format identification field (FID). A field in the TH indicating its format.

formatted system services. A segment of VTAM providing certain services that pertain to receiving field-formatted commands.

gateway NCP. An NCP connecting two or more SNA networks.

half-duplex. Ability to transmit data in one direction at a time only.

IMS/VS. Information Management System/Virtual Storage.

intermediate routing node (IRN). A subarea node with intermediate routing function. A subarea node may also be a boundary node.

local address. Address used by the peripheral node (e.g., cluster controller, terminals). The boundary function of a subarea node translates network address to local address and vice versa.

local attached.   A channel-attached device.

logical unit.   A port through which an end user accesses the SNA network and communicates with another logical unit.

logon mode table.   A VTAM table containing one or many logon modes identified by a logon mode name.

LU.   Logical unit.

major node.   A set of resources (minor nodes) that are given a unique name which can be activated or deactivated by a single command.

minor node.   A resource within a major node.

multiple-domain network.   A network with more than one SSCP.

Multiple Virtual Storage (MVS).   An IBM mainframe operating system.

NAU.   Network addressable unit.

NCCF.   Network Communications Control Facility.

negotiable BIND.   Capability of two LU-LU half-sessions to be able to negotiate the parameters of a session.

NetView.   An IBM network management product consisting of NCCF, NPDA, NLDM, and new enhancements.

network address.   An address consisting of subarea and element fields.

network addressable unit (NAU).   An SSCP, an LU, or a PU.

Network Management Vector Table (NMVT).   A record of information sent to a host by an SNA resource. It contains information about errors, alerts, and line statistics.

Network Terminal Option (NTO).   A program product that runs in the communications controller and allows certain non-SNA devices to have sessions with VTAM application programs.

NMVT.   Network Management Vector Table.

node name.   A symbolic name for a major or minor node.

NTO.   Network Terminal Option.

pacing.   Pertaining to control of data transmission by the receiving station so that the sending station does not cause buffer overrun.

parallel sessions.   Capability of having two or more sessions concurrently active between the same set of two LUs.

path information unit (PIU).   A message consisting of TH and BIU.

physical unit (PU).   A network addressable unit (NAU) that manages attached resources and acts as a routing node for communications between LUs. Examples of PUs are SSCP, communications controllers, and cluster controllers.

PIU. Path information unit.

PLU. Primary logical unit.

primary logical unit (PLU). In an LU-LU session, a PLU is the LU which is responsible for bind, recovery, and control.

PU. Physical unit.

request header (RH). Control information prefixed to the request unit (RU).

request parameter list (RPL). A control block containing parameters pertaining to a data transfer request or session initiation/termination request.

request unit (RU). A message unit containing user data or function management headers (FMH) or both.

return code. A code pertaining to the status of the execution of a particular set of instructions.

RH. Request/response header.

RPL. Request parameter list.

RU. Request/response unit.

SDLC. Synchronous Data Link Control.

session. A logical connection between two network addressable units (NAUs).

single-domain network. A network with one SSCP.

SLU. Secondary logical unit.

SMF. System management facility.

SSCP. System services control point.

System Support Program (SSP). An IBM program product to support NCP.

TAP. Trace analysis program.

TCAS. Terminal control address space.

terminal control address space (TCAS). The address space of TSO/VTAM that provides logon services for TSO user address spaces.

TG. Transmission group.

TH. Transmission header.

TIC. Token-ring interface coupler.

Token-ring interface coupler (TIC). An adapter to connect a communications controller to an IBM Token-Ring network.

trace analysis program (TAP). A program service aid to help in analyzing trace data produced by VTAM and NCP.

transmission group (TG). A group of one or more links between two adjacent subarea nodes that appears as a single logical link.

transmission header (TH).   Information created and used by path control and used as a prefix to a basic information unit (BIU).

transmission priority.   Priority by which the transmission group control component of path control selects a PIU for transmission to the next subarea.

unbind.   Request to terminate a session between two LUs.

unformatted system services (USS).   An SSCP facility that translates a character-coded request (e.g., logon or logoff) into a field-formatted request for processing by formatted system services (FSS).

user exit.   A user-written program which can be given control at a determined point in an IBM program.

USS.   Unformatted system services.

virtual route (VR).   Logical connection between two subareas to provide for transmission priority and underlying explicit routes.

virtual route (VR) pacing.   A technique used by the VR control component of path control to regulate a PIU's flow over a virtual route.

VM SNA console support (VSCS).   A VTAM component for VM providing SNA support and providing for SNA terminals to be VM consoles.

VM/SP.   Virtual Machine/System Product.

VR.   Virtual route.

VSCS.   VM SNA console support.

VSE.   Virtual Storage Extended operating system.

VTAM.   Virtual Telecommunications Access Method.

VTAM application program.   A program that can issue VTAM macro instructions and is known to VTAM through an ACB.

VTAM operator.   A human being or a program authorized to issue VTAM operator commands.

XRF.   Extended recovery facility.

X.21.   CCITT's recommendations for an interface between the DTE and DCE for synchronous communications (e.g., HDLC) over a data network.

X.25.   CCITT's recommendations for an interface between the DTE and packet switching networks.

X.25 NCP Packet Switching Interface (NPSI).   A program product which runs in the communications controller that allows VTAM applications to communicate over an X.25-compliant network to SNA or non-SNA equipment or end users.

Index

ABOUT THE AUTHORS

JAY RANADE is an Assistant Vice-President at Merrill Lynch in New York City. He is also the Editor in Chief of various McGraw-Hill computer series, and the author or coauthor of many McGraw-Hill books, including (with George Sackett) *Advanced SNA Networking: A Professional's Guide to VTAM/NCP*.

GEORGE C. SACKETT is Manager of Network Management and Planning at a leading consumer electronics company. He is also a contributor to *Enterprise Systems Journal* and *IBM Internet Journal*. He has an M.S. in Management of Technology from Polytechic University and is the author of *IBM's Token-Ring Networking Handbook*, available from McGraw-Hill.